4th International Conference on Mechanical and Aeronautical Engineering (ICMAE 2018)

IOP Conference Series: Materials Science and Engineering
Volume 491

Bangkok, Thailand
13 – 16 December 2018

ISBN: 978-1-5108-8331-4
ISSN: 1757-8981

Printed from e-media with permission by:

Curran Associates, Inc.
57 Morehouse Lane
Red Hook, NY 12571

Some format issues inherent in the e-media version may also appear in this print version.

Printed by Curran Associates, Inc. (2019)

For permission requests, please contact the Institute of Physics at the address below.

Institute of Physics
Dirac House, Temple Back
Bristol BS1 6BE UK

Phone: 44 1 17 929 7481
Fax: 44 1 17 920 0979

techtracking@iop.org

Additional copies of this publication are available from:

Curran Associates, Inc.
57 Morehouse Lane
Red Hook, NY 12571 USA
Phone: 845-758-0400
Fax: 845-758-2633
Email: curran@proceedings.com
Web: www.proceedings.com

Table of contents

Volume 491

4th International Conference on Mechanical and Aeronautical Engineering (ICMAE 2018)

13–16 December 2018, Bangkok, Thailand

Published online: 13 March 2019

Preface

Papers

IOP Conf. Series: Materials Science and Engineering **491** (2019) 011001 doi:10.1088/1757-899X/491/1/011001

PREFACE

ICMAE 2018 was held in Bangkok, Thailand during 13-16 Dec., 2018, which is organized by Engineering Science and Hong Kong Society of Mechanical Engineers (HKSME). The conference provides a useful and wide platform both for displaying the latest research and for exchanging research results and thoughts in mechanical engineering, aeronautical engineering, aircraft design, control engineering and other topics. The participants of the conference were from almost every part of the world, with a background in either academia or industry, or even well-known enterprise. The success and prosperity of the conference is reflected in the high level of the papers received.

The proceeding is a compilation of the accepted papers which represents an interesting outcome of the conference. This book covers 4 chapters: Advanced Robotics and Transportation System; Dynamics, Control and Measurement; Innovative Design and Applications & Modelling analysis; Process and Material Innovations and Analysis.

We would like to acknowledge all of those who supported ICMAE 2018. Each individual and institutional help was very important for the success of this conference. Especially we would like to thank the organizing committee for their valuable advices in the organization and helpful peer review of the papers.

We sincerely hope that ICMAE 2018 will be a forum for excellent discussions that will put forward new ideas and promote collaborative researches. We are sure that the proceedings will serve as an important research source of references and the knowledge, which will lead to not only scientific and engineering progress but also other new products and processes.

General Chair
Prof. Y. Lawrence Yao, Columbia University, USA

IOP Conf. Series: Materials Science and Engineering **491** (2019) 011002 doi:10.1088/1757-899X/491/1/011002

COMMITTEES

General Chair

Prof. Y. Lawrence Yao, Columbia University, USA

Conference Committee Co-Chairs

Prof. Dan Zhang,York University, Canada

Prof. Ruxu Du,The Chinese University of Hong Kong, Hong Kong

Prof. Noha Mohamed Hassan, American University of Sharjah, UAE

Conference Program Committee Chair

Prof. Matthew Parkinson, Pennsylvania State University, USA

Advisory Committee

Prof. Andrew Rae, University of the Highlands and Islands, UK

Prof. Kai Cheng, Brunel University, UK

International Technical Committees

Prof. Craig Law, University of the Witwatersrand, South Africa

Dr. Max BAKAEV, Novosibirsk State Technical University, Russia

Dr. Andrew Carruthers, University of Bradford, UK

Dr. Eltayeb Eljack, Aeronautical Engineering Department at King AbdulAziz University, Saudi Arabia

Dr. Redouane Zitoune, Clément Ader Institute of Toulouse University, Toulouse, France

IOP Conf. Series: Materials Science and Engineering **491** (2019) 011002 doi:10.1088/1757-899X/491/1/011002

Dr. Bhupendra Khandelwal, The University of Sheffield, UK

Dr. Michele Ferlauto, Politecnico di Torino, Italy

Dr. Anis Hor, Clément Ader Institute of ISAE - SUPAERO, Toulouse, France

Dr. Vineet Jain, Amity University Haryana, Gurgaon

Peer review statement

All papers published in this volume of *IOP Conference Series: Materials Science and Engineering* have been peer reviewed through processes administered by the proceedings Editors. Reviews were conducted by expert referees to the professional and scientific standards expected of a proceedings journal published by IOP Publishing.

IOP Conf. Series: Materials Science and Engineering **491** (2019) 012001 doi:10.1088/1757-899X/491/1/012001

Design and Optimization of a Blended-Wing-Body Underwater Glider

Pengcheng Ye[1, 2, *] and Guang Pan[1, 2,]

[1]School of Marine Science and Technology, Northwestern Polytechnical University, Xi'an 710072, China
[2]Key Laboratory for Unmanned Underwater Vehicle, Northwestern Polytechnical University, Xi'an, Shaanxi 710072, China

*yepengcheng@nwpu.edu.cn

Abstract. The shape of blended-wing-body underwater glider (BWBUG) is an important factor in determining the hydrodynamic efficiency. In order to reduce the needed computation time and efforts during the search of the optimum shape, a new surrogate-based shape optimization framework is proposed to solve the complicated BWBUG shape design optimization problem in this paper. During the search, seven baseline airfoils are used to build the parametric geometric model of the BWBUG, with the planar surface being fixed. Moreover, a newly proposed ensemble of surrogates based global optimization algorithm using a hierarchical design space reduction method (ESGO-HSR) is employed to optimize each baseline airfoil. Then, the optimum shape of the BWBUG can be determined and rebuilt based on all baseline airfoils that are successful optimized. As the optimization target, the maximum lift to drag ratio of the initial design is increased by 13.72% under the given operating conditions. The results demonstrate that the presented surrogate-based shape optimization framework is efficient and capable in identifying the optimum shape of the BWBUG.

1. Introduction

As a specially-designed flying-wing structure underwater glider, the blended-wing-body underwater glider has the superior hydrodynamic performances compared with the classic cylindrical configurations glider. It's composed of distinct and separate wing structures, and the body is blended and smoothed into the wings [1]. The advantage of this blended-wing-body configuration lies in higher lift to drag ratio and lower wetted area to volume ratio.

The growth of the BWBUG application has increased the significance of optimum shape of the BWBUG. However, it generally involves a larger number of computationally expensive black-box function evaluations to identify the optimal solution compared with the conventional cylindrical AUGs. The hydrodynamic shape optimization is rarely applied in the context of the BWBUG design since the existing optimization methodologies and design tools are not sufficiently fast to be used within a specific optimization target. To relieve the computational burden, surrogate models are regularly used to speed up the search process [2]. Sun et al. use the kriging-based genetic algorithm, called efficient global optimization (EGO) to deal with the hydrodynamic shape optimization of the BWBUG for achieving higher maximum lift to drag ratio [1]. Wang et al. obtain the maximum lift to drag ratio and minimum moment of the BWBUG by using the Gaussian kernel function driven multi-objective optimization method [3]. Sun et al. establish an energy consumption model for a given BWBUG and optimize its shape with a combined computational fluid dynamics (CFD) and EGO method [4]. Zhang et al. introduce a novel optimization approach to find the optimal layout of landing gears for a BWBUG based on which the comprehensive performance of the underwater landing platform is significantly improved [5].

In the authors' previous study [6], a novel ensemble of surrogates based global optimization algorithm using a hierarchical design space reduction method is presented to improve the accuracy, efficiency and robustness for dealing with the complicated engineering design optimization problems. The ESGO-HSR algorithm is used to solve the computation-intensive BWBUG shape optimization problem in this work. It's noted that the 3D BWBUG shape can be considered as a distribution of airfoils across the wing span [1]. That is, the optimum shape of the BWBUG can be obtained through many key section airfoils which are successful optimized. This simplification is beneficial to efficient determination of acceptable approximate solutions as the complex shape of the BWBUG may cause difficulties during optimization. A section parameterization method (SPM) based on the class function/shape function transformation method (CST) is used to parameterize the section airfoils in this study [7, 8]. Then, the ESGO-HSR algorithm is used to optimize the baseline airfoils of the BWBUG according to the optimization target with the maximum lift to drag ratio. Afterwards, the optimum shape of the BWBUG is built based on the optimal baseline airfoils obtained. The open code CFD packages XFOIL is used to simulate the flow around the designed section airfoils of the BWBUG during the search. XFOIL can not only provide the accurate prediction of the flow around the airfoils, but also strike a good balance between computational cost and convergence stability [9, 10]. At last, the hydrodynamic performance of the optimal BWBUG is calculated and analysed to show the accuracy, efficiency and validity of the proposed surrogate-based shape optimization framework.

2. Geometry configuration

The basic design concept of the BWBUG is considered that relates to the development of a hydrodynamics superior vehicle which can be easily launched, operated and recovered underwater without any special handling equipment. An axisymmetric body which is blended and smoothed into the wings is considered in this study, as shown in figure 1.

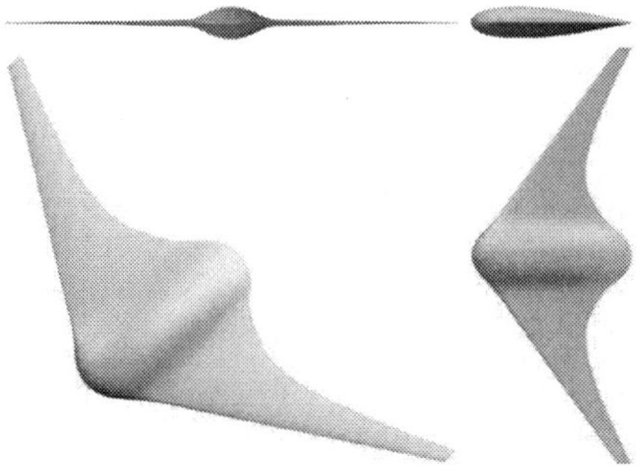

Figure 1. Initial BWBUG shape.

The BWBUG shape is built by Computer Aided Design (CAD) software Unigraphics NX (UG) based on seven typical and symmetric NACA 00XX series airfoils named $XF_1 \sim XF_7$ with the chord lengths of $c_1 \sim c_7$. figure 2 shows an illustration of the parameterization of the hull geometry, including the horizontal distances between seven airfoils and the origin $l_1 \sim l_7$, and the vertical distances between the leading edges of seven airfoils and the origin $d_1 \sim d_7$. The leading edge of XF1 is selected as the origin in this work. The planar surface of the BWBUG is fixed and determined by fourteen control points, i.e. leading and trailing edges of seven baseline airfoils using UG software. Moreover, the baseline airfoils are also employed to determine the thickness of any section of the airfoil by means of UG. That is, the BWBUG shape can be obtained once the geometry configuration of all baseline airfoils is determined. The BWBUG shape shown in figure1 is taken as an initial design to compare with optimal solutions in this paper, which is characterized with a length of $L = 1$ m and a width of $D = 0.35$ m. Its design parameters are listed in table 1.

To parameterize the BWBUG, seven control sections i.e., baseline airfoils are used to determine its hydrodynamic shape. A section parameterization method presented is used to parameterize the baseline airfoils [7, 8]. Details of the problem formulation on the BWBUG shape optimization are discussed in the next section.

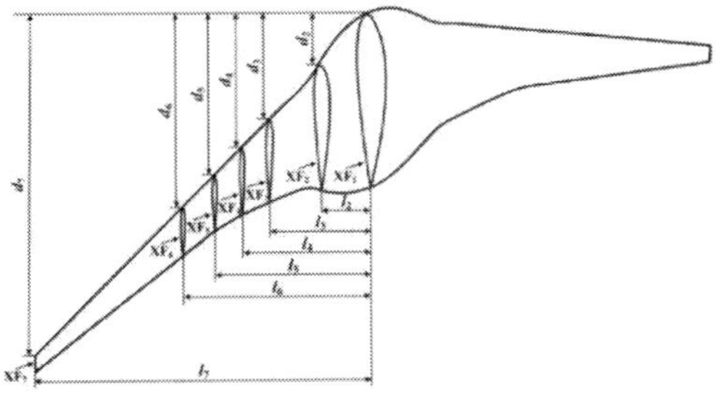

Figure 2. Parameterization of the hull geometry.

Table 1. Parameters of the BWBUG.

Airfoils	XF_1	XF_2	XF_3	XF_4	XF_5	XF_6	XF_7
NACA Series	0022	0017	0008	0008	0008	0008	0008
Chord length c (mm)	350	310	235	165	130	88	30
Horizontal distances l (mm)	0	50	90	150	190	275	500
Vertical distances d (mm)	0	20	55	95	120	175	320

3. Problem formulation

The objective is to maximize the lift to drag ratio of each baseline airfoil at the gliding conditions of the velocity $v = 1$ m/s, the angle of attack $AOA = 7°$ with the constraints on the maximum thickness and cross sectional area of the baseline airfoils. For this case, seven baseline airfoils are parameterized using the section parameterization method of *4th*-order, resulting in 5 design variables for each symmetric section airfoil and 35 in total.

The formulation of the airfoil shape optimization problem can be defined as

$$\begin{aligned} \min \quad & f(X) = -C_L/C_D \\ s.t. \quad & g_1(X) = t_{ini} - t_{opt} \le 0 \\ & g_2(X) = s_{ini} - s_{opt} \le 0 \\ & x_i^l < x_i < x_i^u, \quad i \in 1, 2, ..., 5 \end{aligned} \tag{1}$$

where C_L, C_D represent the drag coefficient and lift coefficient of the airfoil separately, t_{ini}, t_{opt}, s_{ini}, s_{opt} indicate the maximum thickness, cross sectional area of the initial airfoil and optimal airfoil respectively, x_i^l and x_i^u are the lower and upper bounds of the design variables. Considering larger displacement volume increases the loading capacity, the constraints that the maximum thickness and cross sectional area of the optimal airfoil obtained are larger than that of the initial airfoil need to be strictly met. In other words, these two constraints can well balance the weight between the loading capacity and hydrodynamic performance.

According to the parametric design of the BWBUG, the lower and upper bounds of design variables are shown in table 2, which are determined by avoiding abnormal shapes. As shown in figure 1, the empirical design has been completed by using the initial baseline airfoils, i.e. the initial design scheme in the optimization process. A new surrogate-based shape optimization framework used to solve the BWBUG shape optimization problem is introduced in the following section.

Table 2. Parameters ranges of design space.

Parameters	XF_1	XF_2	XF_3	XF_4	XF_5	XF_6	XF_7
x_1^l	-0.0714	-0.0714	-0.0286	-0.0286	-0.0286	-0.0286	-0.0286
x_1^u	0.0714	0.0714	0.0286	0.0286	0.0286	0.0286	0.0286
x_2^l	-0.0627	-0.0627	-0.0251	-0.0251	-0.0251	-0.0251	-0.0251
x_2^u	0.0627	0.0627	0.0251	0.0251	0.0251	0.0251	0.0251
x_3^l	-0.0677	-0.0677	-0.0271	-0.0271	-0.0271	-0.0271	-0.0271
x_3^u	0.0677	0.0677	0.0271	0.0271	0.0271	0.0271	0.0271
x_4^l	-0.0500	-0.0500	-0.0200	-0.0200	-0.0200	-0.0200	-0.0200
x_4^u	0.0500	0.0500	0.0200	0.0200	0.0200	0.0200	0.0200
x_5^l	-0.0702	-0.0702	-0.0281	-0.0281	-0.0281	-0.0281	-0.0281
x_5^u	0.0702	0.0702	0.0281	0.0281	0.0281	0.0281	0.0281

4. Shape optimization framework

The proposed shape optimization framework searches for the optimum shape through four stages: a) build the baseline airfoils parameterization models using the section parameterization method, b) calculate the hydrodynamic performance of baseline airfoils using the XFOIL software, c) explore the design spaces and obtain the optimal baseline airfoils using ESGO-HSR algorithm, d) rebuild the 3D BWBUG shape using all optimal airfoils. The proposed shape optimization framework is summarized into the following steps, and illustrated by the flowchart in figure 3.

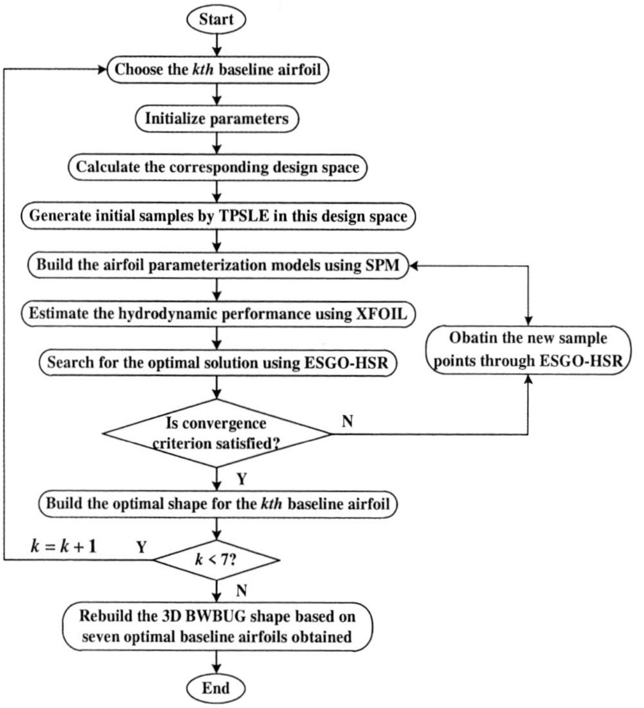

Figure 3. Flowchart of the proposed shape optimization framework.

(1) Set parameter $k = 1$ (k is the index of baseline airfoil).
(2) Choose the *kth* baseline airfoil as the optimization object.
(3) Initialize parameters for ESGO-HSR algorithm [6] and set the operating condition.
(4) Calculate the corresponding design space for the *kth* baseline airfoil.
(5) Generate an initial number of sample points in the obtained design space through sampling method translational propagation and successive local enumeration (TPSLE) [11].
(6) Build the airfoil parameterization models of the new sample points using SPM.
(7) Estimate the hydrodynamic performance for the candidate airfoils using XFOIL software.
(8) Search for the optimal solution using ESGO-HSR algorithm.
(9) If the current best solution satisfies the convergence criterion formulated in equation (2), go to the step 11. Otherwise, go to the next step.

$$\left|\bar{F}_{i+1} - \bar{F}_i\right| \le \varepsilon, \qquad \bar{F}_i = \frac{\sum_{j=1}^{5} f_j}{5} \tag{2}$$

Where ε is a small value given by the designer, f_j is the *jth* smallest function value.
(10) Obtain the new sample points through ESGO-HSR algorithm, and add them to the sample set. Go to the step 6.
(11) Build the optimum shape for the *kth* baseline airfoil and store the relevant optimal parameters.
(12) Check $k < 7$? If it's true, $k = k + 1$ and go to the step 2. Otherwise, go to the next step.
(13) Rebuild the 3D BWBUG shape based on seven optimal baseline airfoils using UG software.

The proposed surrogate-based shape optimization framework starts with a set of initial solutions. Such solutions are generated using the variable bounds and fed into the section parameterization module that generates a design for each of these solutions. The optimal solution for each baseline airfoil is found by using the ESGO-HSR algorithm. The optimization process runs till the convergence criterion is satisfied or the assigned number of function evaluations is completed. The 3D BWBUG shape can be rebuilt through UG software based on seven optimal baseline airfoils obtained.

5. Results and discussions

This study aims to achieve the maximum lift to drag ratio using the proposed surrogate-based shape optimization framework. To verify the viability and effectiveness of this framework, the optimal design of the BWBUG has been completed to compare with the empirical design.

5.1. Shape optimization

The maximum lift to drag ratio is used as the optimization target to complete the design of seven baseline airfoils in the shape optimization process. The inlet velocity is set to be equivalent to the glider velocity 1m/s, and an angle of attack 7° is used for each airfoil performance evaluation. The Reynolds number for each baseline airfoil is given in table 3. It is noted that the chord length of each baseline airfoil is normalized for clearly display. This study focuses on optimizing the hydrodynamic performance of the BWBUG.

The optimal results for each baseline airfoils are given in table 4. In addition, the comparison of hydrodynamic performance between initial airfoils and optimal airfoils with the gliding angle of 7° is also presented in table 5. As shown in table 5, the lift to drag ratio for the optimal airfoils are all larger than the initial airfoils. Thus, the optimal 3D BWBUG model can be rebuilt using the optimal airfoils obtained. Figure 4 shows the comparison of shapes between empirical design and optimal design. It can be found that the initial BWBUG has sharper nose and thinner airfoil of the centerline. On the contrary, the optimal BWBUG has mellower nose and thicker airfoil of the centerline so that it has a larger displacement volume. In actual, the total displacement volume of the optimal BWBUG is 0.0028 m^3, which is 6.64% higher than the initial BWBUG. In a word, the optimal baseline airfoils get both superior hydrodynamic performance that keeps the BWBUG high gliding efficiency and large displacement volume that enables the BWBUG to carry more loads.

Table 3. Reynolds number for each baseline airfoil

No.	XF_1	XF_2	XF_3	XF_4	XF_5	XF_6	XF_7
Re	3.50E+5	3.10E+5	2.35E+5	1.65E+5	1.30E+5	0.88E+5	0.30E+5

Table 4. Optimal results for each baseline airfoil

No.	XF_1	XF_2	XF_3	XF_4	XF_5	XF_6	XF_7
x_1	-0.0616	-0.0456	-0.0456	0.0279	0.0278	0.0274	0.0184
x_2	0.0615	0.0540	0.0540	0.0213	0.0245	0.0228	0.0244
x_3	0.0631	0.0407	0.0407	-0.0216	-0.0186	0.0006	0.0237
x_4	0.0473	-0.0441	-0.0441	0.0150	-0.0129	-0.0102	0.0097
x_5	-0.0630	0.0664	0.0664	-0.0265	-0.0018	-0.0262	-0.0166

Table 5 Comparison of the *L/D* for each baseline airfoil with the gliding angle of 7°

No.	XF_1	XF_2	XF_3	XF_4	XF_5	XF_6	XF_7
Initial *L/D*	46.58	56.09	29.26	23.81	22.07	18.18	11.18
Optimal *L/D*	54.62	63.02	38.60	33.84	30.68	25.47	14.27
Percent	17.25%	12.36%	31.92%	42.13%	39.01%	40.10%	27.68%

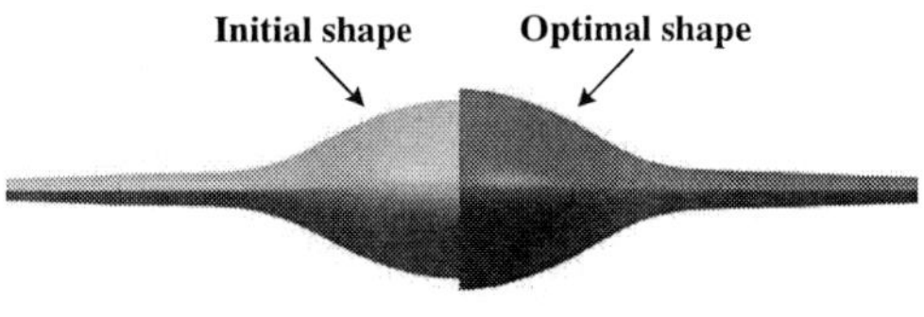

(a) Front view

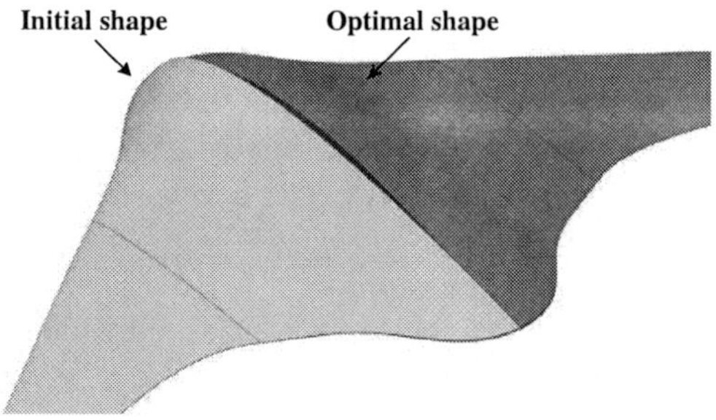

(b) Partial enlarged view

Figure 4. Comparison of shapes between initial design and optimal design.

5.2. Analysis of optimization results

A CFD analysis of the initial shape and optimal shape is also carried out for assessing the performance of the detailed design. As this object is axisymmetric, a half model is used for CFD analysis to reduce the computational costs in this work.

In order to minimize the influence of block caused by the BWBUG, a cuboid domain with a dimension of $25D \times 20D \times 15D$ (D is the width of the BWBUG and is equal to 0.35 m) is chosen. The BWBUG is placed in the mid-plane of the domain and $10D$ to the inlet boundary. Meanwhile, this computational domain is meshed with structured hexahedral grids to obtain more accurate results with smaller number of grids. Here, the non-dimensional y^+ is set to be smaller than 5 for all meshes for the compatibility with the SST k-ω turbulence model [3, 12]. Moreover, the inlet velocity at the front and the bottom of the domain is set to be equivalent to the movement speed of the BWBUG, i.e., 1 m/s with various angels of attack from 1° to 9°. The outlet pressure at the back and the top of the domain is set to be 0 Pa. And, smooth wall condition is imposed at the left side surface, where the shear effects are neglected to minimize the influences of the walls [13]. The standard wall condition is allocated to the surface of the BWBUG. Besides, a symmetry plane through the center of the BWBUG is applied for reducing the computational cost.

The hydrodynamic performance of the preliminary and detailed designs with angels of attack from 1° to 9° is reported in Table 6. Significantly, the optimization results show that the maximum lift to drag ratio of detailed design with the gliding angle of 7° is 13.04, which is 13.72% larger than the preliminary design. Furthermore, the hydrodynamic performance of the optimal design is also comparable to that of the initial design with other angels of attack. The comparative results demonstrate that the proposed surrogate-based shape optimization framework can well enhance the hydrodynamic performance and loading capacity of the BWBUG. Furthermore, it can greatly reduce the computational effort compared with the traditional optimization algorithm.

Table 6 Comparison of hydrodynamic performance between initial and optimal BWBUG

AOA	Initial design			Optimal design			Percent
	C_L	C_D	*L/D*	C_L	C_D	*L/D*	
1°	0.0161	0.0036	4.47	0.0163	0.0037	4.42	-1.02%
2°	0.0321	0.0038	8.41	0.0325	0.0039	8.34	-0.85%
3°	0.0479	0.0042	11.44	0.0485	0.0043	11.37	-0.61%
4°	0.0635	0.0047	13.38	0.0643	0.0048	13.34	-0.27%
5°	0.0785	0.0056	14.08	0.0795	0.0056	14.10	0.11%
6°	0.0925	0.0068	13.60	0.0940	0.0068	13.84	1.74%
7°	0.1033	0.0090	11.46	0.1077	0.0083	13.04	13.72%
8°	0.1127	0.0119	9.44	0.1189	0.0106	11.25	19.19%
9°	0.1210	0.0152	7.93	0.1286	0.0135	9.52	19.99%

6. Conclusions

To handle a complex, black-box engineering design optimization problem, researchers always face the challenge of finding the best tradeoff between the search efficiency and accuracy of optimal solution. In this paper, a surrogate-based shape optimization framework is presented to efficiently solve the computationally expensive BWBUG shape optimization problem. The objective of this work is to further improve the maximum lift to drag ratio of the BWBUG for a given set of user requirements. The primary contributions from the proposed surrogate-based shape optimization framework are summarized as follows.

(1) Introduce a new parametric geometric model for BWBUG. The parameterization model of the BWBUG can be simplistically obtained deriving from the parameterized baseline airfoils and the unaltered planar surface. Meanwhile, the important baseline airfoils are parameterized by using the section parameterization method.

(2) Introduce a novel efficient design optimization approach for the design and optimization of the BWBUG. This approach is developed to greatly reduce the computational time and efforts during optimization. The optimum shape of the BWBUG is completed and the maximum lift to drag ratio of the empirical design has been increased by 13.72%.

The BWBUG has been designed without considering the planar surface. In actual, the planar surface has an effect on the hydrodynamic performance of the BWBUG. The design and optimization of the planar surface will be taken into account in future research.

Acknowledgments

Supports from National Natural Science Foundation of China (Grant No. 61803306), National Key Research and Development Project of China (Grant No. 2016YFC0301300) and Natural Science Basic Research Plan in Shaanxi Province of China (Grant No. 2018JQ5092) are greatly acknowledged.

References

[1] Sun C, Song B and Wang P 2015 Parametric geometric model and shape optimization of an underwater glider with blended-wing-body *International Journal of Naval Architecture and Ocean Engineering* **7** 995-1006

[2] Ye P and Pan G 2017 Global optimization method using adaptive and parallel ensemble of surrogates for engineering design optimization *Optimization* **66** 1135-1155

[3] Wang Z, Yu J, Zhang A Wang Y and Zhao W 2017 Parametric geometric model and hydrodynamic shape optimization of a flying-wing structure underwater glider *China Ocean Engineering 31* 709-715

[4] Sun C, Song B, Wang P and Wang X 2017 Shape optimization of blended-wing-body underwater glider by using gliding range as the optimization target *International Journal of Naval Architecture and Ocean Engineering* **9** 693-704

[5] Zhang B, Song B, Mao Z and Li B 2018 Layout optimization of landing gears for an underwater glider based on particle swarm algorithm *Applied Ocean Research* **70** 22-31

[6] Ye P, Pan G and Dong Z 2018 Ensemble of surrogate based global optimization methods using hierarchical design space reduction. Structural and Multidisciplinary Optimization, https://doi.org/10.1007/s00158-018-1906-6

[7] Liu J, Song W, Han Z and Zhang Y 2017 Efficient aerodynamic shape optimization of transonic wings using a parallel infilling strategy and surrogate models *Structural and Multidisciplinary Optimization* **55** 925-943

[8] Shi L, Han Z, Shahbaz M and Song W 2016 Surrogate-based robust airfoil optimization under aleatory flight condition and geometric uncertainties *54th AIAA Aerospace Sciences Meeting*, San Diego, 2016.1.4-2016.1.8

[9] Hu W, Choi K, Zhupanska O and Buchholz J 2016 Integrating variable wind load, aerodynamic, and structural analyses towards accurate fatigue life prediction in composite wind turbine blades *Structural and Multidisciplinary Optimization* **53** 375-394

[10] Morgado J, Vizinho R, Silvestre M and Pascoa J 2016 XFOIL vs CFD performance predictions for high lift low Reynolds number airfoils *Aerospace Science and Technology* **52** 207-214

[11] Pan G, Ye P and Wang P 2014 A novel latin hypercube algorithm via translational propagation *Scientific World Journal* **2014** 163949 https://doi.org/10.1155/2014/163949

[12] Lu L, Gao Y, Li Q and Du L 2018 Numerical investigations of tip clearance flow characteristics of a pumpjet propulsor *International Journal of Naval Architecture and Ocean Engineering* **10** 307-317

[13] Tian W, Mao Z, Zhao F and Zhao Z 2017 Layout optimization of two autonomous underwater vehicles for drag reduction with a combined CFD and neural network method *Complexity* **2017** 5769794 https://doi.org/10.1155/2017/5769794

IOP Conf. Series: Materials Science and Engineering **491** (2019) 012002 doi:10.1088/1757-899X/491/1/012002

A Method of Foot Trajectory Generation for Quadruped Robots in Swing Phase to Optimize the Joint Torque

Jiaqi Li*, Dacheng Cong and Zhidong Yang

School of mechanical and electrical engineering, Harbin Institute of Technology, China

*E-mail: 120132010@126.com, Telephone number: 0451-86413230-312

Abstract. According to the basic motion requirements of quadruped robots, the mechanism of quadruped robot with 12 degrees of freedom is designed, and the dynamic model of single leg and the description equation of foot position are established. In order to solve the problem of the joint torque optimization of quadruped robots in swing phase, a method of generating the foot trajectory based on dynamics is proposed. Computer simulation results show that the proposed method can reduce the torque output of quadruped robot joints when the quadruped robots are in swing phase and prove the reliability of the proposed method for quadruped robot in the foot trajectory generation. Finally, because the leg structure of quadruped robot has the general characteristics of typical series linkage mechanism, it can also be extended to other series mechanisms with multiple connecting rods.

1. Introduction

Quadruped robots are highly adaptable to unstructured environments compared with wheeled and tracked mobile robots. Compared with biped and other less-legged robots, quadruped robots have more robust balanced gait. Quadruped robots have faster kinematic performance than hexapod and other multi-legged robots. Therefore, quadruped robots have high research and application value. The foot trajectory is the necessary and sufficient condition for the quadruped robot to adapt to the unstructured environment. The foot trajectory of the quadruped robots with better characteristics can make the quadruped robots have good dynamics and kinematics performance on the basis of fully adapting to the unstructured environment.

Many scholars have deeply studied the foot trajectory of quadruped robots, and put forward a series of foot trajectory description equations with certain characteristics. For example, the composite cycloidal foot trajectory proposed by Y.Sakakibara [1] satisfies the requirements of quadruped robot foot trajectory which the velocity and acceleration component of the forward and vertical direction is zero in the moment of landing and leaving the ground. Kyeong Yong Kim and Jong Hyeon Park [2] proposed a quadruped robot foot trajectory based on elliptical trajectory to realize the high-speed galloping of quadruped robot. Rong Xuefen [3] of Neishan University proposed a foot trajectory based on cubic polynomial and straight line combination. This method solves the defect of accelerating and decelerating the joint of quadruped robot in cycloidal trajectory, and reduces the output of joint torque to some extent. The trajectory based on cycloid is the most commonly used as the trajectory of quadruped robot [1,2,4,5,6,7,8]. The above methods do not consider the joint torque output as the constraint condition when they make the description of foot trajectory, thus neglecting the optimization of joint torque output in the process of quadruped robot from leg lifting to landing.

In order to reduce the joint torque output during the period from leg lifting to landing, and to realize more energy saving and more robust movement, a method for generating the foot trajectory is proposed in this paper. Firstly, the simplified dynamic model and foot position equation are established according to Lagrange equations, and the relationship between foot position equation and dynamic model is analysed. The workspace of the foot and the output torque diagram of the joint are drawn. According to the relationship between the position equation of the foot and the dynamic model, the foot trajectory curve is found. The contrasting analysis shows that the trajectory of the foot generated by this method can significantly reduce the joint output torque.

2. Leg Dynamics Model of a Dual-Degree-of-Freedom

On the basis of studying the physiological structure of tetrapods which can run at high speed, such as cheetahs and dogs, the model of quadruped robot with certain bionic structure has been proposed. For example, the 'Big Dog', which developed by Boston Dynamics Corporation, and the bionic cheetah robot developed by MIT University. These quadruped robot models basically satisfy the motion requirements of quadruped animals. According to the basic motion requirements of tetrapods, a 12-DOF quadruped robot model is designed in this paper. Each leg consists of three degrees of freedom, they are the lateral swing degree of hip joint, the front and rear swing freedom of thigh joint and the front and rear swing freedom of knee joint. The three degrees of freedom are responsible for the lateral swing of the quadruped robot, the front and back movement. They are connected by the rotating joints and driven by motors. The mechanism of a quadruped robot is shown in figure 1.

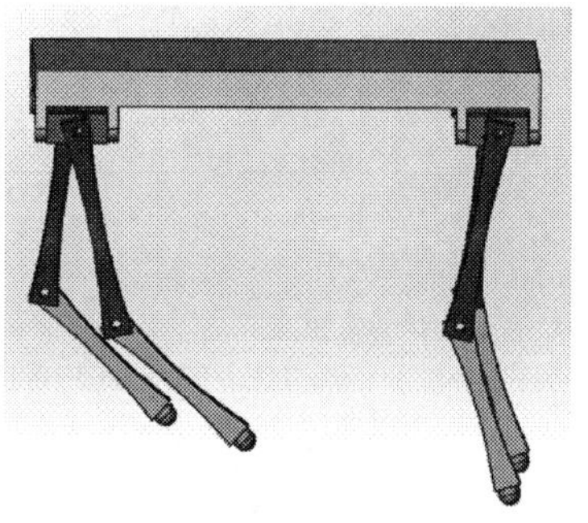

Figure 1. Mechanism diagram of quadruped robot model

The main motions of quadruped robot come from the thigh joint and knee joint, so the hip joint of quadruped robot which adjusts the side pendulum of fuselage is ignored. In addition, in order to simplify the calculation, the legs of quadruped robot are constrained in two-dimensional plane. Consider concentrating the mass of the thigh and calf at the end of the thigh and calf, as shown in figure 2.

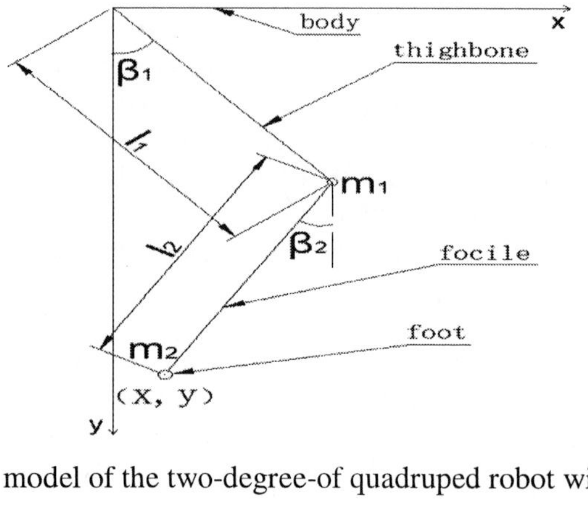

Figure 2. The dynamics model of the two-degree-of quadruped robot with the hip being ignored

In figure 2, $\beta_1, \beta_2$are the rotation angle of the thigh and the calf, $l_1, l_2$are the length of the thigh and the calf, $m_1, m_2$are the assumed concentrated mass of the thigh and the calf, and $(x, y)$is the point of the foot in the coordinate system ,which is also the point on the trajectory curve of the foot. The dynamic equation of the above system can be obtained by Lagrange equation.

The kinetic energy of the system is:

$$E_k = E_{k1} + E_{k2} \tag{1}$$

Where $E_{k1} = \frac{1}{2} m_1 v_1^2, v_1 = l_1 \dot{\beta}_1, E_{k2} = \frac{1}{2} m_2 v_2^2, v_2 = l_1^2 \dot{\beta}_1^2 + l_2^2 \dot{\beta}_2^2 - 2 l_1 l_2 \dot{\beta}_1 \dot{\beta}_2 \cos(\beta_1 + \beta_2)$

The potential energy of the system is:

$$E_p = E_{p1} + E_{p2} \tag{2}$$

Where $E_{p1} = m_1 g l_1 \cos(\beta_1), E_{p2} = m_2 g (l_1 \cos(\beta_1) + l_2 \cos(\beta_2))$

The Lagrange equation is:

$$L = E_k - E_p \tag{3}$$

The torque of the thigh and knee is:

$$\tau_1 = \frac{d}{dt}\left(\frac{\partial L}{\partial \dot{\beta}_1}\right) - \frac{\partial L}{\partial \beta_1} \tag{4}$$

$$\tau_2 = \frac{d}{dt}\left(\frac{\partial L}{\partial \dot{\beta}_2}\right) - \frac{\partial L}{\partial \beta_2} \tag{5}$$

Where

$$\begin{aligned} \tau_1 &= (m_1 l_1^2 + m_2 l_2^2)\ddot{\beta}_1 - m_2 l_1 l_2 \cos(\beta_1 + \beta_2)\ddot{\beta}_2 + m_2 l_1 l_2 \sin(\beta_1 + \beta_2)\dot{\beta}_2^2 + (m_1 + m_2) g l_1 \sin(\beta_1) \\ \tau_2 &= m_2 l_2^2 \ddot{\beta}_2 - m_2 l_1 l_2 \cos(\beta_1 + \beta_2)\ddot{\beta}_1 + m_2 l_1 l_2 \sin(\beta_1 + \beta_2)\dot{\beta}_1^2 + m_2 g l_2 \sin(\beta_2) \end{aligned} \tag{6}$$

The sum of joint torque is:

$$\begin{aligned} \tau &= (m_1 l_1^2 + m_2 l_2^2 - m_2 l_1 l_2 \cos(\beta_1 + \beta_2))\ddot{\beta}_1 + (m_2 l_2^2 - m_2 l_1 l_2 \cos(\beta_1 + \beta_2))\ddot{\beta}_2 \\ &\quad + m_2 l_1 l_2 \sin(\beta_1 + \beta_2)(\dot{\beta}_1^2 + \dot{\beta}_2^2) + (m_1 + m_2) g l_1 \sin(\beta_1) + m_2 g l_2 \sin(\beta_2) \end{aligned} \tag{7}$$

Foot trajectory description point is:

$$\begin{aligned} x &= l_1 \sin(\beta_1) - l_2 \sin(\beta_2) \\ y &= -l_1 \cos(\beta_1) - l_2 \cos(\beta_2) \end{aligned} \tag{8}$$

It can be seen that the variables of equation (6) and (7) can be completely constrained by the point of equation (8). in others words, all variables in equation (6) and (7) can be described by and its derivatives. It can be seen that the output optimization of joint torque can be realized by the foot trajectory optimization.

3. The solution of the equation

In the first section, the equation of foot point and the leg dynamics equation of quadruped robot are established. By comparing the parameters of the foot trajectory description equation (8) with the quadruped robot leg dynamics equation (7), we can see that the foot trajectory position description equation (8) can completely describe the quadruped robot leg dynamics equation (7). In all, equation (8) can describe the joint torque in equation (7). Joint torque output and foot position are closely related. The relationship between joint torque output and foot position can be solved by simultaneous equation (7) and equation (8).

Assume $s_1 = \sin(\beta_1), s_2 = \sin(\beta_2)$

Then we can get

$$\begin{aligned}&\beta_1 = \arcsin(s_1), \beta_2 = \arcsin(s_2)\\&\dot{\beta}_1 = (1-s_1^2)^{-\frac{1}{2}}, \dot{\beta}_2 = (1-s_2^2)^{-\frac{1}{2}}\\&\ddot{\beta}_1 = s_1(1-s_1^2)^{-\frac{3}{2}}, \ddot{\beta}_2 = s_2(1-s_2^2)^{-\frac{3}{2}}\end{aligned} \tag{9}$$

We can get the equation (10) by introducing the equation (9) into the equation (7).

$$\begin{aligned}\tau = &(m_1 l_1^2 + m_2 l_2^2 - m_2 l_1 l_2 \cos(\beta_1+\beta_2)) s_1 (1-s_1^2)^{-\frac{3}{2}}\\&+(m_2 l_2^2 - m_2 l_1 l_2 \cos(\beta_1+\beta_2)) s_2 (1-s_2^2)^{-\frac{3}{2}}\\&+m_2 l_1 l_2 \sin(\beta_1+\beta_2)((1-s_1^2)^{-1}+(1-s_2^2)^{-1})\\&+(m_1+m_2) g l_1 s_1 + m_2 g l_2 s_2\end{aligned} \tag{10}$$

The expression of equation (10) can be obtained by solving equation (8). But the equation (8) is a quadratic equation, it is difficult to solve it. In this paper, a method of foot trajectory generation is proposed, which can avoid solving equation (8). The main ideas are as follows: firstly, according to the leg structure of quadruped robot, the range of joint rotation angle can be obtained, and the parameter range of joint rotation angle is assumed to be here, and the other structural parameters are shown in Table 1. From the Monte Carlo method, we can get the joint torque map and the workspace of the foot. Finally, we can get the trajectory curve of the foot in the workspace of the foot under the condition of the map of joint torque.

Table 1. The main structural parameters

Parameters	Value
Thigh length l_1/m	0.4
Calf length l_2/m	0.6
Thigh mass m_1/kg	2.0
Calf mass m_2/kg	1.5

According to equation (8), the workspace of the foot is drawn by using Monte Carlo method in MATLAB. As shown in figure 3, the mechanism workspace in this paper is asymmetrical. The maximum operating range in the y direction is 0 m to -1 m, and in the x direction is -0.6 m to 0.4 m, which fully meets the requirements of the motion space of the quadruped robot.

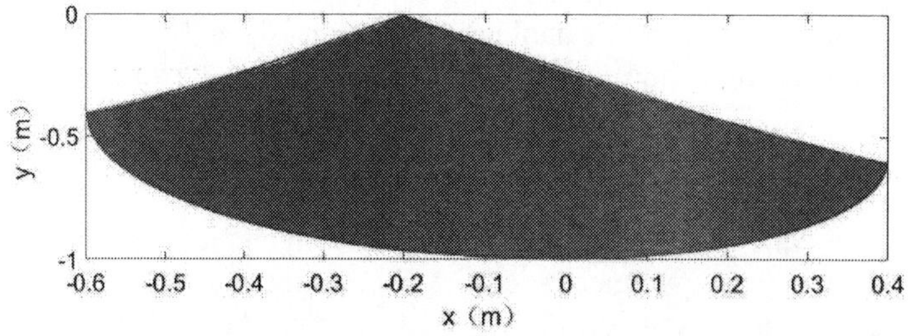

Figure 3. The work space of foot

According to equation (10), the joint torque figure is drawn. In figure 4, it can be found that the torque output of the joint is infinite at some points, which shows that the mechanism is singular in the vicinity of these points. Because the actuator output force is limited, this part of the region can`t be used as the foot trajectory design area, it should be sacrificed, and finally the nonsingular workspace is obtained, as shown in figure 5.

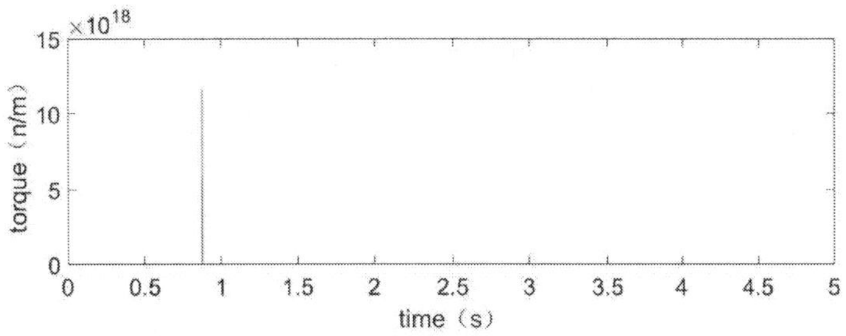

Figure 4. The output torque of joints

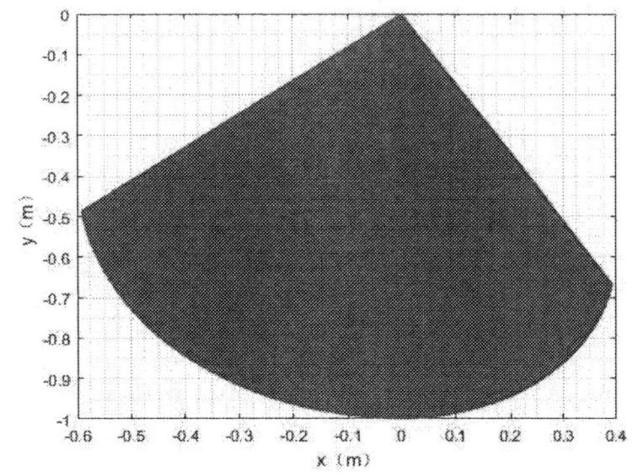

Figure 5. Non-singular workspace

By comparing figure5 and figure 3, we can find that the nonsingular workspace shown in figure 5 forgets the boundary region in figure 3. It is shown that the boundary region of the singular workspace is the singular region of the mechanism. These regions should be avoided in the design of foot trajectory. In others words, the foot trajectory cannot be contained in the region which is near the boundary in the singular workspace. In order to obtain a large step size and a certain height, the trajectory design is carried out in $x = -0.5 \sim 0.3, y = -0.8 \sim -0.65$.

4. Foot trajectory generation method

In the second section, the nonsingular workspace of the quadruped robot leg mechanism is determined, and the design area of the foot trajectory in the non-singular workspace is determined according to the requirements of the step size and the height of the foot. The foot trajectory will be generated in the foot trajectory design area. Firstly, the foot trajectory design area and the torque of the joint in the foot trajectory design area are given, as shown in figure 6. The joint output torque in figure 6 and the joint torque distribution in this area can be seen to be more uniform, which is more favorable to the joint force.

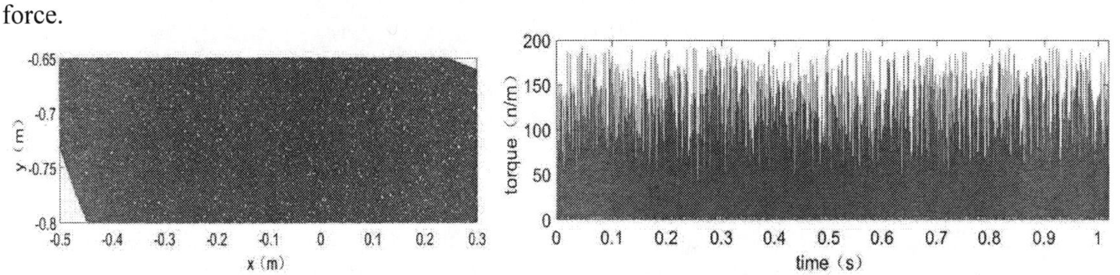

Figure 6. Foot trajectory design area and the output torque of joints

In the foot trajectory design area shown in figure 6, the trajectory curve is searched according to the principle of joint output torque optimization. The fundamental problem is to find a series of points satisfying the constraints. In the circular region with a certain step size as the radius, the minimum position of the joint torque is searched in the circular region, and the track point of the foot corresponding to the minimum position of the moment obtained in the next step is searched again. The algorithm flow chart is shown in figure 7 until the boundary of the foot trajectory design area is reached.

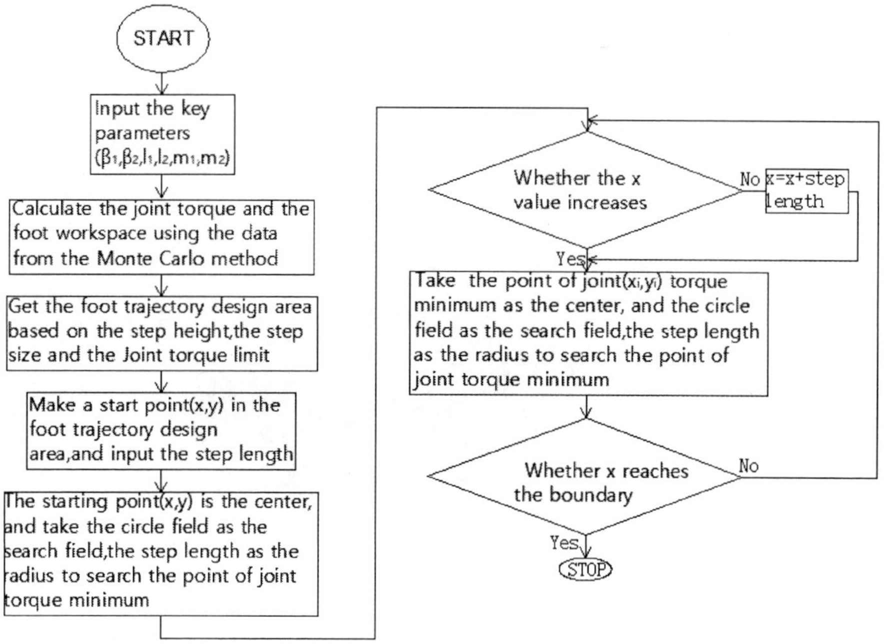

Figure 7. The flow chart of foot trajectory algorithm

The trajectory searched by the above process is actually a series of points, not an equation description. When the joints are moving, they can be directly obtained according to the inverse kinematics operation to obtain the required motion angles of the joints. It is no longer necessary to solve the foot trajectory description equation. Finally, the search points are drawn into the foot trajectory design area, as shown in figure 8. In figure 8, it can be seen that the track has a large step length of 0.65 m, the leg height is 0.06 m, and the leg height is slightly insufficient. The further work, that is, increasing the constraint of the leg lift height for further optimization, can meet the requirements of the leg lift height.

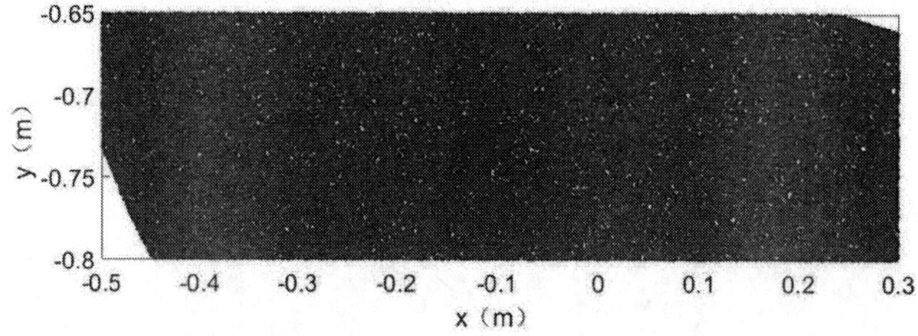

Figure 8. Foot trajectory curve

Figure 9 shows the joint output torque in the whole foot trajectory. Compared with the joint output torque in figure 6, it can be seen that the joint torque output after optimization search in the nonsingular region is greatly reduced. In figure 9, we can see that the maximum of the foot trajectory generated by the proposed method is 28 N / m, the minimum output torque is 10 N / m, the average output torque is 19.5 N / m, and the torque output is smooth and continuous. The maximum torque output is near the end of the foot trajectory.

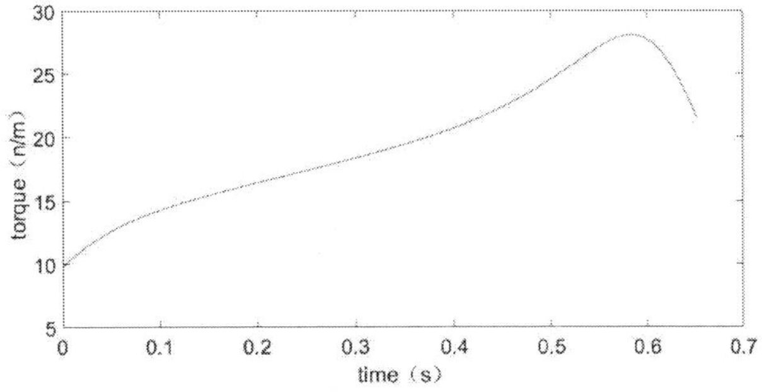

Figure 9. The output torque of joints

5. Comparative analysis and extended application

The foot trajectory of quadruped robot based on cycloid is the most commonly used trajectory at present. In reference [5], an improved cycloidal foot trajectory is proposed, its equation is (11), (12). With the same step and leg height, the foot trajectory is shown in figure 10.

$$x = \frac{0.65}{2\pi}(\theta - \sin(\theta)) \tag{11}$$

$$y = \begin{cases} 0.12(\frac{\theta}{2\pi} - \frac{1}{4\pi}\sin(2\theta)), \theta = (0, \pi) \\ 0.12 - 0.12(\frac{\theta}{2\pi} - \frac{1}{4\pi}\sin(2\theta)), \theta = (\pi, 2\pi) \end{cases} \tag{12}$$

Where θ is independent variable, and (x, y)is the point of foot trajectory.

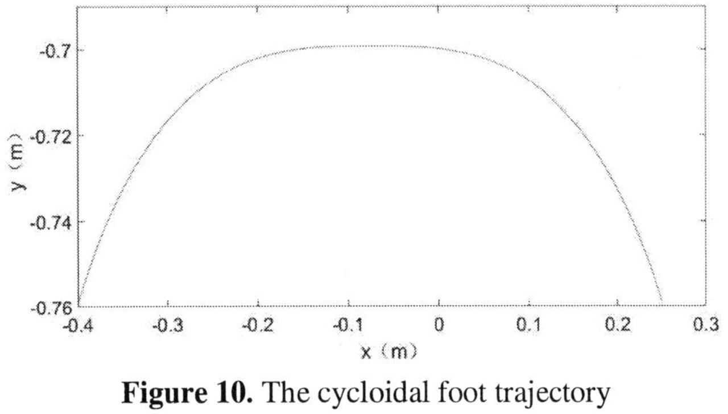

Figure 10. The cycloidal foot trajectory

The joint output torque in the improved cycloidal foot trajectory proposed in [5] is shown in figure 11. As can be seen in figure 11, the output maximum torque of the improved cycloidal foot trajectory proposed in reference [5] is 39 N / m, the minimum output torque is 13.4 N / m, the average output torque is 22.4 N / m, and the joint torque output is smooth and continuous. The maximum torque output from the joint is near the beginning of the trajectory.

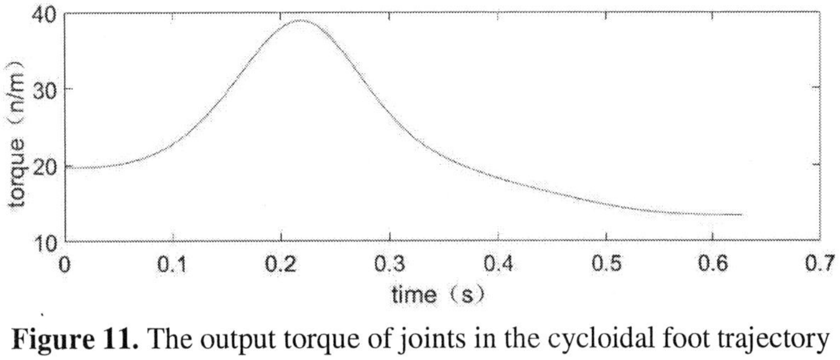

Figure 11. The output torque of joints in the cycloidal foot trajectory

Finally, by the comparison of the above data, shown in the table 2, we can see that the maximum output torque from the method proposed in the paper is 11 N / m smaller than the improved cycloidal foot trajectory proposed in reference [5], the minimum output torque is 3.4 N / m smaller, the average output torque is 2.9 N / m smaller, which shows that the method proposed in this paper can significantly reduce joint torque and output.

Table 2. Comparative analysis of the two methods

Item	Value(N/m)
The maximum of joint torque	11
The maximum of joint torque	3.4
The average of joint torque	2.9

The dynamic model established in this paper is a typical two-degree-of-freedom series connecting rod system, and its dynamic model structure is the same for the multi-degree-of-freedom series multi-linkage system. The end position and pose description equation of multi-link can also be established according to the principle of coordinate transformation. After obtaining the dynamic model of the multi-link system and the terminal position and pose description equation, the method proposed in this paper can be used to design the end trajectory so that the output torque and the joint torque of the multi-linkage system can be optimized.

6. Conclusion

In this paper, a method for generating the foot trajectory of quadruped robot based on dynamic constraints is proposed. The relationship between joint rotation angle and foot position is established by kinematics, and then the leg dynamic equation of quadruped robot is established by Lagrange equation. A foot trajectory is generated by constraining the kinematics equation and dynamic equation of quadruped robot leg. Compared with the cycloidal trajectory, the output torque of joint is reduced under the same structure parameters. At the same time, the method can be extended to the multi-linkage series-parallel mechanism system, such as the 6-DOF manipulator system, which has a great application prospect.

Reference

[1] Sakakibara Y, Kan K, Hosoda Y, et al. Low-impact foot trajectory for a quadruped walking machine[J]. Advanced Robotics, 1990, 7(4):343-360.
[2] Kim K Y, Park J H. Ellipse-based leg-trajectory generation for galloping quadruped robots[J]. Journal of Mechanical Science & Technology, 2008, 22(11):2099-2106.
[3] Rong X, Li Y, Ruan J, et al. Design and simulation for a hydraulic actuated quadruped robot[J]. Journal of Mechanical Science & Technology, 2012, 26(4):1171-1177.
[4] Liu Hongyi, Song Weigang, Peng Zhaoxing. A method of controlling the movement of foot and foot on foot. [J]. Robot, 1994(6):350-356.
[5] Wang Lipeng, Wang Junzheng, Wang Shoukun, et al. Gait Control Strategy of hydraulic quadruped Robot based on foot trajectory Planning algorithm[J]. Journal of Mechanical Engineering, 2013, 49(1):39-44.
[6] Yang Chaofeng. Study on gait planning of hydraulic quadruped robot[D]. Beijing Institute of Technology, 2015.
[7] He Dongqing. Study on trot gait Motion characteristics of JTUWM- III quadruped Robot[D]. Shanghai Jiao Tong University, 2006.
[8] Zhuang Ming. Gait Planning and Simulation of hydraulic driven quadruped Robot based on Virtual prototype Technology[D]. NUAA, 2012.

IOP Conf. Series: Materials Science and Engineering **491** (2019) 012003 doi:10.1088/1757-899X/491/1/012003

Development of a Heavy Duty Commercial Vehicle Dynamic Model with a Hydraulic-Mechanical Transmission Model Based on Field Test Results of a Vehicle with Automatic Transmission

Beomjoon Pyun[1], Chulwoo Moon[1,*], Changhyun Jeong[1] and Dohyun Jung[1]

[1]Korea Automotive Technology Institute, 31214 Chungnam, Republic of Korea

*Corresponding author's e-mail: cwmoon@katech.re.kr

Abstract. The modeling of dynamic systems is advantageous for saving time and effort when developing new systems. In the automobile industry, making a model is beneficial for predicting the feasibility and performance of a conceptual design. Hence, in this study, a heavy duty commercial vehicle with an Automatic Transmission (AT) was tested, and the model was developed and verified. A conceptual design model of Hydraulic-Mechanical Transmission (HMT) was then developed. The commercial vehicle model and the HMT model were integrated and compared to evaluate the feasibility and the difference in performance.

1. Introduction

Building a vehicle model is advantageous for saving time and effort when it comes to developing a new transmission module in the automobile industry. A transmission model should be considered in a vehicle model because the operational environment of transmission is affected by vehicular reaction forces. For this reason, a vehicle model with a commercial transmission model needs to be compared to a vehicle model with a newly developed transmission model to verify the performance [1]. Furthermore, in the conceptual design stage of a vehicle's development, a conceptual transmission model should be compared to the commercial transmission as a vehicle model in order to predict its feasibility and performance. A conceptual transmission should then be verified with a commercial transmission to predict its feasibility and performance.

In this paper, a heavy duty commercial vehicle model was developed based on vehicle tests, and a Hydraulic-Mechanical Transmission (HMT) model was developed using conceptual specifications and integrated into the vehicle model to predict the feasibility and performance of the HMT. Section 2 describes how a heavy duty commercial vehicle (AT vehicle) model was developed and verified with the vehicle test results. Section 3 introduces the concept of HMT, and explains how the model developed. Section 4 explains how the commercial vehicle model was integrated with the HMT model and how the models compare.

2. Heavy Duty Commercial Vehicle Modeling Based on Vehicle Tests

Vehicle dynamics is basically considered in order to develop a simulation environment of a heavy duty commercial vehicle model based on vehicle test results [2]. In vehicle dynamics, longitudinal dynamics is the predominant factor for developing a transmission, and the longitudinal model is comprised of a powertrain model and a vehicle and road loads model [3].

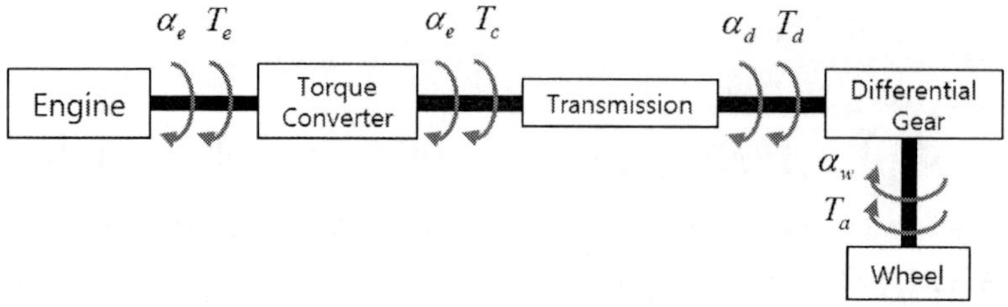

Figure 1. Schematic of a powertrain power flow

$$I_e\alpha_e = T_e - T_c \tag{1}$$

$$I_t\alpha_e = T_c - \frac{1}{N_t}T_d \tag{2}$$

$$I_d\alpha_d = T_d - \frac{1}{N_f}T_a \tag{3}$$

$$I_w\alpha_w = T_a - rF_x \tag{4}$$

$$F_x = [-I_w\alpha_w - N_{tf}^2(I_e + I_t)\alpha_w + N_{tf}T_e - N_f^2 I_d\alpha_w]/r \tag{5}$$

Where, $\alpha_e = N_t\alpha_d$, $\alpha_d = N_f\alpha_w$, $\alpha_e = N_{tf}\alpha_w$, and I_e is the engine inertia (kgm^2), α_e is the angular acceleration of the engine (rad/s), T_e is the engine torque (Nm), T_c is the torque of the torque converter (Nm); I_t is the transmission inertia (kgm^2), N_t is the gear ratio of transmission (-), T_d is the differential torque (Nm), I_d is the differential gear inertia (kgm^2), α_d is the angular acceleration of the differential gear (rad/s), N_f is the gear ratio of the differential gear (-), T_a is the wheel torque (Nm), I_w is the wheel inertia (kgm^2), α_w is the angular acceleration of the wheel (rad/s), r is the wheel effective radius (m), and F_x is the traction force (N).

Figure 1 shows the schematic of a powertrain power flow. The power flow can be described by the following formulas: Formula (1) describes the power transfer of the engine, formula (2) describes the power transfer of the transmission, formula (3) describes the power transfer of the differential gear, and formula (4) describes the power transfer of the wheel. In formula (5), engine torque and angular acceleration are transferred to the wheel as a tractive force [4]. Many other aspects should also be considered for powertrain system modeling: Engine map, torque converter properties, transmission scheduling map, type of transfer case, gear ratio of the differential gear, wheel torque, and so on. In this modeling of a powertrain system, these are obtained and updated to the powertrain model.

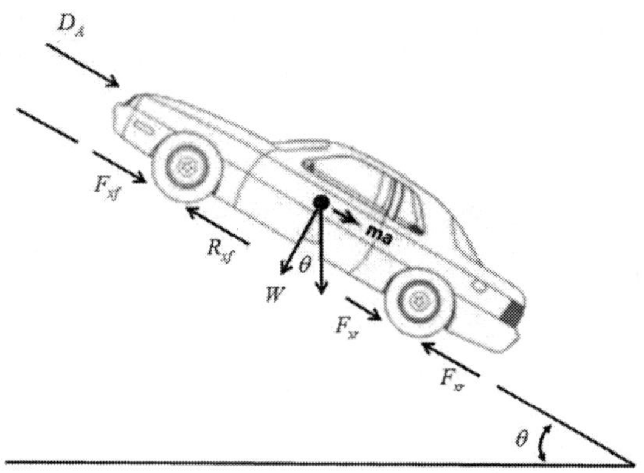

Figure 2. Schematic of a vehicle and road load model

$$F_x = F_{xf} + F_{xr} \tag{6}$$

$$R_x = R_{xf} + R_{xr} \tag{7}$$

$$D_A = \frac{1}{2}\rho C_d A V^2 \tag{8}$$

$$\frac{W}{g} a_x = F_{xf} + F_{xr} - D_A - (R_{xf} + R_{xr}) - W \sin\theta \tag{9}$$

Where, F_x is the traction force (N), R_x is the rolling resistance (N), D_A is the drag force (N), a_x is the vehicle acceleration (m/s^2), θ is the road slope (rad), ρ is the atmospheric pressure (Pa), C_d is the drag coefficient (-), and A is the frontal area (m^2).

Figure 2 shows a vehicle and road loads model composed of traction force and opposite forces. Formula (9) is the traction force of a vehicle, and formulas (10) and (11) are the opposite forces. Formula (12) is a longitudinal dynamics formula considering these forces. These formulas can be described as the schematic shown in Figure 2. With these formulas, a vehicle and road loads model can be developed.

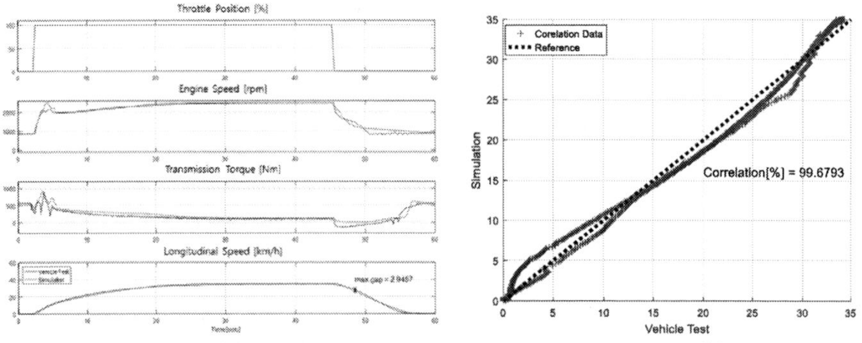

(a) (b)

Figure 3. Heavy duty commercial vehicle (AT vehicle) model verification

To verify a longitudinal vehicle model, maximum error and correlation (Pearson's r) [5] of the longitudinal speed are generally used. Hence, using the Throttle Position input in Figure 3 (a) from the vehicle test, the Longitudinal Speed outputs could be compared. The correlation result is shown in Figure 3 (b). The AT heavy vehicle model was developed and verified with a maximum error of 2.95km/h and correlation of 99.68%.

3. Hydraulic-Mechanical Transmission Modeling

Hydraulic-Mechanical Transmission (HMT) is a transmission that combines mechanical transfer parts and a Hydrostatic Unit (HSU) or a Hydrostatic Transmission (HST) part.

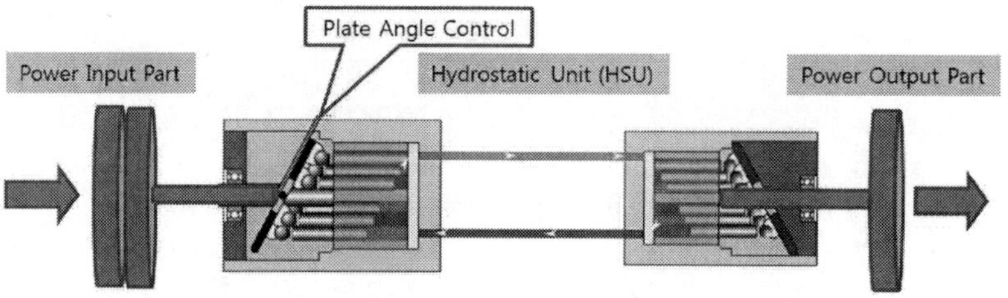

Figure 4. Schematic of a Hydrostatic Unit

Figure 4 shows an HSU operated by hydraulic power. The flow rate in the HSU is changed by Plate Angle Control, and the gear ratio is changed by that.

HMT is a transmission that has an HSU part integrated with mechanical parts because an HSU has the following disadvantages: low efficiency at high speed and narrow range of gear ratio. An HMT is more advantageous than Continuously Variable Transmission (CVT) because an HMT has better instantaneous response, durability, and efficiency in cruising speed. Additionally, an HMT is more advantageous than Automatic Transmission (AT) in terms of its instantaneous response, and it also has better efficiency at low speed. However, an HMT has disadvantages in that it has high processing costs, and low efficiency at high speed. As a result, increasing powertrain efficiency and fuel efficiency is the ultimate goal using a control algorithm because an HMT can provide good performance with a control algorithm. Section 3 describes how an HMT model became developed from a conceptual design.

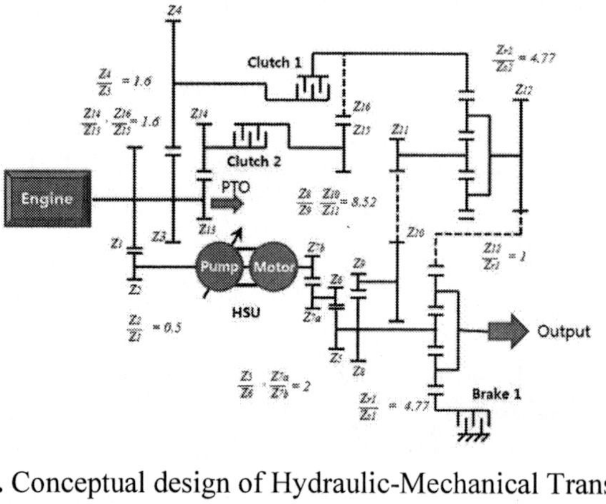

Figure 5. Conceptual design of Hydraulic-Mechanical Transmission

Figure 5 shows a conceptual design of an HMT. In Figure 5, Z_k indicates that the k-th gear's name and formulas including division are the gear ratio. The HMT is composed of two clutches, two planetary gears, one brake, and one HSU part. From this conceptual design, an HMT model was developed using the commercial simulation tool, AMESim.

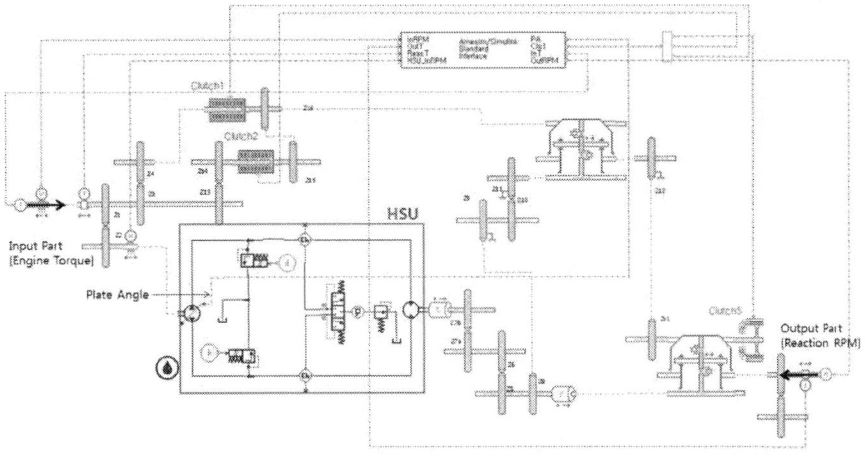

Figure 6. Simulation model of Hydraulic-Mechanical Transmission

Figure 6 shows the simulation model of HMT, developed from the conceptual design. Torque from the engine is transferred to the Input Part, and reaction rotational speed is transferred from the drive shaft to the Output Part. With the power transfer, Plate Angle is controlled to control the continuous gear ratio, and Clutches 1/2/3 are controlled to control discrete gear ratio.

4. Integrated Simulation Environment of an AT vehicle model and an HMT model

A heavy duty commercial vehicle model was developed with an Automatic Transmission model from the vehicle test results. The model comprised of a powertrain model and a vehicle and road loads model because the major consideration is longitudinal dynamics [6]. Based on the AT vehicle model, the HMT model from Section 3 was integrated into the AT vehicle model.

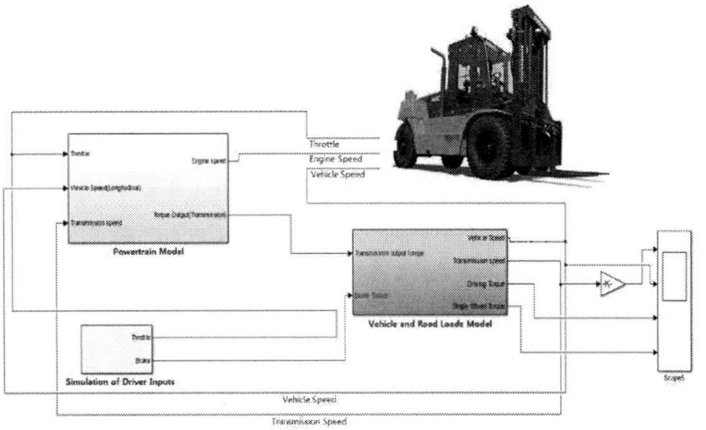

Figure 7. Simulation environment of an AT vehicle model in Matlab/Simulink

Figure 7 shows the AT vehicle model using Matlab/Simulink. In Figure 7, the gray-colored block is a powertrain model that includes an engine map, a torque converter model, and a transmission shifting algorithm from the test results, and the blue-colored block is a vehicle and road loads model that includes driving environment and vehicle information.

IOP Conf. Series: Materials Science and Engineering **491** (2019) 012003 doi:10.1088/1757-899X/491/1/012003

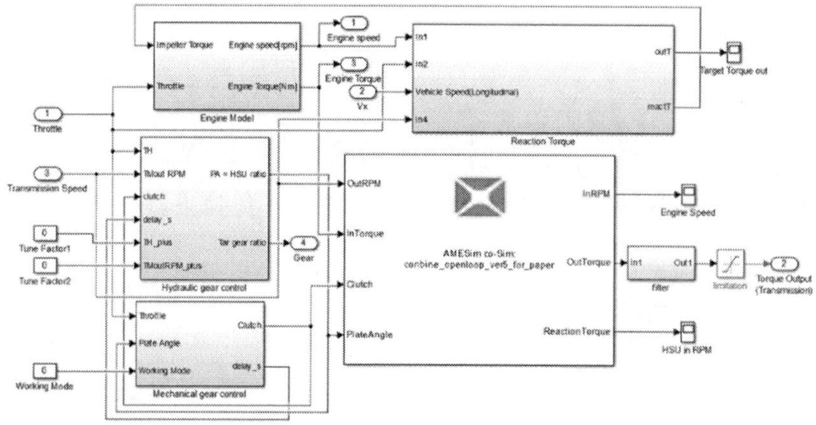

Figure 8. Hydraulic-Mechanical Transmission model in an integrated vehicle model

Based on the AT vehicle model in Figure 7, the AT model was replaced with the HMT model using AMESim. In Figure 8, the 'AMESim co-Sim' block is the HMT model using AMESim, and it is integrated in the vehicle model [7]. Because the HMT model was developed from a conceptual design, the HMT model and the AT model had different gear ratio ranges (AT gear ratio: 4.714, 2.341, 0.974; HMT gear ratio: 10~0.9). The HMT gear ratio was controlled manually. That is, the plate angle was controlled with a 'revised transmission map' from the Automatic Transmission map [8], and the clutches were controlled with commands of engagement and disengagement time.

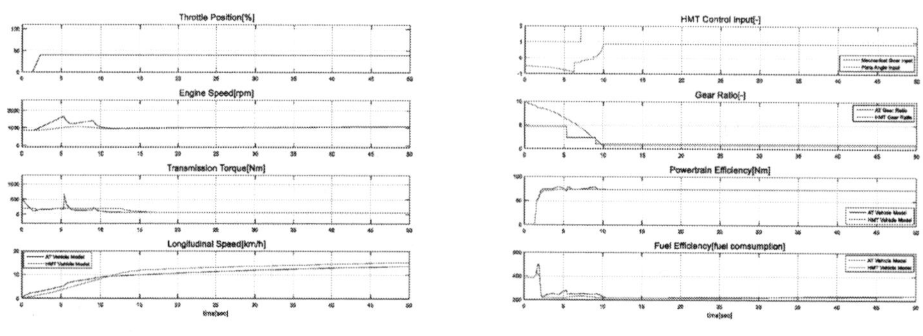

(a) **(b)**

Figure 9. Simulation comparison results of the AT vehicle model and HMT vehicle model

The Figure 9 graphs show the comparison results between the AT vehicle model and the HMT vehicle model. Figure 9 (a) shows that the input values are the same as the vehicle models. In the simulation comparison in Section 2, the Throttle Position 100% test data are used. This is because a driver's stable accelerator control is possible with full acceleration mode. However, in this section, Throttle Position 40% was used for describing the general input of a driver. Figure 9 (b) shows that the HMT Gear Ratio (red line) changed continuously and discretely. This is because the HSU caused the smooth change, and HMT discrete gears caused the stepped change at 2 seconds. From the control input, the HMT vehicle model showed better fuel efficiency and worse powertrain efficiency than the AT vehicle model. Therefore, the HMT vehicle model was developed to adapt a control algorithm.

5. Conclusions

In this paper, a heavy duty commercial vehicle model was developed and integrated with a Hydraulic-Mechanical Transmission (HMT) model that was in conceptual design stage. From the test results of a vehicle with Automatic Transmission (AT), an AT vehicle model was developed and verified. An HMT model was developed based on a conceptual design. The AT vehicle model and the HMT model were integrated, and the integrated model was compared with the AT vehicle model to consider the feasibility and difference in performance.

In future work, the conceptual design will be tested. The specifications of the HMT will be confirmed, and the HMT model will be developed and tested. The HMT model will be verified based on the model test, and the model correlation will be increased.

References

[1] C. Senatore, C. Sandu, *Torque distribution influence on tractive efficiency and mobility of off-road wheeled vehicles* (Journal of Terramechanics, 2011)

[2] T. D. Gillespie, *Fundamentals of Vehicle Dynamics,* (SAE, 1992)

[3] T. E. Scott, Power Transmission (Prentice Hall, 2000)

[4] J. Reimpell, H. Stoll, J. W. Betzler, The Automotive Chassis: Engineering Principles 2nd Edition (Butterworth-Heinemann, 2001)

[5] J. D. Gibbons, S. Chakraborti, Nonparametric Statistical Inference, 4th Edn, CRC Press/Taylor & Francis, 2014

[6] K. Ogata, System Dynamics (Prentice Hall, 2004)

[7] B. J. Pyun, C. W. Moon, C. H. Jeong, D. H. Jung, *Development of High Precision Vehicle Dynamic Model with an Intelligent Torque Transfer System (All-Wheel Drive System)*, MATEC Web of Conferences, Volume 167, 2018

[8] U. Kiencke, L. Nielsen, Automotive Control Systems: For Engine, Driveline and Vehicle 2nd Edition (Springer, 2004)

A Real-time Moving Target Following Mobile Robot System with Depth Camera

Xueling Luo[1], Dan Zhang[1,*] and Xiaodong Jin[1,2]

[1] York University, Toronto, Canada, M3J 1P3
[2] Beijing Jiaotong University, Beijing, China, 100044

[*]dzhang99@yorku.ca

Abstract. An intelligent system that could obtain a certain moving target human and could follow the target in real-time by a mobile robot could be widely applied in manufacturing as assistant or service industry to provide a human robot interaction experience. In this work, an improved efficient human following algorithm for a mobile robotic system is proposed. The system consists by a mobile robot and a depth camera which overcomes target scale variation, missing and occlusion. An improved three-dimensional tracking method is presented with the depth camera. Besides, the depth camera also provides information to avoid obstacles in robot's path. The omnidirectional mobile robot in the system provides a flexible and fast motion responding solution. In the experiments, the proposed system is tested in different scenes and the results show the reliability of the 3D tracking system in terms of accuracy and robustness to the environment.

1. Introduction

Human tracking is one of the most important and promising application for a robotic system in the service industry and manufacturing. A human tracking robot could play an important role in daily life such as a cargo transportation assisting robot at the workshop or an individual luggage carrying robot at any transportation terminal [1]. Therefore, a human following robotic system should able to detect the target continuously and keep follow movement of the target while could acknowledge surrounding environment to find its own path. Unlike surveillance systems, cameras are not static in human tracking robotic system. Challenges for specified moving target localization in a mobile robotic system are brought by environment changing during the movement such as camera position changing, illumination variation, changing of the target shape and target occlusion. [2]

In recent years, researches have put effort on moving target detection and tracking for human robot interaction application [3]. Several researches have been presented to achieve human following for a robotic system with different sensor sets approach such as using laser scanner to discriminate human legs with other objects [4] and moving targets tracking with integrated range sensors [5]. Also, several studies present on path planning and localization to achieve moving target following [6]. Applying visual system for moving target tracking is one the most popular research area among them. Texture features usually been applied on targets localization in a series of RGB images and those features could be generally classified into two categories: global features and local features. To consider the efficiency requirement on real-time tracking system, a few remarkable work have been present by using local feature in visual tracking filed [7]-[9]. In reality world, certain moving target tracking is a three dimensional problem and it is different from general pedestrians tracking problem that all human in the picture are targets. Discrete Fourier Transform (DFT) is one of the typical methods have been used on solving correlation filter to achieve high calculating efficiency [10]. However, such methods are using for two dimensional image processing.

In this paper, we present an efficient and accurate moving target detecting and tracking algorithm for a mobile robot which equip with an RGB-D Camera and a laptop. An efficient modified Kernelized Correlation Filter (KCF) is applied as visual tracker. Information obtained from depth camera is used to calculate 3D position of the target and provides environmental information for the robot to decide its path. The system has great performance on the target detecting, tracking and following and the system is able to moving with avoiding obstacles if there are some in the scene.

2. Proposed method

In this section, we will briefly introduce the KCF tracker which is the base of our tracking algorithm. A lot of KCF trackers are designed for searching target in a window without drift and tracking without failure when the target is moving with changing speed. To achieve that more calculation is needed which highly reduce computing speed of the algorithm which is not applicable in real-time tracking system. [11] We take variation velocity of the target in to consideration to modify the KCF tracker and to use the same vector of the movement for the mobile robot to predicting target trajectory. Besides, we also include the depth camera information in the algorithm for the mobile robot to avoid obstacles in its path.

2.1. Kernelized correlation filter

The goal of Kernelized Correlation filter (KCF) tracker [7] is to solve the ridge regression which is to minimize error between input image patches $t_{w,h}$ (w × h) and its regression labels $y_{w,h}$.

$$\min \sum_{w,h} \left| \langle f(t_{w,h}), W \rangle - y_{w,h} \right|^2 + \gamma \|W\|^2 \tag{1}$$

Where f represent the mapping to the nonlinear space which introduced by kernel, $t_{w,h}$ is the image patch of the target, the γ is regular parameter to control overfitting, W So, the target has to be selected manually at beginning. The shifts $t_{w,h}$ $((w,h) \in \{0, \dots, W-1\} \times \{0, \dots, H-1\})$ around the target are considered as training sample. W is the solution of $W = \sum_{w,h} \alpha_{w,h} f(t_{w,h})$.

The variable is presented as calculating discrete Fourier transformation and its inverse of following expression:

$$\alpha = F^{-1}\left[\frac{F(y)}{F\left(K(t_{w,h},t)\right)+\gamma}\right] \tag{2}$$

Where $K(t_{w,h}, t)$ represents the kernel correlation of the target model t. Then the target position within a pitch z could be found by the maximum response of

$$\hat{f}_z = F^{-1}\left(F\left(K(t_{w,h}, t)\right) F(\alpha)\right) \tag{3}$$

2.2. Target tracking

In order to tracking a target without constant moving speed and to predict the motion, the system variable vector is defined as: $x = [p_x, p_y, v_x, v_y]^{\mathrm{T}}$. Where (p_x, p_y) is the centre location of the target and (v_x, v_y) is the velocity of the target. The initial position of the target need to be selected manually and initial velocities are zero. To use a-second order autoregressive model for target status estimating [11]:

$$x^{k+1} = Ax^k + N(\tau) \tag{4}$$

Where A is the state matrix and $N(\tau)$ is the noise matrix which is Gaussian distributed.

To improve tracking accuracy, the system is described as follow for three dimensional position prediction as $x_p = [p_x, p_y, p_z, v_x, v_y, v_z]^T$, where $(p_y)^2 = (d_x)^2 - (p_x)^2 + (p_z)^2$ is the depth position of the target respect to the mobile robot which is obtained by the depth camera and d_x is the

IOP Conf. Series: Materials Science and Engineering **491** (2019) 012004 doi:10.1088/1757-899X/491/1/012004

depth of position x. Position offset from x direction to the z direction could be represent by $\sin\theta_r = d_y/p_x$ where θ_r is the rotation vector for mobile robot control. And $v_y = d\ (p_y)$ is the moving velocity of the target in depth direction. We applied extended Kalman filter for target position predicting and following. The target state and measurement equations could be written as

$$\hat{X}_{k+1} = \hat{X}_k \quad (5)$$

$$P_{k+1} = \phi_{k+1} P_k \phi_{k+1}^T + Q \quad (6)$$

Where P_{k+1} is the covariance matrix and ϕ_{k+1} is the . Q is the noise of the system. The state of system would be updated by

$$K_{k+1} = P_{k+1} H_{k+1}^T \left(H_{k+1} P_{k+1} H_{k+1}^T + R\right)^{-1} \quad (7)$$

$$\hat{X}_{k+1} = \hat{X}_{k+1} + K_{k+1}\left(Z_{k+1} - \hat{h}(X_{k+1})\right) \quad (8)$$

$$P_{k+1} = [I - K_{k+1} H_{k+1}] P_{k+1} \quad (9)$$

Where $\hat{h}(X_{k+1})$ is the predicted measurement model of the system and Z_{k+1} is the measurement result from the system.

2.3. Obstacle avoiding

In this work, only relative position of obstacles in the camera view are need to be determinated. The depth camera we used in this work is Kinect which could provide RGB and depth images synchronously. When there is no obstacle in the scene, the mobile robot moves in vector direction $[X_r, Y_r, \theta_r, V_r]^T$ with $[X_r, Y_r, \theta_r]$ as moving direction which are obtained from prediction in 2.2 and V_r as moving velocity which is calculated from v_x *and* v_y since the robot is not able to move in vertical direction. In figure 1, the module of tracking system is shown when $p_x = 0$.

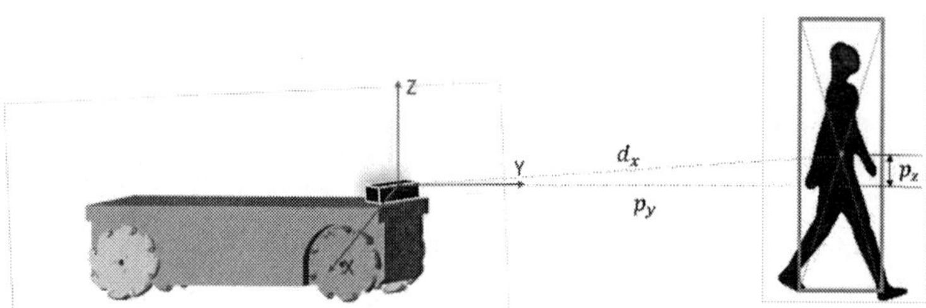

Figure 1. The system coordinate

When there are obstacles in the path of the robot. In order to receive abnormal notice timely during the mobile robot moving, we only consider information from the depth camera in the maximum height of the robot $H_{r,max}$. In figure 2, a detection result is presented with depth camera in robot height. All marked point in the picture shows the edges of obstacles in the scene for accessibility judgement. We assume the robot could pass through an alley with minimum width W_{min} or it will stop. So the robot will have an adjusting vector $[x_r, y_r, \rho_r, v_r]^T$. Where $x_r = -x_o$ and x_o is the location in x direction of the obstacle, y_r will be the same as Y_r, $\sin\rho_r = f_1(X_r - x_r, Y_r)$ and $v_r = f_2(X_r - x_r, Y_r)$

Figure 2. The detection result from depth camera with robot height $H_{r,max}$

2.4. Mobile robot

Omnidirectional robot is able to move in more direction comparing with normal wheeled robot. Different motions of the system are decoupled for a simple motion like translation or rotation which provide flexibility and fast responsive performance for the system. [12] The mecanum wheeled robot is a typical omnidirectional system which is shown in Figure 3.

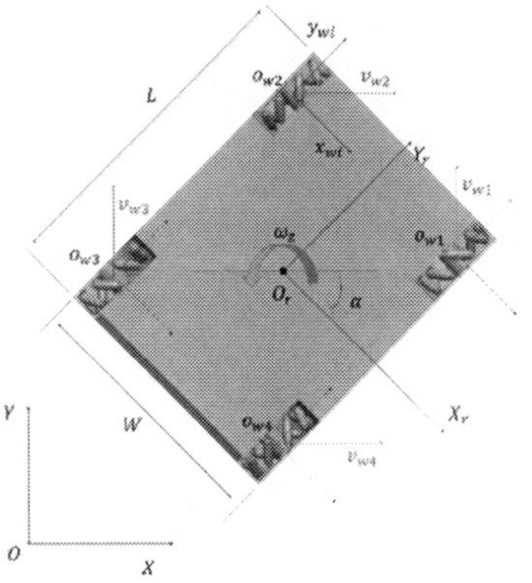

Figure 3. The configuration of the robot

Where, the coordinates OXY is the fixed coordinate and the $O_rX_rY_r$ is the moving coordinates for the robot. Besides, the coordinates o_{wi} (i=1, 2, 3, 4) are coordinate frame for each wheel.

Let the position of each wheel to be described as $S_{wi} = [x_{wi}, y_{wi}, \alpha_i]^T$ (i=1, 2, 3, 4), $\dot{\theta}_{xi}$ is the angular velocity around the hub, $\dot{\theta}_{ri}$ is the angular velocity of the roller and $\dot{\theta}_{zi}$ is the angular velocity about the contact point. R is the radius of the wheel and r is the roller radius. φ_i is the slope angler for the roller and in the robot, $\varphi_1 = \varphi_3 = 45^{\circ}$ and $\varphi_2 = \varphi_4 = -45^{\circ}$. $\dot{S}_{wi}$ could be represented as equation (10).

$$\dot{S}_{wi} = \begin{bmatrix} \dot{x}_{wi} \\ \dot{y}_{wi} \\ \dot{\alpha}_i \end{bmatrix} = \begin{bmatrix} 0 & r\sin\varphi_i & 0 \\ R & -r\cos\varphi_i & 0 \\ 0 & 0 & 1 \end{bmatrix} \begin{bmatrix} \dot{\theta}_{xi} \\ \dot{\theta}_{ri} \\ \dot{\theta}_{zi} \end{bmatrix} \tag{10}$$

The coordinates translation matrix from wheel coordinates to robot coordinates T_{wi} is given as

$$T_{w2ri} = \begin{bmatrix} cos\alpha_{w2ri} & -sin\alpha_{w2ri} & d_{xw2ri} \\ sin\alpha_{w2ri} & cos\alpha_{w2ri} & d_{yw2ri} \\ 0 & 0 & 1 \end{bmatrix} \tag{11}$$

Where α_{w2ri} (i=1,2,3,4) is the rotation angle of wheel to the $O_rX_rY_r$ coordinates, d_{xw2ri} and d_{yw2ri} are the translational distance between two coordinates. For the position of the robot, $S_r = [x_r, y_r, \alpha_r]^T = T_{w2ri}S_{wi}$. In figure 2, α_{w2ri} could be found equal to 0.
Then, the inverse kinematics is obtained as

$$\begin{bmatrix} \dot{x}_r \\ \dot{y}_r \\ \dot{\alpha}_r \end{bmatrix} = \frac{R}{4} \begin{bmatrix} -1 & 1 & -1 & 1 \\ 1 & 1 & 1 & 1 \\ \frac{W+L}{2} & -\frac{W+L}{2} & -\frac{W+L}{2} & \frac{W+L}{2} \end{bmatrix} \begin{bmatrix} \dot{\theta}_1 \\ \dot{\theta}_2 \\ \dot{\theta}_3 \\ \dot{\theta}_4 \end{bmatrix} \tag{12}$$

Where $[\dot{\theta}_1, \dot{\theta}_2, \dot{\theta}_3, \dot{\theta}_4]$T is the angular velocity of each wheel.
In the control of the robot, the signal could be sent in as the coordinate$[\dot{x}_r, \dot{y}_r, \dot{\alpha}_r]$.

3. Experiment

In figure 4, detection and tracking results are shown. During the tracking, the algorithm is able to track the target when the target is occluded and lost for a short time. When the target is partially occluded and fully occluded, the target box is still located on the target path. With relatively long tracking period with complex interference, the target could still be tracked.

Figure 4. Test result in Dataset BAHNHOF[13]

In the mobile robot system used in the experiment is shown in figure 5 which consists by an omnidirectional mobile robot, a Kinect camera system and a laptop. The camera system is already made into wireless power supply. The experiment is run in MATLAB 2018a.

Figure 5. This is a figure with setup of the system which includes a RGB-D camera

In figure 6, few pictures from the experiment is shown.

Figure 6. Test result in the experimental scene

4. Conclusion

In the work, we use a relatively simple system to achieve certain human tracking algorithm for a mobile robot in 3D. To improve efficiency and accuracy of the system, a modified KCF is applied as the target tracker and a EKF is also used to update information from different dimensional. The system is verified in a real system and achieves functions as detecting, tracking the target and control the mobile robot following the target. However, when the target is moving too fast with changing speed the system is not perfectly worked for the mobile robot following part. We will keep put our effect on it in our future work.

Acknowledgement

The authors would like to thank the financial support from the Natural Sciences and Engineering Research Council of Canada (NSERC). The authors gratefully acknowledge the financial support from Kaneff Research Chairs program.

IOP Conf. Series: Materials Science and Engineering **491** (2019) 012004 doi:10.1088/1757-899X/491/1/012004

References

[1] Hirose N, Tajima R, and Sukigara K, "Personal robot assisting transportation to support active human life—Human-following method based on model predictive control for adjacency without collision," in Proc. IEEE Int. Conf. Mechatronics (ICM), Nagoya, Japan, 2015, pp. 76–81.

[2] Gupta M, Kumar S, Behera L, and Subramanian V. K., "A Novel Vision-Based Tracking Algorithm for a Human-Following Mobile Robot," IEEE Trans. on Systems, Man, and Cybernetics: Systems, vol. 47, no. 7, 2017, pp. 1415-1427.

[3] Bellotto N and Hu H, "Multisensor-based human detection and tracking for mobile service robots," IEEE Trans. Systems, Man, and Cybernetics, B, Cybernetics., vol. 39, no. 1, 2009, pp. 167–181.

[4] Xiao Q, Sun F, Ge R and et al., "Human Tracking and Following of Mobile Robot with a Laser Scanner," 2017 2nd Int. Conf. on Advanced Robotics and Mechatronics (ICARM), Anhui, China, 2017, pp. 675-680

[5] Hoshino F and Morioka K, "Human Following Robot based on Control of Particle Distribution with Integrated Range Sensors," IEEE/SICE Int. Symposium on System Integration, Kyoto, Japan, 2011, pp. 212-217

[6] Yoshimi T, Nishiyama M, Sonoura T, et al., "Development of a Person Following Robot with Vision Based Target Detection," IEEE/RSJ International Conference on Intelligent Robots and Systems, Beijing, China, 2006, pp. 5286-5291.

[7] Henriques J. F., Caseiro R, Martins P, and et al., "High-speed tracking with kernelized correlation filters," IEEE Trans. on Pattern Analysis and Machine Intelligence, vol. 37, no. 3, 2015, pp. 583–596.

[8] S. Liu, T. Zhang, X. Cao, and C. Xu, "Structural correlation filter for robust visual tracking," in Proc. IEEE Conf. Computer Vision Pattern Recognition (CVPR), Les vegas, United States, 2016, pp. 4312–4320.

[9] T. Liu, G. Wang, and Q. Yang, "Real-time part-based visual tracking via adaptive correlation filters," in Proc. IEEE Conf. Computer Vision Pattern Recognition (CVPR), Boston, United States, 2015, pp. 4902–4912.

[10] Wang M, Liu Y, Su D and et al., "Accurate and Real-Time 3-D Tracking for the Following Robots by Fusing Vision and Ultrasonar Information," IEEE/ASME Trans. On Mechatronics, vol. 23, no. 3, 2018, pp. 997-1006

[11] Wang X, O'Brien M, Xiang C and et al., "Real-time visual tracking via robust Kernelized Correlation Filter," IEEE Int. Conf. on Robotics and Automation (ICRA), Singapore, 2017, pp. 4443-4448

[12] Taheri H, Qiao B and Ghaeminezhad N, "Kinematic Model of a Four Mecanum Wheeled Mobile," International Journal of Computer Applications, vol. 113, no.3, 2015, pp. 6-9

[13] Ess a, Leibe B and Gool L. V., "Depth and Appearance for Mobile Scene Analysis," IEEE 11th International Conference on Computer Vision, Rio de Janeiro, Brazil, 2007

IOP Conf. Series: Materials Science and Engineering **491** (2019) 012005 doi:10.1088/1757-899X/491/1/012005

Improvement of Modeling Consistency for Vehicle Dynamic Analysis Using Design of Experiment

Doyeong Kim[1,2], Chulwoo Moon[1,*], Changhyun Jeong[1] and Dohyun Jung[1]

[1]Korea Automotive Technology Institute, 31214 Chungnam, Republic of Korea
[2]KoreaTech, 31253 Chungnam, Republic of Korea

[*]Corresponding author: cwmoon@katech.re.kr

Abstract. In the present study, a development method for high-accuracy dynamics models using Design of Experiments (DOE) was proposed based on experimental results obtained in full-vehicle dynamic testing. In vehicle dynamics modeling, the design parameters of each model need to be experimentally obtained by measuring their characteristic values or referring to detailed information regarding those models so that these design parameters can be reflected in the modeling process; however, in practice, it is difficult to gain access to detailed information for most design parameters. To analyze the effects on vehicles of design parameters for which relevant data have not been acquired, DOE tools available in Minitab, a statistical analysis program, were employed. As a result, it was possible to identify major design parameters that affected the behavior of vehicle models, and a method to improve the consistency of vehicle models by preferentially modifying those major design parameters was proposed.

1. Introduction

In attempts to improve automotive fuel efficiency, global car manufacturers are continuing to develop high-efficiency powertrain systems, including AMT and DCT; at the same time, the scope of applications is now extending to not only passenger vehicles but also commercial vehicles. Most 3.5-ton small- and medium-sized commercial vehicles on the road in Korea are fitted with manual transmissions and are currently making a gradual shift to automatic transmissions.

Developing a new powertrain system requires pretesting that occurs when the system is equipped into vehicles. Considering safety issues and costs, however, vehicle dynamics simulations are widely carried out on developed products as pre-tests to improve their integrity. Again, high-accuracy system and vehicle modeling are required as prerequisites for those vehicle dynamics simulations.

In general, vehicle dynamics simulations are performed based on characteristics test results of essential auto parts, such as dampers, springs, and bushing. Commercial vehicles, however, are equipped with parts that are more robustly designed compared to those used in passenger vehicles, and their relevant characteristics data, including Center of Gravity (COG) positions and tire property data, are not as general as those of passenger vehicles. As a result, high-accuracy modeling of commercial vehicles is more challenging overall.

In the present study, a method of using Design of Experiments (DOE) to efficiently implement vehicle modeling was proposed based on test results of full-vehicle dynamic testing. In Section 2, details on full-vehicle testing to assess vehicle dynamics and existing modeling methods are described. In Section 3, a method of using DOE to improve the consistency of vehicle modeling is proposed.

Finally, details about vehicle models with improved consistency as a result of the proposed method are described, and major findings of the present study are specified in the Conclusion.

2. Vehicle Modeling for Vehicle Dynamics Analysis

2.1 Full-vehicle Testing to Developing Vehicle Models

Full-vehicle dynamic testing of a target vehicle was conducted to develop corresponding vehicle models. Here, a 3.5-ton commercial vehicle manufactured by the H motor company was selected as the target vehicle, and special experimental environments for full-vehicle tests were created. In the meantime, the vehicle was fitted with various instrumentation devices, including a telemetry system to measure the torque of the propeller shaft, a Global Positioning System/Inertial Navigation System (GPS/INS) to estimate the dynamic behavior of the vehicle, and a data acquisition system to store measurement data. Experimental environments were configured as shown in Fig. 1.

Figure 1. Test Vehicle Setting

A total of ten full-vehicle dynamic tests were performed, including five lateral dynamics tests and five longitudinal dynamics tests. Considering the special operating conditions of commercial vehicles, the same test was conducted twice in succession: once without live load and the other under live load[1]. By doing so, the effect of live load on the dynamic characteristics of the vehicle was determined and analyzed. Testing scenarios are shown in Table 1 below.

Table 1. Vehicle Dynamics Test Scenarios

Test Classification	Test Scenario	reference
Longitudinal Dynamics Test	Acceleration Test – APS 50%, 100%	KATECH
	Deceleration Test – DCC 0.3g, 0.5g	KATECH
	Coast-down Characteristic Test	KATECH
	Hill Climbing Characteristic Test	KATECH
	Downhill Characteristic Test	KATECH
Lateral Dynamics Test	Double Lane Change Test	ISO-3888
	Slalom Test	KATECH
	Steady-State Cornering Test	ISO-4138
	Continuous Sinusoidal steering test	KATECH
	J-turn test	ISO-7401

2.2 Vehicle Modeling

The development process of vehicle modeling in the present study is shown in Fig. 2. Trucksim, a commercial modeling tool made by Mechanical Simulation, was employed for commercial-vehicle modeling, and data acquired from characteristics testing of essential auto parts were used in the modeling. Subsequently, the consistency improvement process for the developed lateral-direction vehicle model was carried out during 107 iterations of trial and error by varying each of the design parameters based on experience and know-how accumulated during previous vehicle modeling studies. Determination coefficients used for verification of modeling consistency are shown in Eq. (1).

$$\rho(A,B) = \frac{1}{N-1}\sum_{i=1}^{N}\overline{(\frac{A_i-\mu A}{\sigma A})}(\frac{B_i-\mu B}{\sigma B}) \tag{1}$$

μA and σA are the average and standard deviation, respectively, of A values.

μB and σB are the average and standard deviation, respectively, of B values.

The lateral-direction consistency improvement process was conducted based on the results of double-lane change testing; as a result, a vehicle model was developed to provide lateral-correlation properties, as follows: Lateral-Acceleration at 96.40% and Yaw-Rate at 99.70%. The result set of vehicle consistency were configured as shown in Fig. 3.[2]

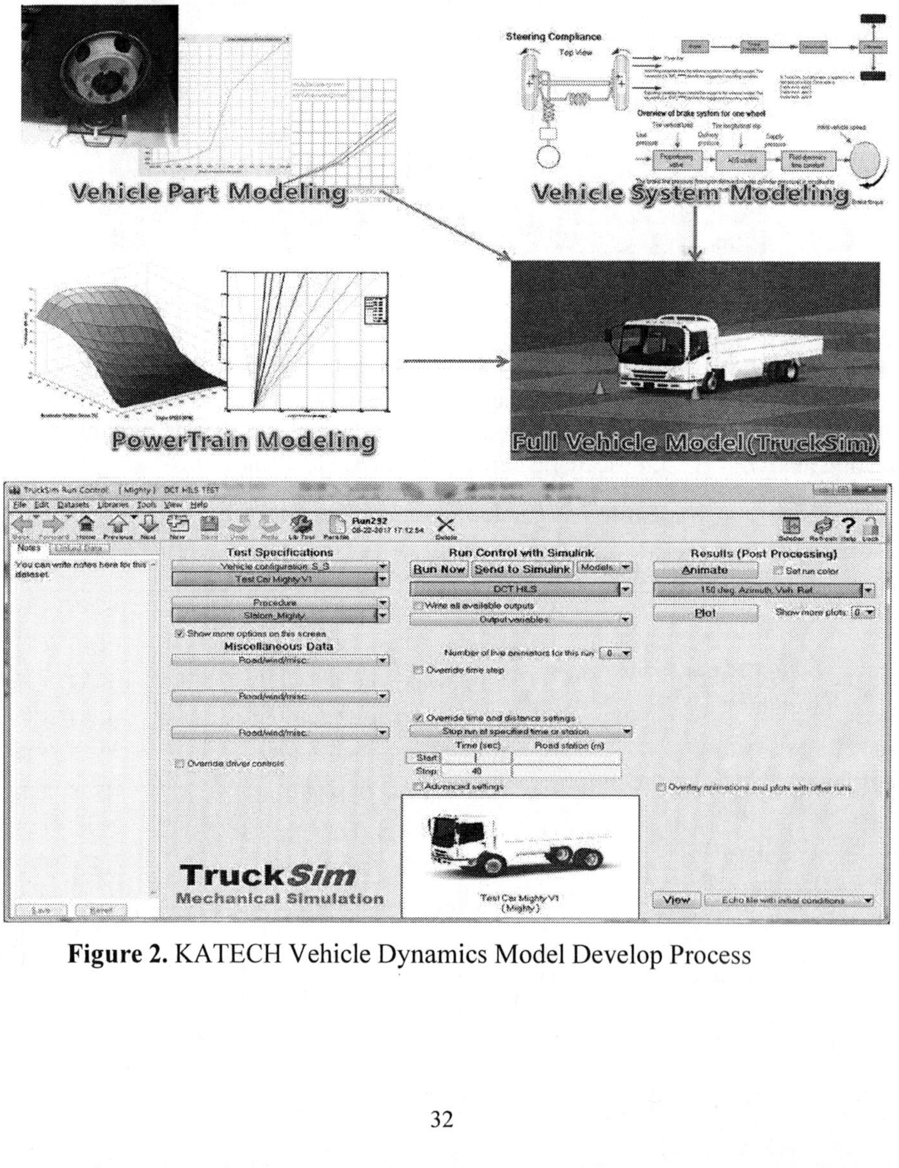

Figure 2. KATECH Vehicle Dynamics Model Develop Process

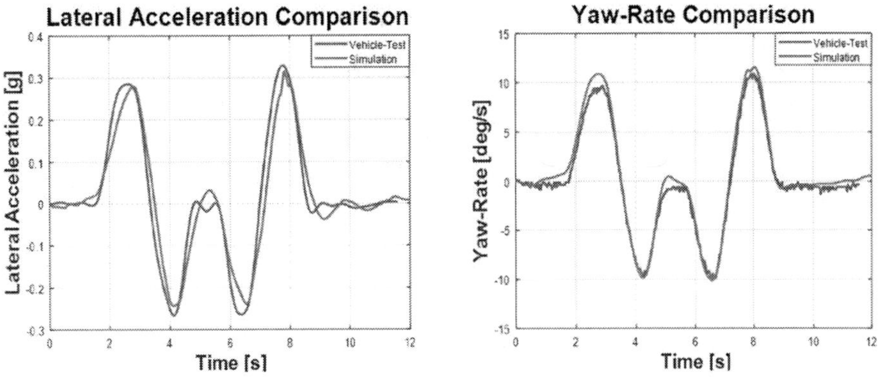

Figure 3. Vehicle Test and Simulation Result comparison of Lateral Acceleration & Yaw-Rate

3. Improvement of Vehicle Model Consistency Using DOE

3.1 Parameter Selection for Simulation

Among various factors other than spring constant and damper characteristics curves, which were acquired from the essential auto part testing, eight factors were selected as design parameters for the simulation, including the COG height, COG position with respect to front tires (weight ratio between the front and rear wheels), front tire stiffness, rear tire stiffness, stabilizer constant, camber, toe value, and caster.[3] To analyze the effects of adjusted combinations of the eight parameters on the lateral-direction behavior of the vehicle, DOE tools available in Minitab, a statistical analysis program, were employed[4]. A design parameter table for DOE was developed, as shown below, to ensure that each arrangement set the resolution to a value of 4, so that only the combination of primary effects of the design parameters without interaction components could be reflected and analysed[5]. Also, another 16 simulations were run based on a dataset that was generated by adding weighted values of +1 and -1 to the basic dataset of the vehicle model developed in 2.2.

Table 2. Design Parameter Combination for DOE.

Run Order	C.O.G. Height	FR : RR Weight Ratio	Front Tire Stiffness	Rear Tire Stiffness	Stabilizer Constant	camber	Toe	Caster
1	800	1700	1.1	0.9	600	1	-1	1
2	800	1300	0.9	1.1	800	1	-1	1
3	840	1700	0.9	0.9	800	1	-1	-1
4	800	1300	0.9	0.9	600	-1	-1	-1
5	840	1700	1.1	1.1	800	1	1	1
6	800	1300	1.1	0.9	800	1	1	-1
7	800	1700	1.1	1.1	800	-1	-1	-1
8	840	1300	1.1	0.9	800	-1	-1	1
9	840	1300	0.9	1.1	800	-1	1	-1
10	800	1300	1.1	1.1	600	-1	1	1
11	840	1300	1.1	1.1	600	1	-1	-1
12	840	1700	0.9	1.1	600	-1	-1	1
13	800	1700	0.9	0.9	800	-1	1	1
14	840	1300	0.9	0.9	600	1	1	1
15	800	1700	0.9	1.1	600	1	1	-1
16	840	1700	1.1	0.9	600	-1	1	-1

IOP Conf. Series: Materials Science and Engineering **491** (2019) 012005 doi:10.1088/1757-899X/491/1/012005

3.2 Analysis of Primary Effects Based on Simulation Results

Based on the results of the 16 simulation runs, as shown in Table 2, the effect of each design parameter on the behavior of the vehicle was analyzed. Here, the DOE tools of Minitab were used, while the response was set to be the lateral correlation of the vehicle's Lateral-Acceleration. As a result, the strongest determining factor for the vehicle's Lateral-Acceleration was found to be the COG position with respect to the front tires (weight ratio between the front and rear wheels), followed by the COG height and the stabilizer constant. The Yaw-Rate was determined to be most significantly affected by the COG position with respect to the front tires (weight ratio between the front and rear wheels), followed by the toe values. The standardized effect for Lateral-Acceleration and Yaw-rate were configured as shown in Fig. 4 and 5.

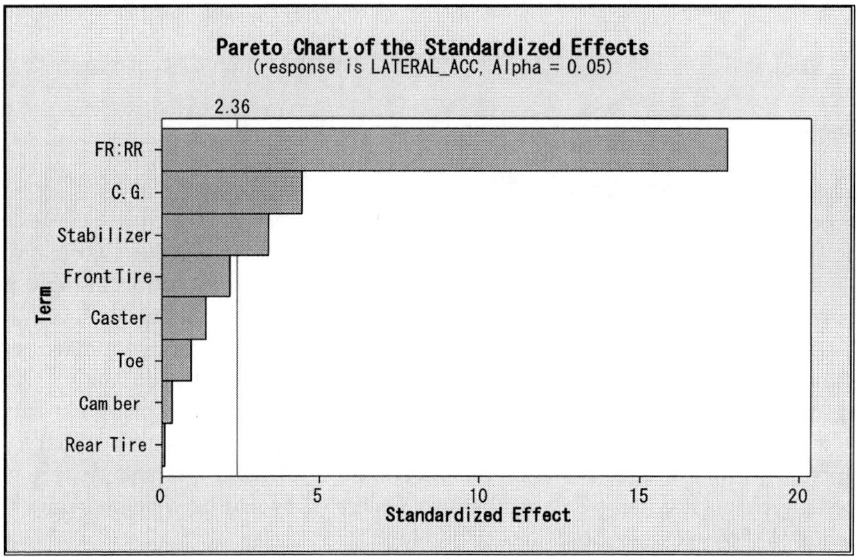

Figure 4. Pareto Chart of the Standardized Effect : Lateral Acceleration

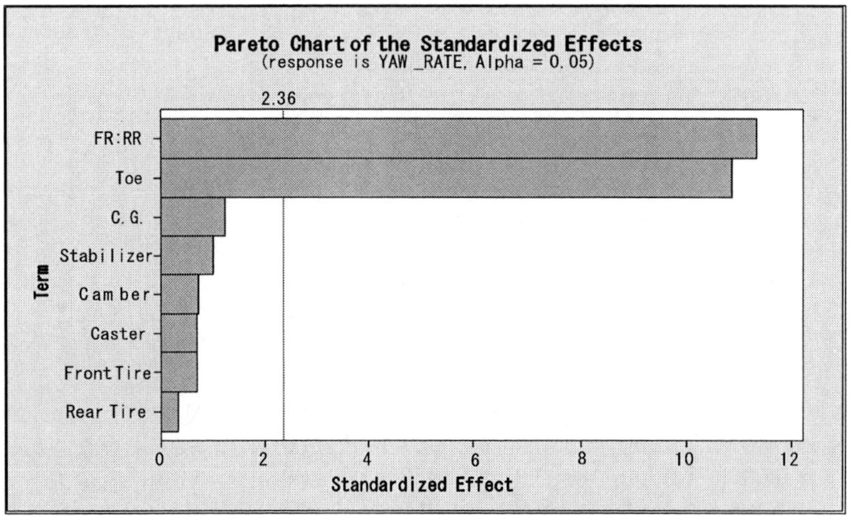

Figure 5. Pareto Chart of the Standardized Effect : Yaw-Rate

3.3 Improvement of Simulation Consistency by Changing Primary Effects

As can be seen in the 16 simulation-run results shown in Table 2, Run Order 6 had the highest consistency in both Lateral-Acceleration and Yaw-rate. All Run Orders recorded over 99% consistency in the Yaw-Rate, and thus additional simulations were conducted by setting as a design parameter the weight ratio between the front and rear wheels, the primary determining factor of Lateral-Acceleration. The results are shown in Table 3.

Table 3. Results of Additional Simulations

FR:RR Weight Ratio	1100	1200	1300	1400	1500
Lateral-Acceleration	97.54	97.38	97.17	96.98	96.74
Yaw-Rate	99.75	99.77	99.76	99.75	99.73

Additional simulation results showed that the smaller the weight ratio became, the higher the consistency was in the Lateral-Acceleration. The consistency in the Yaw-Rate, however, was the highest when the COG position with respect to the front tires was set to 1200. Therefore, the optimized design parameter for the vehicle was determined to be a COG position with respect to the front tires of 1200.

4. Conclusions

In the present study, eight determining parameters to improve the consistency of vehicle models were selected, including COG height, COG position with respect to front tires (weight ratio between front and rear wheels), front tire stiffness, rear tire stiffness, stabilizer constant, camber, toe value, and caster. Using these parameters, a Design of Experiments (DOE) method was employed to determine the optimized modeling consistency conditions, as follows:

1) In the present study, most determining factors for the consistency of vehicle modeling, which are difficult to experimentally characterize, were confirmed. By doing so, the effects of these parameters on the lateral-direction characteristics of the test vehicle were demonstrated and analyzed; as a result, the strongest determining factor was confirmed.

2) A consistency improvement process was performed by adjusting the design parameter that had been determined to most significantly affect the modeling consistency, leading to improvement as follows: consistency in the Lateral-Acceleration improved from 96.40 to 97.38% and consistency in the Yaw-Rate improved from 99.70 to 99.77%. Improvement data results were configured as shown in Fig. 6.

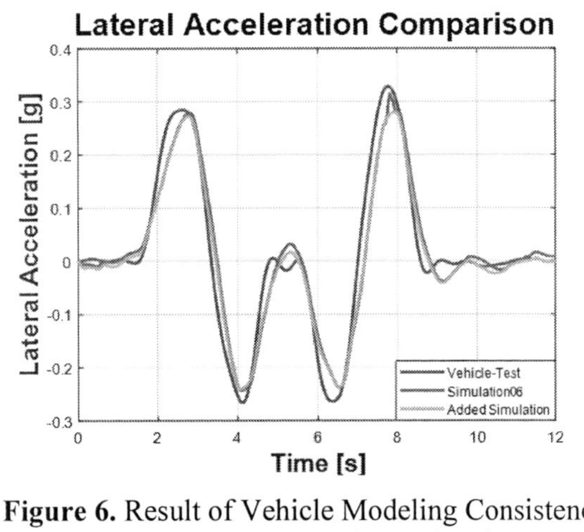

Figure 6. Result of Vehicle Modeling Consistency Improvement : Lateral-Acceleration

3) Through these procedures, a vehicle model with improved modeling consistency was developed. Also, it was confirmed that the application of this DOE based method significantly reduced the time needed to implement consistency-improving procedures in vehicle models.

Reference

[1] Gyoojae Choi, Kihan Noh and Yungmyun Yoo 2001 Real-time Dynamic Simulation Using Multibody Vehicle Model *Transactions of the Korean Society of Mechanical Engineer.* **A 25(3)**, 2001.3, 486-494 (9pages)
[2] Jungho Kim, Taeuk Kim and Youngsuk Kim 2012 A Study of Multi-Disciplinary Optimization for Vehicle Simulation *KASE 2012 Autumn Conference*. **2012.5**, 965-971 (7 pages)
[3] Sung-hyun Kim, Yong-sik Bang, Dae-Oh Kang, Yong-Hoon Lee, Seung-Jin Heo and Hong-Jae Yim 2010 Robust Design Optimization of Vehicle Body Structure Considering Multi-disciplinary *KASE 2010 Autumn Conference*. **2010.11**, 2511-2516 (6 pages),
[4] Ostle 2000 Engineering Statistics : the industrial experience (SciTechmedia)
[5] Wan Suk Yoo, Sang Do Na and Kwang Suk Kim 2012 Real-Time Dynamic Analysis of Vehicle with Experimental Vehicle Model *Transactions of the Korean Society of Mechanical Engineers*. **A 36(9)**, 2012.9, 1003-1008 (6 pages)

IOP Conf. Series: Materials Science and Engineering **491** (2019) 012006 doi:10.1088/1757-899X/491/1/012006

Novel Design of a 3-RRUU 6-DOF Parallel Manipulator

Zhongxing Yang[1,a] **and Dan Zhang**[1,b,*]

[1]Department of Mechanical Engineering, Lassonde School of Engineering, York University, Toronto, Canada

E-mail addresses: [a]zxyang@yorku.ca; [b,*]dan.zhang@lassonde.yorku.ca

Abstract. A novel 3-RRUU 6-DOF parallel manipulator is designed based on the multi-plane design principle, that provides larger workspace, less chance of link interference and simplified kinematic analysis. The link interference detection by pentagon identification and cross-product is demonstrated. Workspace is optimized by different orientations. The Jacobian matrix by global cartesian coordinate system is developed. A transformation matrix could convert that to Jacobian matrix of a new coordinate system.

1. Introduction

A parallel manipulator has a moving platform that is actuated and connected to the base by more than one limb. This structure has higher stiffness and precision but smaller workspace [1].

Six limbed 6-DOF parallel manipulators, such as 6-UPS [2] and 6-RUS [3], are common designs. There are some topology methods [4] [5] [6] for chain design of unique applications, that include three limbed 6-DOF manipulators with serial chain for each limb [7] [8] [9] [10]; three limbed 6-DOF manipulators with parallel actuation in each limb [11] [12] [13] [14]. Compared to the six limbed counterparts, three limbed 6-DOF manipulators usually have larger workspace and less link interference [7] [8] [9] [10] [13]. Although the chance of interference is lower, some three limbed 6-DOF manipulators [12] [14] still need to detect such conditions, generally by determining the link to link distance. The R̲HHR̲ chain in [13] is free from interference, while the spherical joints connected to the moving platform might limit the motion range (e.g. the rotation around z axis).

This paper discusses the multi-plane design principle. Different from [11] where the links connected to the platform have their motions restricted in the planes perpendicular to the platform plane, this paper suggests deploying these links in three intersecting planes that coincide at a line which is perpendicular to the horizontal base plane and passes platform equilateral triangle center. The links are designed similarly as to the PRU link in HALF and HANA structure [15], this avoids using spherical joints with limited motion range [16]. The five-bar robot [17] with large workspace is employed for the limb movement. The actuating five-bar links are on the horizontal base plane. This manipulator also has the lie-flat function [18] [19] that protects it from outdoor weather damages.

2. Mechanical Design

The 3-RRUU parallel manipulator has 6 DOFs. Figure 1 shows (a) the structure and (b) the kinematic chains represented by topology diagram [20]. A structure could also be expressed by a notation [14]. Therefore the structure is denoted as 3-[(2-R̲R)UU], where underlined joints are actuated. Each limb has two motors $A_{[Api]}$ and $A_{[Aqi]}$ in hybrid chain. This increases the precision and reduces the bending of link $B_{[pi]}$ and $B_{[qi]}$. The revolute joints $R_{[Api]}$ and $R_{[Aqi]}$ are driven. The revolute joint $R_{[Ci]}$

connects to the universal joint $U_{[Di]}$ which consists two revolute joints and one of these is colinear to $R_{[Ci]}$.

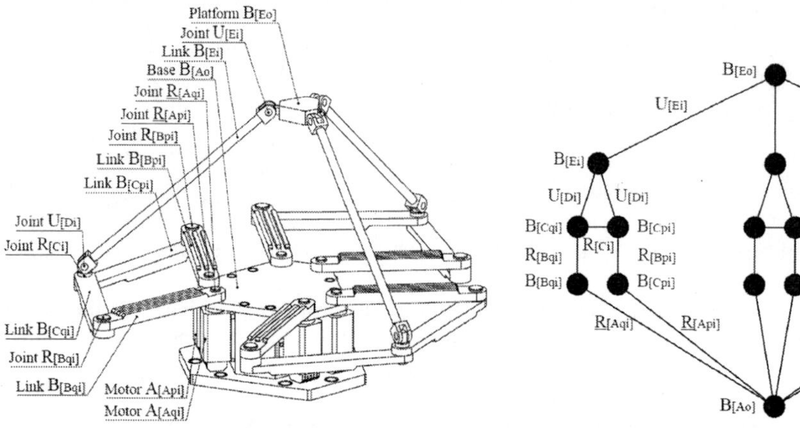

(a) (b)

Figure 1. Mechanical design.

3. Inverse Kinematics

This chapter gives the calculation of link positions, and detection of link interference.

3.1. Link Positions

The global cartesian coordinate is at base center $\boldsymbol{A_o}$: $[0\ 0\ 0]^T$ (figure 2. (a)).

$$\boldsymbol{X_o} = [1\ 0\ 0]^T, \boldsymbol{Y_o} = [0\ 1\ 0]^T, \boldsymbol{Z_o} = [0\ 0\ 1]^T \tag{1}$$

$$\boldsymbol{A_{pi}} = r_b \cdot [\cos(\alpha_i - \beta)\ \ \sin(\alpha_i - \beta)\ \ 0]^T \tag{2.a}$$

$$\boldsymbol{A_{qi}} = r_b \cdot [\cos(\alpha_i + \beta)\ \ \sin(\alpha_i + \beta)\ \ 0]^T \tag{2.b}$$

The pose of the moving platform (an equilateral triangle) is defined by its orientation $\boldsymbol{\theta_e}$ and position $\boldsymbol{E_o}$ from global coordinate.

$$\boldsymbol{\theta_e} = [\theta_x\ \ \theta_y\ \ \theta_z]^T \text{ and } \boldsymbol{E_o} = [x_e\ \ y_e\ \ z_e]^T \tag{3}$$

Let $\theta_e = \sqrt{\theta_x{}^2 + \theta_y{}^2 + \theta_z{}^2}$, so that $u_x = \theta_x/\theta_e$, $u_y = \theta_y/\theta_e$, $u_z = \theta_z/\theta_e$.

With orientation matrix R, the positions $\boldsymbol{E_i}$ are calculated, where the universal joints are located with one rotational axis along symmetry lines $\boldsymbol{E_iE_o}$ and the other rotational axis along $\boldsymbol{Y_{di}}$. The platform has a normal vector $\boldsymbol{N_e}$ that points upward.

$$R = \begin{bmatrix} \mathrm{c}\,\theta_e + u_x{}^2(1-\mathrm{c}\,\theta_e) & u_xu_y(1-\mathrm{c}\,\theta_e) - u_z\,\mathrm{s}\,\theta_e & u_xu_z(1-\mathrm{c}\,\theta_e) + u_y\,\mathrm{s}\,\theta_e \\ u_xu_y(1-\mathrm{c}\,\theta_e) + u_z\,\mathrm{s}\,\theta_e & \mathrm{c}\,\theta_e + u_y{}^2(1-\mathrm{c}\,\theta_e) & u_yu_z(1-\mathrm{c}\,\theta_e) - u_x\,\mathrm{s}\,\theta_e \\ u_xu_z(1-\mathrm{c}\,\theta_e) - u_y\,\mathrm{s}\,\theta_e & u_yu_z(1-\mathrm{c}\,\theta_e) + u_x\,\mathrm{s}\,\theta_e & \mathrm{c}\,\theta_e + u_z{}^2(1-\mathrm{c}\,\theta_e) \end{bmatrix} \tag{4}$$

$$\boldsymbol{E_i} = R \cdot r_e \cdot [\cos(\alpha_i)\ \ \sin(\alpha_i)\ \ 0]^T + \boldsymbol{E_o} \tag{5}$$

$$\boldsymbol{N_e} = \frac{\boldsymbol{E_oE_1}\times\boldsymbol{E_oE_2}}{|\boldsymbol{E_oE_1}\times\boldsymbol{E_oE_2}|} \tag{6}$$

$$\boldsymbol{Y_{di}} = \frac{\boldsymbol{Z_o}\times\boldsymbol{E_oE_i}}{|\boldsymbol{Z_o}\times\boldsymbol{E_oE_i}|} \text{ and } \boldsymbol{X_{di}} = \boldsymbol{Y_{di}} \times \boldsymbol{Z_o} \tag{7}$$

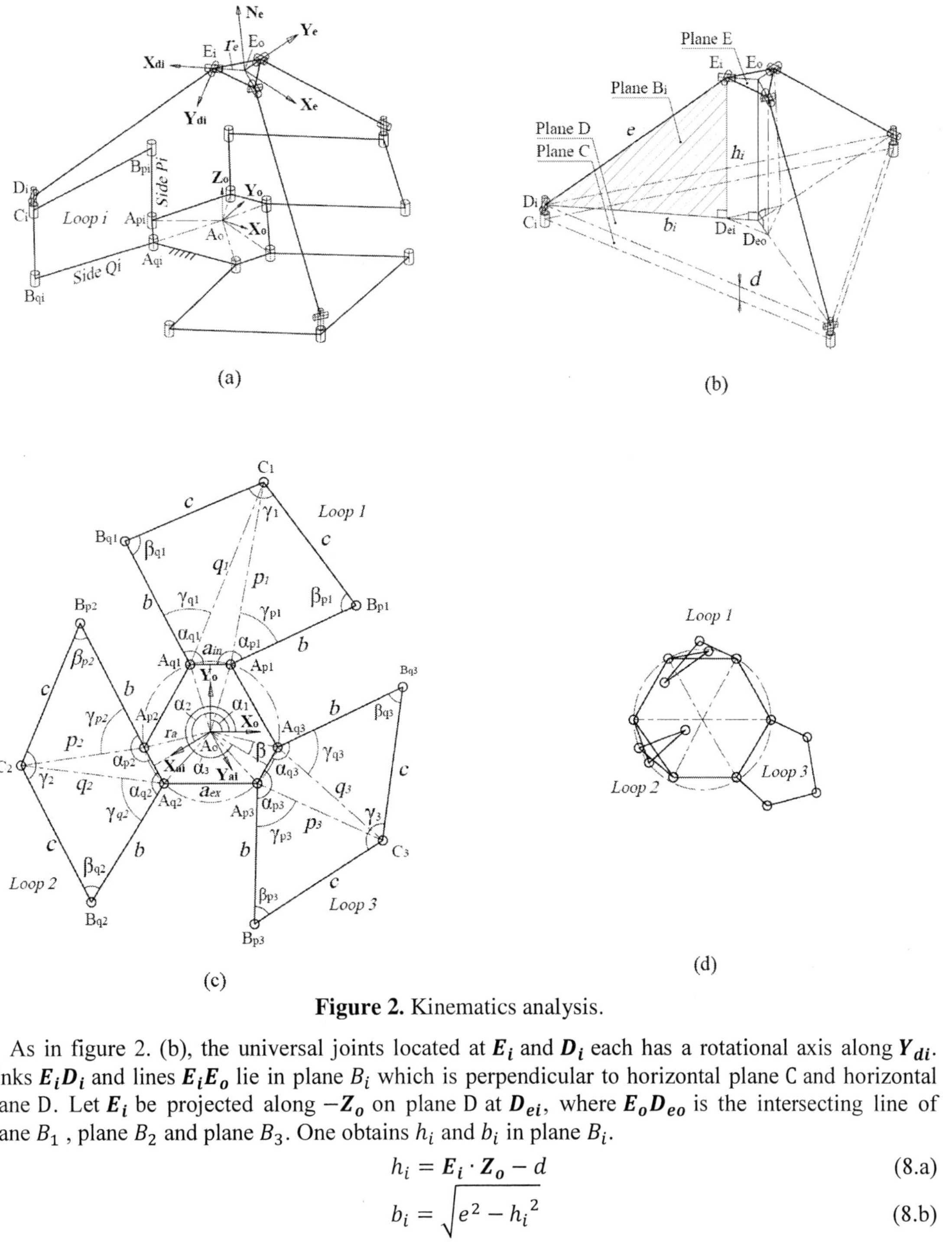

Figure 2. Kinematics analysis.

As in figure 2. (b), the universal joints located at $\boldsymbol{E_i}$ and $\boldsymbol{D_i}$ each has a rotational axis along $\boldsymbol{Y_{di}}$. Links $\boldsymbol{E_i D_i}$ and lines $\boldsymbol{E_i E_o}$ lie in plane B_i which is perpendicular to horizontal plane C and horizontal plane D. Let $\boldsymbol{E_i}$ be projected along $-\boldsymbol{Z_o}$ on plane D at $\boldsymbol{D_{ei}}$, where $\boldsymbol{E_o D_{eo}}$ is the intersecting line of plane B_1 , plane B_2 and plane B_3. One obtains h_i and b_i in plane B_i.

$$h_i = \boldsymbol{E_i} \cdot \boldsymbol{Z_o} - d \tag{8.a}$$

$$b_i = \sqrt{e^2 - {h_i}^2} \tag{8.b}$$

Position $\boldsymbol{D_i}$ indicates where the universal joint is located. The universal joint has one rotational axis along $\boldsymbol{Z_o}$ and the other rotational axis along $\boldsymbol{Y_{di}}$. As in figure 2. (c), let the five-bar mechanisms be in the plane C where the limbs are actuated. The position $\boldsymbol{C_i}$ indicates the location of revolute joint. The revolute joints rotate around $\boldsymbol{Z_o}$, and colinear with the universal joints at $\boldsymbol{D_i}$.

$$\boldsymbol{D_i} = \boldsymbol{E_i} + b_i\boldsymbol{X_{di}} - h_i\boldsymbol{Z_o} \tag{9}$$
$$\boldsymbol{C_i} = \boldsymbol{D_i} - d\boldsymbol{Z_o} \tag{10}$$

The $\boldsymbol{B_{pi}}$ and $\boldsymbol{B_{qi}}$ represent the locations of two revolute joints, rotational axis along $\boldsymbol{Z_o}$.

$$p_i = |\boldsymbol{A_{pi}C_i}| \text{ and } q_i = |\boldsymbol{A_{qi}C_i}| \tag{11.a}$$
$$\gamma_{pi} = \cos^{-1}\left(\frac{b^2+p_i{}^2-c^2}{2bp_i}\right) \text{ and } \gamma_{qi} = \cos^{-1}\left(\frac{b^2+q_i{}^2-c^2}{2bq_i}\right) \tag{11.b}$$
$$R_{pi} = \begin{bmatrix} \cos\gamma_{pi} & \sin\gamma_{pi} & 0 \\ -\sin\gamma_{pi} & \cos\gamma_{pi} & 0 \\ 0 & 0 & 1 \end{bmatrix} \text{ and } R_{qi} = \begin{bmatrix} \cos\gamma_{qi} & -\sin\gamma_{qi} & 0 \\ \sin\gamma_{qi} & \cos\gamma_{qi} & 0 \\ 0 & 0 & 1 \end{bmatrix} \tag{11.c}$$
$$\boldsymbol{B_{pi}} = R_{pi} \cdot \boldsymbol{A_{pi}C_i} \cdot \frac{b}{p_i} + \boldsymbol{A_{pi}} \text{ and } \boldsymbol{B_{qi}} = R_{qi} \cdot \boldsymbol{A_{qi}C_i} \cdot \frac{b}{q_i} + \boldsymbol{A_{qi}} \tag{11.d}$$

One could calculate actuation angles α_{pi} and α_{qi} by $\boldsymbol{A_{pi}B_{pi}}$ or $\boldsymbol{A_{qi}B_{qi}}$.

3.2. Link Interference

There are some literatures [12] [14] [21] [22] [23] discussing the link interference detection, where the main algorithm is to keep safe distance between links. One needs to check that three five-bar mechanisms have no link interference. The interference detection in this analysis has three sections.

Section 1 is to limit α_{pi} and α_{qi} ranges so that the moving sides of the link will not collide with the base arc of a neighboring link. Such as in figure 3. (a), the side of $\boldsymbol{A_{q1}B_{q1}}$ interferes with the arc of $\boldsymbol{A_{p1}}$. Let $a_{in} = 2r_a \sin\beta$ and $a_{ex} = 2r_a \sin\left(\frac{\pi}{3} - \beta\right)$ that are the lengths of hexagon sides. One needs to verify that $a_{in} > 2r_m$ and $a_{ex} > 2r_m$ so that it has enough distance for two motors with radius r_m.

$$a_{min} = \sqrt{b^2 + 4r_b{}^2} \text{ and } a_{max} = b + 2r_b \tag{12.a}$$
$$\gamma_{in} = g(a_{in}) \text{ and } \gamma_{ex} = g(a_{ex}) \tag{12.b}$$

Table 1. Angles offsets.

Conditions	$g(x)$
$x \in (0, a_{min}]$	$\sin^{-1}\left(\frac{2r_b}{a_x}\right)$
$x \in [a_{min}, a_{max}]$	$\cos^{-1}\left(\frac{b^2 + a_x{}^2 - 4r_b{}^2}{2 \cdot a_x \cdot b}\right)$
$x \in [a_{max}, \infty)$	0

Table 1 gives γ_{in} and γ_{ex} as offsets on α_{pi} and α_{qi}, to be interference free for section 1.

$$\alpha_{pi} \; and \; \alpha_{qi} \in \left[\gamma_{in}, \frac{240}{180}\pi - \gamma_{ex}\right] \tag{12.c}$$

Section 2 is to wipe out the conditions where α_{pi} and α_{qi} are within ranges but there exists intersection of links in a loop (see loop 1 and loop 2 in figure 2. (d)). An algorithm of pentagon identification could easily find intersection. For concave or convex pentagons, the sum θ_i should be 3π to be interference free for section 2.

$$\theta_i = \alpha_{pi} + \beta_{qi} + \alpha_{pi} + \beta_{qi} + \gamma_i \tag{13}$$

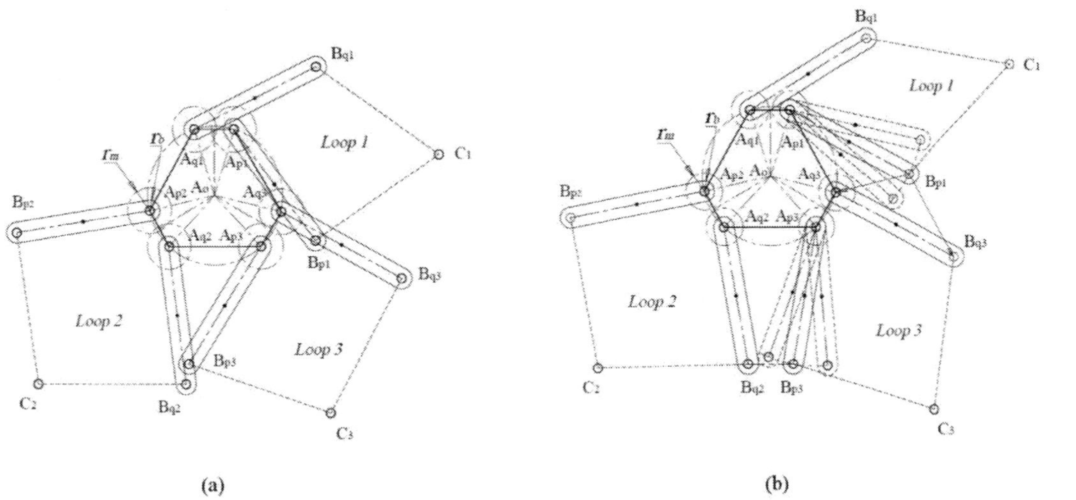

Figure 3. Interference detection.

Section 3 is to avoid the moving arc of a link interfering with the moving side of a link in another loop. Such as in figure 3. (a), the arc of $\boldsymbol{B_{p1}}$ interferes with the side of $\boldsymbol{A_{q3}B_{q3}}$. For when $\alpha_{pi} + \alpha_{qk} > 5\pi/3$ (i and k index two loops in clockwise order), the cross-product method could conveniently calculate the distance between the two. The link $\boldsymbol{B_{pi}C_i}$ or $\boldsymbol{B_{qi}C_i}$ are in different levels, thus are exempted form examination. The criteria in table 2 need to be met to be interference free for section 3.

Table 2. Cross product criteria.

Conditions	Criteria
$\alpha_{pi} > \alpha_{qk}$	$\boldsymbol{Z_o} \cdot (\boldsymbol{B_{pi}A_{qk}} \times \boldsymbol{B_{pi}B_{qk}}) - 2r_b b > 0$
$\alpha_{pi} < \alpha_{qk}$	$\boldsymbol{Z_o} \cdot (\boldsymbol{B_{qk}B_{pi}} \times \boldsymbol{B_{qk}A_{pi}}) - 2r_b b > 0$

4. Workspace Optimization

The workspace is evaluated by setting three orientations $\boldsymbol{\theta_{ej}}$ ($j = 1,2,3$) of the platform, and then searching for all eligible positions of the $\boldsymbol{E_o}$ within the searching ranges. The number n_{ej} of the eligible positions indicate the volume of the workspace by orientation $\boldsymbol{\theta_{ej}}$.

Considering the efficiency of workspace (largest workspace made by minimum limb sizes), cost function f_j is taken. Pareto method [24] is used to search for the best result of each cost function with non-sacrifice to another cost function.

$$f_j = -\frac{n_{ej}}{\left(\sum_{i=1}^{5} x_i\right)^3} \tag{14}$$

The basic modeling parameters are given in table 3. Table 4 gives the ranges of the optimization variables x_1~x_6, considering that each revolute or universal joint is at least 0.05 m from the center of a body it connects to. Table 5 gives the three orientations, and the $\boldsymbol{E_o}$ search ranges by each orientation.

Table 6 shows the results of the optimization and the cost function value, where $F = f_1 + f_2 + f_3$. The four results are selected from the final optimization solutions, with result 1 for the best of f_1; result 2 for the best of f_2; result 3 for the best of f_3; and result 4 for the best of F overall. The result 4 is chosen for the full workspace plotting in $\boldsymbol{\theta_{e1}}$, $\boldsymbol{\theta_{e2}}$ and $\boldsymbol{\theta_{e3}}$ in figure 4.

Table 3. Modeling parameters.

Parameters	Values	Units
d	0	m
r_b	0.015	m
r_c	0.010	m
r_m	0.025	m
α_i	$(4i-1)\cdot\pi/6$	rad

Table 4. Optimization variables.

	x_1	x_2	x_3	x_4	x_5	x_6
Variables	e	r_e	r_a	b	c	β
Ranges	[0.1, 0.7]	[0.05,0.7]	[0.05,0.7]	[0.1, 0.7]	[0.1,0.7]	$[\pi/18, 5\pi/18]$
Units	m	m	m	m	m	rad

Table 5. Motion and search ranges.

Orientations and positions	Values	Units
$\boldsymbol{\theta_{e1}}$	$[0 \quad 0 \quad 0]$	rad
$\boldsymbol{\theta_{e2}}$	$[\pi/6 \quad 0 \quad 0]$	rad
$\boldsymbol{\theta_{e3}}$	$[0 \quad 0 \quad \pi/6]$	rad
x_e	-0.1: 0.05: 0.1	m
y_e	-0.1: 0.05: 0.1	m
z_e	0.1: 0.05: 0.3	m

Table 6. Optimization results.

	x_1	x_2	x_3	x_4	x_5	x_6	f_1	f_2	f_3	F
Result 1	0.361	0.051	0.102	0.207	0.218	0.296	-126.574	-89.204	-94.026	-309.804
Result 2	0.345	0.051	0.102	0.208	0.219	0.303	-119.825	-119.825	-99.644	-339.294
Result 3	0.386	0.051	0.079	0.245	0.232	0.344	-120.724	-85.939	-111.52	-318.179
Result 4	0.348	0.051	0.102	0.205	0.219	0.300	-121.277	-117.488	-101.07	-339.829

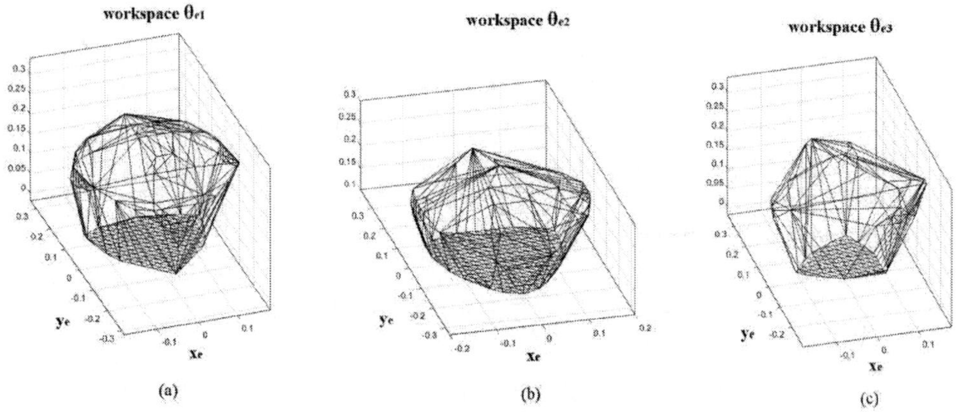

Figure 4. Full workspace in $\boldsymbol{\theta_{e1}}, \boldsymbol{\theta_{e2}}, \boldsymbol{\theta_{e3}}$ of result 4.

5. Jacobian-based Stiffness

In figure 5. (a), from the global cartesian coordinate at $\boldsymbol{A_0}$, the moving platform center $\boldsymbol{E_0}$ has a linear velocity $\boldsymbol{v_{eo}}$ and an angular velocity $\boldsymbol{\omega_{eo}}$, that contribute to the linear velocity at $\boldsymbol{E_i}$. The linear velocity at $\boldsymbol{D_i}$ could be divided to $\boldsymbol{v_{xdi}}$ and $\boldsymbol{v_{ydi}}$, which are along $\boldsymbol{X_{di}}$ and $\boldsymbol{Y_{di}}$ respectively.

A method of dot multiplication [25] [26] can be used since $\boldsymbol{v_{xdi}}$ is not perpendicular to $\boldsymbol{E_iD_i}$.

IOP Conf. Series: Materials Science and Engineering **491** (2019) 012006 doi:10.1088/1757-899X/491/1/012006

$$\boldsymbol{E_iD_i} \cdot (\boldsymbol{v_{eo}} + \boldsymbol{\omega_{eo}} \times \boldsymbol{E_oE_i}) = \boldsymbol{E_iD_i} \cdot \boldsymbol{v_{xdi}} \tag{15.a}$$

So that

$$\boldsymbol{v_{xdi}} = \frac{\boldsymbol{E_iD_i} \cdot (\boldsymbol{v_{eo}} + \boldsymbol{\omega_{eo}} \times \boldsymbol{E_oE_i})}{\boldsymbol{E_iD_i} \cdot \boldsymbol{X_{di}}} \cdot \boldsymbol{X_{di}} \tag{15.b}$$

The angular velocity $\boldsymbol{\omega_{eo}}$ could be divided to $\boldsymbol{\omega_{ei}}$, $\boldsymbol{\omega_{yi}}$ and $\boldsymbol{\omega_{zi}}$, that are along $\boldsymbol{E_oE_i}$, $\boldsymbol{Y_{di}}$ and $\boldsymbol{Z_o}$ respectively. Components $\boldsymbol{\omega_{ei}}$ and $\boldsymbol{\omega_{yi}}$ have no contribution to $\boldsymbol{v_{ydi}}$, except $\boldsymbol{\omega_{zi}}$.

$$\boldsymbol{v_{eo}} \cdot \boldsymbol{Y_{di}} \cdot \boldsymbol{Y_{di}} + \boldsymbol{\omega_{zi}} \times \boldsymbol{E_oD_i} = \boldsymbol{v_{ydi}} \text{ , where } \boldsymbol{\omega_{zi}} = \left(\boldsymbol{\omega_{eo}} \cdot \boldsymbol{Z_o} - \frac{\boldsymbol{E_oE_i} \cdot \boldsymbol{Z_o}}{\boldsymbol{E_oE_i} \cdot \boldsymbol{X_{di}}} \cdot \boldsymbol{\omega_{eo}} \cdot \boldsymbol{X_{di}}\right) \cdot \boldsymbol{Z_o} \tag{16.a}$$

So that

$$\boldsymbol{v_{ydi}} = \boldsymbol{v_{eo}} \cdot \boldsymbol{Y_{di}} \cdot \boldsymbol{Y_{di}} + \left(\boldsymbol{\omega_{eo}} \cdot \boldsymbol{Z_o} - \frac{\boldsymbol{E_oE_i} \cdot \boldsymbol{Z_o}}{\boldsymbol{E_oE_i} \cdot \boldsymbol{X_{di}}} \cdot \boldsymbol{\omega_{eo}} \cdot \boldsymbol{X_{di}}\right) \cdot \boldsymbol{Z_o} \times \boldsymbol{E_oD_i} \tag{16.b}$$

The angular velocity $\omega_{pi} \cdot \boldsymbol{Z_o}$ and $\omega_{qi} \cdot \boldsymbol{Z_o}$ are the actuation angular velocities at $\boldsymbol{A_{pi}}$ and $\boldsymbol{A_{qi}}$.

$$\boldsymbol{B_{pi}C_i} \cdot (\boldsymbol{v_{xdi}} + \boldsymbol{v_{ydi}}) = \boldsymbol{B_{pi}C_i} \cdot (\boldsymbol{Z_o} \times \boldsymbol{A_{pi}B_{pi}}) \cdot \omega_{pi} \tag{17}$$

So that,

$$\omega_{pi} = \boldsymbol{J_{pvi}}^T \cdot \boldsymbol{v_{eo}} + \boldsymbol{J_{pwi}}^T \cdot \boldsymbol{\omega_{eo}} \tag{18.a}$$

where,

$$\boldsymbol{J_{pvi}} = \frac{\boldsymbol{B_{pi}C_i} \cdot \boldsymbol{X_{di}}}{\boldsymbol{B_{pi}C_i} \cdot (\boldsymbol{Z_o} \times \boldsymbol{A_{pi}B_{pi}})} \cdot \frac{\boldsymbol{E_iD_i}}{\boldsymbol{E_iD_i} \cdot \boldsymbol{X_{di}}} + \frac{\boldsymbol{B_{pi}C_i} \cdot \boldsymbol{Y_{di}} \cdot \boldsymbol{Y_{di}}}{\boldsymbol{B_{pi}C_i} \cdot (\boldsymbol{Z_o} \times \boldsymbol{A_{pi}B_{pi}})} \tag{18.b}$$

$$\boldsymbol{J_{pwi}} = \frac{\boldsymbol{B_{pi}C_i} \cdot \boldsymbol{X_{di}}}{\boldsymbol{B_{pi}C_i} \cdot (\boldsymbol{Z_o} \times \boldsymbol{A_{pi}B_{pi}})} \cdot \frac{\boldsymbol{E_iD_i} \times \boldsymbol{E_iE_o}}{\boldsymbol{E_iD_i} \cdot \boldsymbol{X_{di}}} + \frac{\boldsymbol{B_{pi}C_i} \cdot (\boldsymbol{Z_o} \times \boldsymbol{E_oD_i}) \cdot \boldsymbol{Z_o}}{\boldsymbol{B_{pi}C_i} \cdot (\boldsymbol{Z_o} \times \boldsymbol{A_{pi}B_{pi}})} - \frac{\boldsymbol{E_oE_i} \cdot \boldsymbol{Z_o}}{\boldsymbol{E_oE_i} \cdot \boldsymbol{X_{di}}} \cdot \frac{\boldsymbol{B_{pi}C_i} \cdot (\boldsymbol{Z_o} \times \boldsymbol{E_oD_i}) \cdot \boldsymbol{X_{di}}}{\boldsymbol{B_{pi}C_i} \cdot (\boldsymbol{Z_o} \times \boldsymbol{A_{pi}B_{pi}})} \tag{18.c}$$

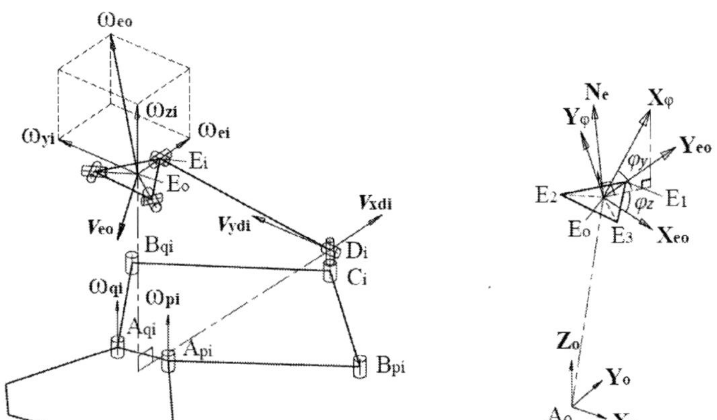

Figure 5. Jacobian matrix analysis.

One could have the similar for ω_{qi}, $\boldsymbol{J_{qvi}}$ and $\boldsymbol{J_{qwi}}$. The Jacobian matrix is obtained as below.

$$\boldsymbol{J} = \begin{bmatrix} \boldsymbol{J_{pv1}} & \boldsymbol{J_{pv2}} & \boldsymbol{J_{pv3}} & \boldsymbol{J_{qv1}} & \boldsymbol{J_{qv2}} & \boldsymbol{J_{qv3}} \\ \boldsymbol{J_{pw1}} & \boldsymbol{J_{pw2}} & \boldsymbol{J_{pw3}} & \boldsymbol{J_{qw1}} & \boldsymbol{J_{qw2}} & \boldsymbol{J_{qw3}} \end{bmatrix}^T \tag{19.a}$$

So that,

$$\begin{bmatrix} \boldsymbol{\omega_p} \\ \boldsymbol{\omega_q} \end{bmatrix} = \boldsymbol{J} \cdot \begin{bmatrix} \boldsymbol{v_{eo}} \\ \boldsymbol{\omega_{eo}} \end{bmatrix} \tag{19.b}$$

The stiffness of manipulator at $\boldsymbol{E_o}$ by the global cartesian coordinate at $\boldsymbol{A_o}$ is given as $\boldsymbol{K_a}$ which is related to the stiffness of actuator $\boldsymbol{K_q} = k_q \cdot \boldsymbol{I_6}$ through Jacobian matrix $\boldsymbol{J}$ [27].

$$\boldsymbol{K_a} = \boldsymbol{J}^T \cdot \boldsymbol{K_q} \cdot \boldsymbol{J} \tag{19.c}$$

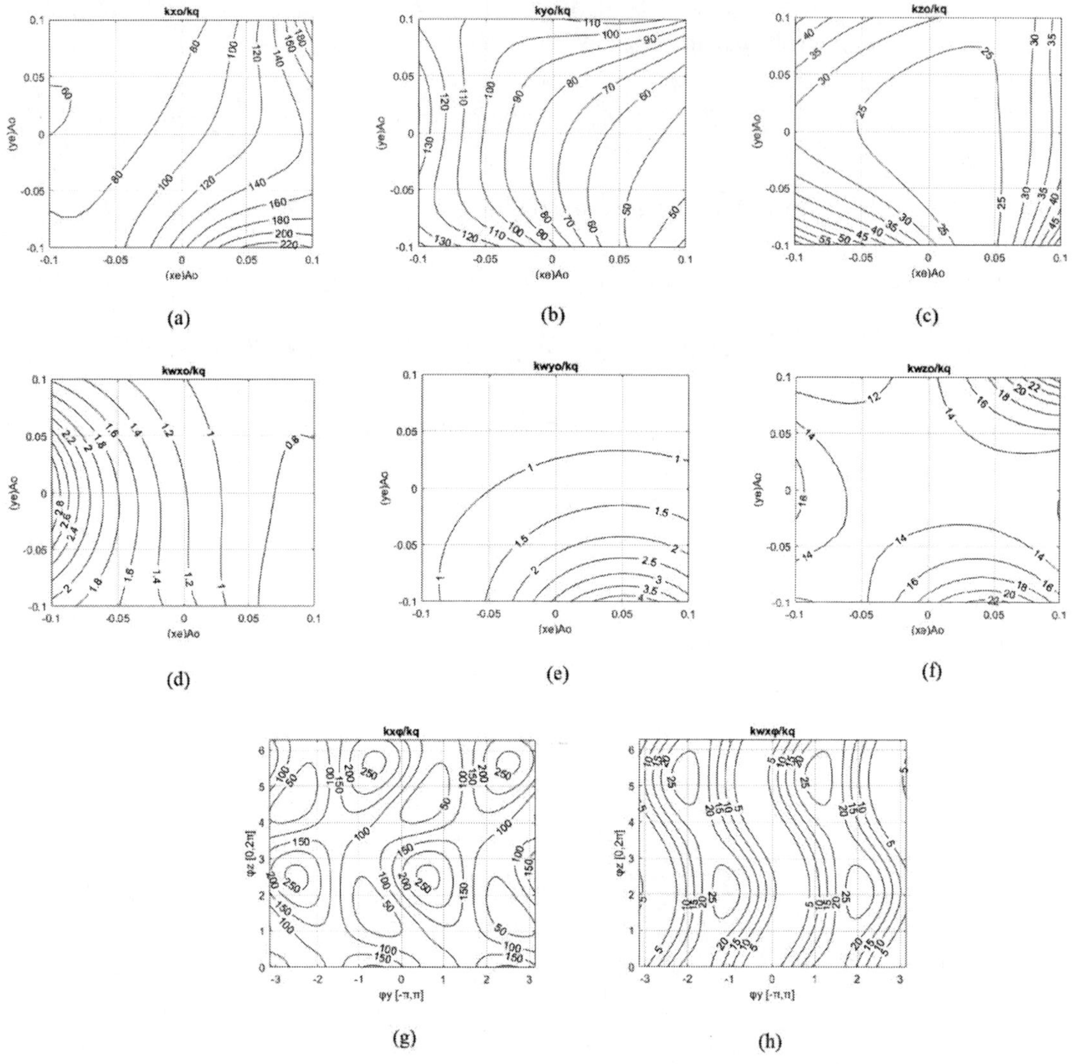

Figure 6. Stiffness mapping $\boldsymbol{\theta_e} = \left[15 \cdot \frac{\pi}{180} \quad 20 \cdot \frac{\pi}{180} \quad 10 \cdot \frac{\pi}{180}\right]^T$.

A transformation matrix is needed [19] to convert to another coordinate system. In figure 5. (b), the coordinate system is established at $\boldsymbol{E_o}$ with unit vectors $\boldsymbol{X_{eo}}$, $\boldsymbol{Y_{eo}}$ and $\boldsymbol{N_e}$. Vector $\boldsymbol{X_o}$ from global coordinate $\boldsymbol{A_o}$ equals the $\boldsymbol{X_{eo}}$ from the platform coordinate at $\boldsymbol{E_o}$. One could transform $\boldsymbol{J}$ to $\boldsymbol{J_e}$.

$$(\boldsymbol{X_o})_{Ao} = (\boldsymbol{X_{eo}})_{Eo} \tag{20.a}$$

$$(\boldsymbol{X_{eo}})_{Ao} = R \cdot (\boldsymbol{X_o})_{Ao} = R \cdot (\boldsymbol{X_{eo}})_{Eo} \tag{20.b}$$

$$\boldsymbol{J_e} = \boldsymbol{J} \cdot \begin{bmatrix} R & 0_{3,3} \\ 0_{3,3} & R \end{bmatrix} \tag{20.c}$$

Similarly, In figure 5. (b) a coordinate of Φ has a vector $\boldsymbol{X_\varphi}$ that rotates from $\boldsymbol{X_{eo}}$ by φ_y and φ_z.

$$(\boldsymbol{X_o})_{Ao} = (\boldsymbol{X_\varphi})_\Phi \tag{21.a}$$

$$(\boldsymbol{X_\varphi})_{Ao} = R \cdot (\boldsymbol{X_\varphi})_{Eo} = R \cdot R_z \cdot R_y \cdot (\boldsymbol{X_\varphi})_\Phi \tag{21.b}$$

$$R_z = \begin{bmatrix} \cos\varphi_z & -\sin\varphi_z & 0 \\ \sin\varphi_z & \cos\varphi_z & 0 \\ 0 & 0 & 1 \end{bmatrix} \text{ and } R_y = \begin{bmatrix} \cos\varphi_y & 0 & \sin\varphi_y \\ 0 & 1 & 0 \\ -\sin\varphi_y & 0 & \cos\varphi_y \end{bmatrix} \tag{21.c}$$

A vector in coordinate $\boldsymbol{A_o}$ could be converted to that in coordinate Φ. One could transform $\boldsymbol{J}$ to $\boldsymbol{J_\varphi}$.

$$\boldsymbol{J_\varphi} = \boldsymbol{J} \cdot \begin{bmatrix} R \cdot R_z \cdot R_y & 0_{3,3} \\ 0_{3,3} & R \cdot R_z \cdot R_y \end{bmatrix} \tag{21.d}$$

$$\boldsymbol{K_\varphi} = \boldsymbol{J_\varphi}^{\boldsymbol{T}} \cdot \boldsymbol{K_q} \cdot \boldsymbol{J_\varphi} \tag{21.e}$$

As one could calculate the stiffness in or around $\boldsymbol{X_\varphi}$. The polar stiffness of the end effector center $\boldsymbol{E_o}$ at any pose could be plotted about rotation angle φ_y and φ_z. The unit stiffness in and around vectors $\boldsymbol{X_o}, \boldsymbol{Y_o}, \boldsymbol{Z_o}$ from global coordinate at $\boldsymbol{A_o}$ are mapped over an area of $\boldsymbol{E_o} = [x_e \quad y_e \quad 0.2]^T$ in figure 6. (a)~(f). The unit stiffness in and around $\boldsymbol{X_\varphi}$ from given coordinate of Φ is mapped about angle $[\varphi_y \quad \varphi_z]$ when $\boldsymbol{E_o} = [0.1 \quad 0.1 \quad 0.2]^T$ in figure 6. (g) and (h).

6. Conclusion

A 3-[(2-RR)UU] 6-DOF parallel manipulator is designed with multi-plane principle. The pentagon identification and cross-product method are developed for link interference detection. The workspace boundaries of different orientation angles are plotted. The unit stiffness in and around the global coordinate vectors are mapped. With a transformation matrix this stiffness could be converted to indicate stiffness in or around any vector that is a rotation angle from the global coordinate system.

Acknowledgement

The authors would like to thank the financial support from the Natural Sciences and Engineering Research Council of Canada (NSERC). The authors gratefully acknowledge the financial support from Kaneff Research Chairs program.

References

[1] D. Zhang, Chapter 1 Introduction, Parallel Robotic Machine Tool, Springer 2010
[2] D. Zhang, Chapter 5 Spatial Parallel Robotic Machines with Prismatic Actuators, Parallel Robotic Machine Tool, Springer 2010
[3] D. Zhang, Chapter 6 Spatial Parallel Robotic Machines with Revolute Actuators, Parallel Robotic Machine Tool, Springer 2010
[4] D. Zhang, Chapter 3 Architectures of Parallel Robotic Machine, Parallel Robotic Machine Tool, Springer 2010
[5] Q. Zeng and K. F. Ehmann, 2014, "Design of Parallel Hybrid-loop Manipulators with Kinematotropic Property and Deployability", Mechanism and Machine Theory, 71 (2014) 1–26
[6] C. Tian, Y. Fang, Q. J. Ge, 2017, "Structural Synthesis of Parallel Manipulators with Coupling Sub-chains", Mechanism and Machine Theory, 118 (2017) 84–99
[7] Kim, J., Park, F. C., Ryu, S. J., Kim, J., Hwang, J. C., Park, C., and Iurascu, C. C., 2001, "Design and Analysis of a Redundantly Actuated Parallel Mechanism for Rapid Machining," IEEE Trans. Rob. Autom., 17(4), pp. 423–434.
[8] Yang, G., Chen, I. M., Chen, W., and Lin, W., 2004, "Kinematic Design of a Six-DOF Parallel-Kinematics Machine With Decoupled-Motion Architecture," IEEE Trans. Rob., 20(5), pp. 876–

884.
[9] Chen, C., Gayral, T., Caro, S., Chablat, D., Moroz, G., and Abeywardena, S., 2012, "A Six Degree of Freedom Epicyclic-Parallel Manipulator," ASME J. Mech. Rob., 4(4), p. 041011.
[10] R. Lin, W. Guo, F. Gao, 2016, "Type Synthesis of a Family of Novel Four, Five, and Six Degrees-of-Freedom Sea Lion Ball Mechanisms With Three Limbs", Journal of Mechanisms and Robotics Transaction of ASME, APRIL 2016, Vol. 8
[11] Tahmasebi, F, Tsai, L-W, 1992, "Jacobian and Stiffness Analysis of a Novel Class of Six-DOF Parallel Minimanipulators", 22nd ASME Biennial Mechanisms Conference, September 1992.
[12] Monsarrat, B., and Gosselin, C. M., 2003, "Workspace Analysis and Optimal Design of a 3-Leg 6-DOF Parallel Platform Mechanism," IEEE Trans. Rob. Autom., 19(6), pp. 954–966.
[13] W. Li, Jorge Angeles, 2017, "A Novel Three-Loop Parallel Robot with Full Mobility: Kinematics, Singularity, Workspace, and Dexterity Analysis", Journal of Mechanisms and Robotics, ASME, OCTOBER 2017, Vol. 9
[14] J. Landure and C. Gosselin, 2018, "Kinematic Analysis of a Novel Kinematically Redundant Spherical Parallel Manipulator", Journal of Mechanisms and Robotics Transactions of ASME, Volume 10, April 2018
[15] X. Liu, J. Wang, C. Wu, and J. Kim, 2009, "A New Family of Spatial 3-DOF Parallel Manipulators with Two Translational and One Rotational DOFs", Robotica 27(2):241-247
[16] D. Talaba, 2012, "The Angular Capacity of Spherical Joints Used in Mechanisms with Closed Loops and Multiple Degrees of Freedom", Robotics and Computer-Integrated Manufacturing, Volume 28, Issue 5, October 2012
[17] L. Campos, F. Bourbonnais, I. Boney, P. Bigras, 2010,"Development of a Five-Bar Parallel Robot With Large Workspace", Proceedings of the ASME 2010 International Design Engineering Technical Conferences, IDETC2010 August 2010, Montreal, QC, Canada
[18] Z. Yang and D. Zhang, 2017, "Reconfigurable 3-PRS Parallel Solar Tracking Stand", Proceedings of the ASME 2017 International Design Engineering Technical Conferences & Computers and Information in Engineering Conference, IDETC2017 August 6-9, 2017, Cleveland, USA
[19] Z. Yang and D. Zhang, 2018, "Hybrid Harvester 3-RPS Robotic Parallel Manipulator", 2018 2nd International Conference on Artificial Intelligence Applications and Technologies, AIAAT 2018, August 8-10, 2018, Shanghai, China
[20] C. Gosselin, "Chapter 3 Analysis of Complex Kinematic Chain", Kinematic Analysis, Optimization and Programming of Parallel Robotic Manipulators, 1988
[21] J-P. Merlet, 1992, "Geometrical Determination of the Workspace of a Constrained Parallel Manipulator", In ARK, pages 326-329, Ferrare September 7-9, 1992
[22] J-P. Merlet, 1995, "Determination of the Orientation Workspace of Parallel Manipulators", Journal of Intelligent and Robotic Systems June 1995, Volume 13, Issue 2, pp 143–160
[23] D. Ferrari and H. Giberti, 2014, "A Genetic Algorithm Approach to the Kinematic Synthesis of a 6-DoF Parallel Manipulator", 2014 IEEE Conference on Control Applications (CCA) Part of 2014 IEEE Multi-conference on Systems and Control October 8-10, 2014. Antibes, France
[24] "Multi-Objective Optimization", from www.polymtl.ca
[25] Y. Li and Q. Xu, 2004, "Kinematics and Stiffness Analysis for a General 3-PRS Spatial Parallel Mechanism", ROMANSY Conference 2004, June 14-18, 2004, Montreal, Canada
[26] S. R. Babu, V. R. Raju and K. Ramji, 2013, "Design for Optimal Performance of 3-RPS Parallel Manipulator Using evolutionary Algorithms", Transactions of the Canadian Society for Mechanical Engineering, Vol. 37, No. 2, 2013
[27] C. Gosselin, "Stiffness Mapping for Parallel Manipulators", IEEE Transactions on Robotics and Automation, Vol. 6, Issue 3, Jun 1990

Non-Destructive Testing and Diagnostic of Rotating Machinery Faults in Petrochemical Processing Plant

Ong Zhi Chao[1,*], Mohd Bakar Mohd Mishani[1], Khoo Shin Yee[1] and Zubaidah Ismail[2]

[1]Department of Mechanical Engineering, University of Malaya, 50603 Kuala Lumpur, Malaysia
[2]Department of Civil Engineering, University of Malaya, 50603 Kuala Lumpur, Malaysia

[*]alexongzc@um.edu.my, (6012)4192911

Abstract. Machinery fault detection as part of predictive maintenance based on the machine vibration has been practiced comprehensively in oil and gas industry and power generation plant for maintaining the reliability of the machineries. Despite the advantages, the efforts of identifying machinery fault earlier may lead to inaccurate interpretation and diagnosis due to lack of knowledge, skill and experience. Conventional vibration data acquisition and analysis using single axis accelerometer are time consuming. Therefore, an approach which relies on the visualisation of machine vibration motion technique is proposed. The visualisation of machine vibration motion was performed through a simplified Operating Deflection Shape (ODS) analysis. ODS analysis on a 4 measurement points is enough to show the motion of general machinery arrangement which consist of 2 bearings on the drive side and another 2 bearings on the driven side. The effectiveness of the simplified technique was tested in a laboratory condition before being applied in oil and gas industry. The results show vibration data acquisition is more efficient and time saving by using tri-axis accelerometer and relative phase technique. Furthermore, machinery fault diagnosis can be identified easily, faster and more accurate through visualising the machine vibration motion using a simplified 4- point ODS. This simplified technique is seen as practical approach to be proposed and integrated into conventional machinery fault detection and predictive maintenance based on the machine vibration.

1. Introduction

Vibration is defined as a movement in rotational, swinging or linear direction. Most machines in operation produce certain amount of vibration affecting the associated components and systems including pipelines and pipe systems. The systems are of great importance in many industries such as water supply, oil and gas, and nuclear power generating. Structural reliability of structural systems such as beams and beam-like structures, frame structures, and bridges can be affected by vibration and has to be continually inspected [1–3]. Liu and Kleiner [4] have reviewed the state-of-the-art of inspection techniques and technologies towards condition assessment of water distribution and transmission mains. Research works of fault diagnosis on structural systems and machinery fault were also been studied [5,6]. After a continuous run, a machine and its associated systems will have their downtime due to the fatigue and malfunction of its critical parts. Failure of the systems has to be avoided since it can have disastrous effects, leading to injuries and fatalities, as well as substantial cost to industry and the environment [7,8]. Therefore, it is very crucial to perform maintenance activity to ensure the continuation of the machine operation.

General machinery maintenance can be categorised into three which are breakdown maintenance, scheduled (preventive) maintenance, and predictive maintenance. Oil and gas industry and power generation plant are two sectors that comprehensively conducting predictive maintenance on maintaining machinery reliability. Predictive maintenance program can predict developing problems before failure and damage can occur by trending and analysing the machinery vibration. The most desirable way to maintain machinery condition is by performing the inspection while machine under operation. Early detection of minor defects and machine failure characteristic can be identified while the machine runs and it will not affect the operation performance [9].

Despite the advantages of predictive maintenance program, conventional machinery fault detection which is part of the predictive maintenance program is still relying on using single axis accelerometer for vibration data acquisition and using a tachometer as a mean of gathering the phase data. To get an absolute phase using a tachometer, a machine needs to be shut-down and re-run again. This conventional practice is seen as time consuming and can be unsuccessful due to lack of skilled and experienced manpower. Conventional machinery fault detection involves expensive and complicated instruments and software and it requires competent personnel to perform machinery fault detection and analyse the output [10].

The primary problem in oil and gas industry and power generation plants around the world is rotor imbalance and misalignment. They cause a lot of problems that lead to other secondary damage [11]. As a result, parts of a machine will suffer, bearings will wear out, foundations will crack, and machines will have a shorter live. Conventionally, maintenance personnel can detect imbalance and misalignment problems from vibration spectra. However, a research found that analysis by depending only to vibration spectrum does not provide the best information for detecting machinery imbalance and misalignment. There are cases where imbalance, misalignment, and looseness are producing the same vibration characteristic, which may lead to false data interpretation and diagnosis [12]. Research works and studies were carried out to explore better method to accurately identify machines part failure. Other alternatives using current signal and Operating Deflection Shape (ODS) analysis were proposed to detect rotor imbalance and misalignment [13]. ODS analysis, which can be performed by using both frequency domain and time domain, is a measurement technique of determining the motion of a structure while it is in operation. The motion of a machine in actual condition at certain frequencies can be obtained from ODS analysis and it helps the personnel to determine the cause of the motion using some digital signal processing technique [14]. Time domain ODS which requires a lot of channels and accelerometers for data acquisition is not practical for predictive maintenance program. On the other hand, a full ODS analysis from frequency domain is more practical to be implemented into machinery fault detection compared to time domain ODS as it reduced the numbers of the accelerometers and number of channels required in the DAQ device. Two accelerometers are adequate to measure the vibration signal with one of the accelerometers remains fixed while the other unit is moving throughout the selected points on the structure [5].

A routine predictive maintenance program requires the machine vibration to be monitored while the machine operation is not interrupted. To date, no research has been done to implement the frequency domain of ODS technique into machinery fault detection and predictive maintenance program. This concept is utilised and was presented by the authors in [14], but the application on real industrial machinery was not covered. The proposed technique is able to gather both vibration and phase data concurrently by using two accelerometers. The usage of two accelerometers is able to provide the phase between two measurement points through relative phase where interruption on the machine operation can be avoided. The phase data is crucial for identifying rotor imbalance and coupling misalignment problem. The proposed technique will eventually enhance current practice of machinery fault detection by reducing time required for performing vibration acquisition and analysis.

2. Background Theories

An ODS depends on the applied forces/loads. Therefore, the forced vibration response of a linear structure for a general system.. For a system which subjected to a harmonic excitation with harmonic force defined by $\{F\} = \{f\}$, a solution form can be derived as $\{x\} = \{X(\omega)\}$, where ω is the excitation frequency, t is the time and $\{X(\omega)\}$ is the location amplitude vector defined as:

$$X(\omega) = \sum_{r=1}^{n} \frac{\{\phi\}_r^T\{F\}\{\phi\}_r}{\omega_r^2 - \omega^2 + 2\sigma_r \omega i} \quad (1)$$

Which represents the ODS of the system subject to a harmonic excitation.

In this research, the operating condition on steady-state machinery was not changing during the measurement, therefore, the cross-power spectrum between each roving point DOF and the reference point DOF was adequate. The cross-power spectrum can be written as in:

$$G_{ij}(\omega) = X_i(\omega)X_j^*(\omega) \quad (2)$$

Where X^* is the complex conjugate, i is the measured response and j is the reference response. Only the phase difference between the reference point and roving point is important. Therefore, the Auto-Power Spectrum can be written as in [15]:

$$G_{jj}(\omega) = X_j(\omega)X_j^*(\omega) \quad (3)$$

However, the variation in amplitude part of the auto-power spectrum output is used to measure the absolute amplitude of all points. Combining both the phase differences and amplitudes for all points when linked to the geometry will give the operating deflection. The data can be animated and visualised using DASYLab® software [14].

3. Methodology

In this research, a common accelerometer was used as the reference. The response signal was captured using a tri-axis accelerometer by roving it all around the measurement points. The advantage of using the tri-axis accelerometer is it can cover three measurement acquisitions concurrently, as compared to conventional machinery fault detection practice which relies on single axis accelerometer. Generally, machines consist of a driver and driven machine with two bearings each to support the shaft. Therefore, a 4-point ODS should be enough to simulate the vibration on a machine. This technique requires only four measurement points in providing 12 vibration data. Every measurement points provide the data of vibration amplitude and phase. Both vibration amplitude and phase data are important in displaying machine vibration visualisation.

In the scope of this research, machinery fault detection program is developed in DASYLab®, an icon-based, data acquisition, control, and analysis software package. The program will capture the vibration in acceleration as well as the phase of each measurement point. The relation between this vibration and phase value will gives the visualisation of a 4-point ODS. Same method and procedure were developed by the authors in previous study to prove the concept of the proposed technique [14].

The simplified 4-point ODS technique was then applied on industrial rotating equipment to show the effectiveness of this technique in the implementation in conventional machinery fault detection practice. There were four different types of machine which are suitable in demonstrating the machinery fault such as static imbalance, dynamic imbalance, parallel misalignment and angular misalignment. These machines in petrochemical plant located in east coast of Peninsular Malaysia are vertical centrifugal pump, horizontal centrifugal pump, horizontal centrifugal blower and belt driven centrifugal blower.

Figure 1(a) shows a vertical centrifugal pump. The pump only consists of one bearing which is represented by point 3, while measurement point 4 is located on the pump base. Figure 1(b) shows a horizontal centrifugal pump while Figure 1(c) shows a horizontal centrifugal blower. Both Figure 1(b) and Figure 1(c) represent a common machinery arrangement with four bearings. Figure 2 shows the arrangement of belt driven centrifugal blower. The bearings of the blower are located on top of motor.

Both the motor and the blower shaft are placed inside a cover for safety reason.

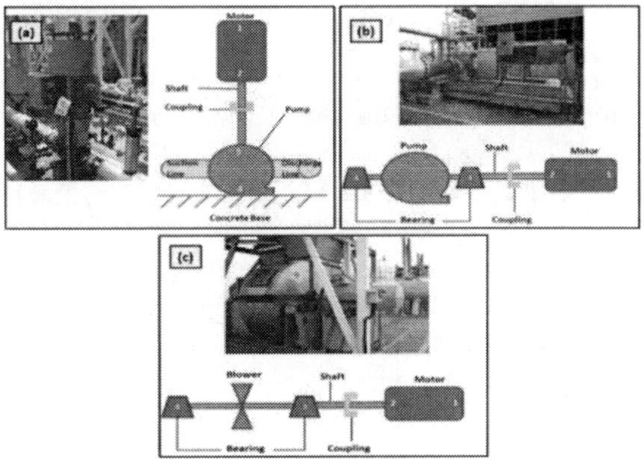

Figure 1. Industrial machinery picture and diagram (a) Vertical centrifugal pump with static imbalance problem (b) Horizontal centrifugal pump with dynamic imbalance problem (c) Horizontal centrifugal blower with parallel misalignment problem

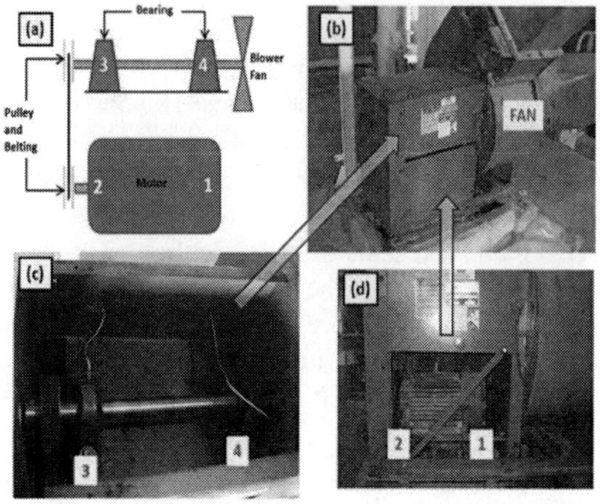

Figure 2. Industrial belt driven centrifugal blower (a) Equipment diagram (b) Equipment photo (c) Blower shaft (d) Motor location

4. Results and Discussion

The testing results on the real industrial machinery using the 4-point ODS technique to identify the machinery fault easily and effectively is deliberated and discussed in this section.

The effectiveness of ODS visualisation was first demonstrated from the industrial vertical centrifugal pump with static imbalance problem. The ODS as shown in Figure 3(a) shows point 1 and point 2 which are both at motor side are moving in relatively higher amplitude and in phase indicates a static imbalance problem. Conventionally, the vibration spectra as in Figure 5(a) were analysed. It shows a harmonic pattern where it may mislead to looseness problem. Further investigation found that the static imbalance problem occurs due to the pipe strain. It is crucial to ensure that both the suction and discharge of the pump are on the same level with the corresponding pipe line to avoid this problem. The proposed simplified 4-point ODS was then tested on an industrial horizontal centrifugal pump which undergoes dynamic imbalance problem. Figure 3(b) shows the ODS of point 3 and point 4 which are both at pump side are moving in relatively higher amplitude and out of phase which indicates the dynamic imbalance problem. Conventionally, the complex vibration spectra as shown in Figure 5(b) were analysed. It could not indicate the imbalance problem with the spectra only. It only shows a frequency component corresponding to the blade passing frequency (BPF). BPF with low vibration amplitude is considered as acceptable. However, due to low discharge pressure, further investigation was conducted. The problem occurred due to severe pump cavitation which causes the pump impeller to worn out. Furthermore, the pump cavitation problem was characterised by the raised noise floor which can be seen in the vibration spectra.

The ODS visualisation of parallel misalignment was then demonstrated from the industrial horizontal centrifugal blower. The ODS is as shown in Figure 4(a) shows that point 2 and point 3, i.e. points across coupling are horizontally out of phase indicating parallel misalignment problem. Conventionally, vibration spectra as shown in Figure 6(a) were obtained where harmonic patterns can be observed in the spectra. Again, this finding may mislead the personnel to diagnose it as looseness problem. The investigation was conducted due to abnormal sound. Although the vibration spectra show relatively lower vibration amplitude, the 4-point ODS is still able to show the motion of parallel misalignment. Further investigation found that the coupling shim was damaged. In addition, the simplified 4-point ODS was tested on a belt driven centrifugal blower with angular misalignment problem. Figure 4(b) shows the ODS that point 2 and point 3, i.e. points across pulley and belting are axially out of phase. Conventionally, vibration with the running speed, 2-time running speed components and multiple harmonic patterns spectra as shown in Figure 6(b) were observed and analysed. The harmonic pattern may mislead to belting looseness. Further investigation found that the vibration remain high after the spare motor was installed. The motor base plate was found flexible and was not properly aligned. Due to uneven level of the motor base plate, the pulley was axially misaligned. By using the 4-point ODS, both parallel and angular misalignment can be detected earlier and easier.

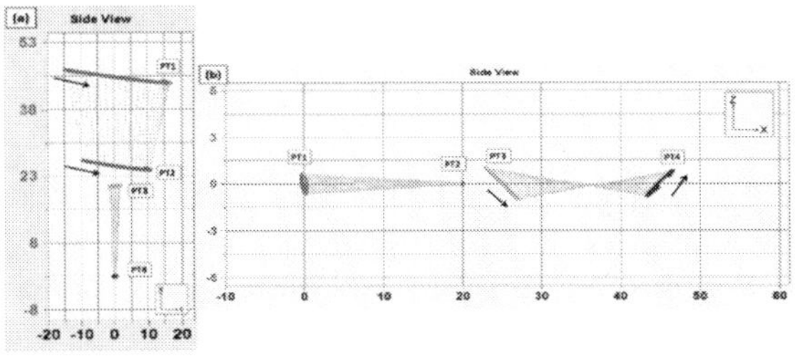

Figure 3. ODS of (a) Static imbalance (b) Dynamic imbalance

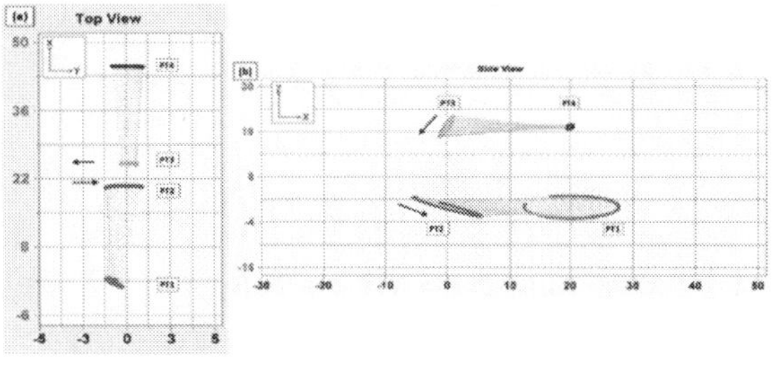

Figure 4. ODS of (a) Parallel misalignment (b) Angular misalignment

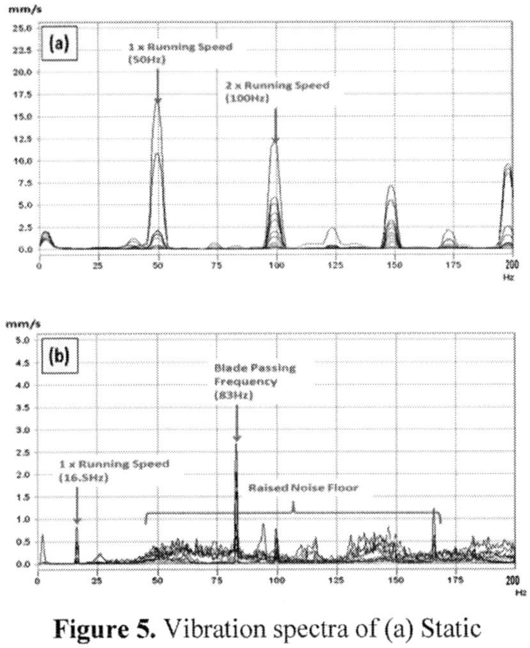

Figure 5. Vibration spectra of (a) Static imbalance (b) Dynamic imbalance

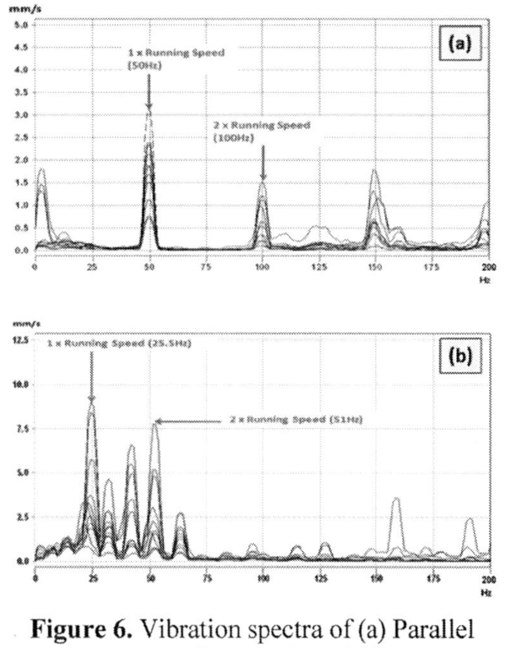

Figure 6. Vibration spectra of (a) Parallel misalignment (b) Angular misalignment

5. Conclusion

Conventionally, machinery failure diagnosis can be performed through transient analysis, steady-state analysis and modal analysis where the machine is totally shut-down. In this research, a simplified machinery fault detection approach using a simplified 4-point ODS shows that a steady-state machine vibration can be analysed without shutting down the machine to acquire the phase data. While other existing analysis method using Nyquist plot and Bode polar are plot only applicable for transient machine vibration which also requires a permanent tachometer or key-phasor as the reference for phase data. Hence, the proposed method shows that imbalance and misalignment can be distinguished easily through machinery motion visualisation without any interruption on the machine operation. Machine motion visualisation is crucial and should be integrated into conventional machinery fault detection practice which relies on vibration spectrum only. From the machine motion visualisation, personnel can visualise the motion of the machine. Therefore, primary machine problem such as imbalance and misalignment can be detected earlier and easier, hence avoiding any secondary failure such as bearing damage, worn impeller, broken rotor bar, and cracked gear teeth in shorter time without analysing every single spectrum. Furthermore, advanced or specialised knowledge is unnecessary for maintenance personnel to apply this technique into machinery fault detection and predictive maintenance program. The integration of this technique will eventually enhance the conventional practice of machinery fault detection by reducing the time required for performing vibration data acquisition and analysis and may possibly reduce downtime cost with early and effective machinery fault detection.

Acknowledgements
This research was financially supported by University of Malaya Research Grant (RP022D-2013AET), Post-graduate Research Fund (PG087-2014A). The authors wish to acknowledge the advice given by Advanced Shock and Vibration Research (ASVR) Group of University of Malaya and other project collaborators.

References
[1] M.M. Fayyadh, H.A. Razak, Z. Ismail, Combined modal parameters-based index for damage identification in a beamlike structure: theoretical development and verification, Arch. Civ. Mech. Eng. 11 (2011) 587–609.
[2] H. Monajemi, H.A. Razak, Z. Ismail, Damage detection in frame structures using damage locating vectors, Measurement. 46 (2013) 3541–3548.
[3] Z. Ismail, Z. Ibrahim, A.Z.C. Ong, A.G.A. Rahman, Approach to reduce the limitations of modal identification in damage detection using limited field data for nondestructive structural health monitoring of a cable-stayed concrete bridge, J. Bridg. Eng. 17 (2012) 867–875.
[4] Z. Liu, Y. Kleiner, State of the art review of inspection technologies for condition assessment of water pipes, Measurement. 46 (2013) 1–15.
[5] A.G. Abdul Rahman, S. Noroozi, M. Dupac, S.M.S.M. Al-Attas, J.E. Vinney, A hybrid approach for nondestructive assessment and design optimisation and testing of in-service machinery, Nondestruct. Test. Eval. 28 (2013) 44–57.
[6] A. Widodo, B. Yang, E.Y. Kim, A.C.C. Tan, Nondestructive Testing and Evaluation Fault diagnosis of low speed bearing based on acoustic emission signal and multi-class relevance vector machine, Nondestruct. Test. Eval. 24 (2009) 313–328.
[7] Z. Ismail, R. Karim, Some technical aspects of spills in the transportation of petroleum materials by tankers, Saf. Sci. 51 (2013) 202–208.
[8] Z. Ismail, J.C. Tai, K.K. Kong, K.H. Law, S.M. Shirazi, R. Karim, Using data envelopment analysis in comparing the environmental performance and technical efficiency of selected companies in their global petroleum operations, Measurement. 46 (2013) 3401–3413.
[9] R.K. Mobley, An Introduction to Predictive Maintenance, Butterworth-Heinemann, 2002.
[10] C. Scheffer, P. Girdhar, Practical Machinery Vibration Analysis and Predictive Maintenance, Newnes, 2004.
[11] R.K. Mobley, Root Cause Failure Analysis, Newnes, 1999. doi:10.1007/978-1-4419-6018-4.
[12] P. Pennacchi, A. Vania, S. Chatterton, Nonlinear effects caused by coupling misalignment in rotors equipped with journal bearings, Mech. Syst. Signal Process. 30 (2012) 306–322.
[13] M.A. Saleem, G. Diwakar, Detection of unbalance in rotating machines using shaft deflection measurement during its operation, J. Mech. Civ. Eng. 3 (2012) 8–20.
[14] M.B. Mohd Mishani, Z.C. Ong, W.T. Chong, S.Y. Khoo, A.G. Abdul Rahman, A simplified technique to distinguish unbalance and misalignment using frequency domain Operating Deflection Shape, J. Teknol. 76 (2015) 1–4.
[15] B. Schwarz, M.H. Richardson, Measurements required for displaying Operating Deflection Shapes, in: Proceeding Int. Modal Anal. Conf., Dearborn, MI, 2004: pp. 1–6.

The Effect of the Low Frequency Mode of the Machine Tool Structure on the Milled Surface Generation

Yili Peng, Bin Li, Xinyong Mao*, Hongqi Liu, Xuchu Jiang and Ling Yin

National NC System Engineering Research Centre, Huazhong University of Science and Technology, Wuhan 430074, PR China

*E-mail addresses: maoxyhust@hust.edu.cn, maoxyhust@163.com (X. Mao).

Abstract: The prediction of machining process stability and surface generation is primarily based on the dynamics of the tool and work piece. However, the vibrations caused by the low frequency of machine tool structure also affect the stability of machining process and the surface integrity in the condition of low speed and heavy load. Therefore, this paper presents a milling time domain simulation model based on the low frequency mode of machine tool structure to study the effect of the low frequency mode of the machine tool structure on the surface generation. Based on this model, the correlation between the multimode natural frequencies of machine tool structure and the surface waviness is studied and the effect of the inertia force caused by the change of cutting speed on the surface shape is studied. Finally, a set of cutting tests is performed to verify the surface generation prediction.

1. Introduction

The milled surface topography could be predicted through the time domain simulation of the machining process. The accurate prediction of surface generation is beneficial to the optimization of processing parameters. Montgomery et al. [1] proposed the model of the mechanics and kinematics of dynamic milling to predict the cutting force and the topology of a finished work piece surface. Campoanes et al. [2] presented an improved time domain simulation model for small radial immersions. Paris et al. [3] presented a machined surface generation model for high speed milling and studied the effect of machining parameters on the surface shape and roughness of the machined surface. Eksioglu et al. [4] presented a discrete time modeling of dynamic milling system that could be used for variable pitch and helix angles to predict the surface location errors, chatter stability and cutting forces. Yang et al. [5] studied the surface generation mechanism in peripheral milling with variable pitch end mills and predicted the surface topography considering the machining error sources from tilting, run-out and the deflection of the tool and work piece displacement. Kiran et al. [6] presented a two degree-of-freedom frequency-domain solution for surface location error prediction.

To predict the surface shape accurately, the study of the relationship of vibration during cutting and dynamics of machine tool and the surface integrity is significant and meaningful. de Aguiar MM et al. [7] studied the correlation between the frequency response function and both the tool wear and workpiece surface roughness. The research indicated that high amplitude vibrations occur when the harmonics of tooth passing frequencies are close to the system's natural frequency. Good workpiece surface roughness related to long tool life of long slender tools can be achieved provided that the tooth passing frequency used does not produce high frequency response function values. Peigne et al. [8] studied the impacts of cutting vibration on the roughness of the machined surface and reported a different finding, claiming that during the stable cutting, the forced vibrations have only a very slight impact, and the unstable cutting primarily impacts roughness. Shimana et al. [9] studied the relationship between the vibration displacement and the machining error and observed that the machining error caused by the deflection of the tool at the cutting point in end-milling processes at a spindle speed of 1000rpm could be closely estimated from the static stiffness of the end mill system and the quasi-static cutting force but the machining error caused by the vibration must be considered with the increase of the spindle speed. Zhang et al. [10] studied the effect of the multimode high frequency vibration on the surface generation in ultra-precision diamond turning. Schmitz et al. [11] experimentally quantified the relative contributions of geometric, thermal, contouring, and cutting force errors to the machined part dimensional errors and observed that the cutting force error contribution is greater than the combined effect of the other error sources for certain choices of spindle speeds but may become insignificant at slightly different spindle speeds.

The current prediction model was mainly based on the dynamics of the tool and the work piece. This model adapts the condition of high speed and light load, however, under the conditions of low speed and heavy load, the effect of the direction and position dependent dynamics of machine tool structure must be considered. Law et al. [12] studied the effect of the position dependent dynamics of machine tool structure on the machining stability. Kono et al. [13] studied the direction dependency of tool-workpiece compliance of machine tool and reported the large discrepancy of compliance in different directions in the frequency range lower than 250Hz; the direction dependency was small in the frequency range higher than 250Hz which was primarily from the spindle and tool. The current prediction model of surface topography and the study of the effect of vibration on the surface topography in the machining process only considered the vibration caused by the cutting force. The vibration caused by the inertia force due to the change of cutting speed was not considered. The research of Wang et al. [14] showed that the high acceleration and braking forces cause the elastic deformation and vibration of a machine tool.

This paper establishes a time domain simulation model considering the influence of the low frequency mode of machine tool structure and the inertia force to study the effect of the dynamics of the machine tool structure on the surface generation. The principle of the model is provided in Section 2. In Section 3, a simulation case is conducted and the link between the surface waviness and the dynamics of machine tool structure is analysed. Section 4 provides the experimental verification of the result of the simulation. Finally, the study's conclusions is presented in Section 5.

2. Theoretical background

The prediction of machined surface generation could be realized through the time domain simulation of the machining process. The presented time domain simulation model is based on the model of "Cycloidal tool path, Regenerative Force, Dynamic Deflection Model" [15]. The effect of inertia force caused by the change of cutting speed is considered in the model of this paper.

2.1. Calculation of the vibration displacement caused by the inertia force

The inertia force caused by the cutting speed can be approximated by a rectangular wave.

$$g(t)=\begin{cases}A_0, t\in[0,\tau]\\ 0, t\notin[0,\tau]\end{cases} \tag{1}$$

The dynamic equations of the machine tool structure under the excitation of inertia force are as follows:

$$m_{x,r}\ddot{u}_{x,r}+c_{x,r}\dot{u}_{xx,r}+k_{x,r}u_{x,r}=g_x \tag{2}$$

$$m_{y,r}\ddot{u}_{y,r}+c_{y,r}\dot{u}_{y,r}+k_{y,r}u_{y,r}=g_y \tag{3}$$

Where $m_{x,r}$. $c_{x,r}$ and $k_{x,r}$ are the r-th order modal parameters in the x direction. $r=1,2,...,n$; $u_{x,r}$ is the r-th order modal displacement in the x direction; g_x is the inertia force in the x direction. $m_{y,r}$.

$c_{y,r}$, $k_{y,r}$, $u_{y,r}$, and g_y are the corresponding parameters in the y direction. The total displacement caused by the inertia force is the sum of the multiple modal displacements.

$$x_i = \sum_{r=1}^{n} u_{x,r} \text{、} y_i = \sum_{r=1}^{n} u_{y,r} \tag{4}$$

2.2. *Calculation of the vibration displacement caused by the cutting force*

The instantaneous cutting force can be determined as follows:

$$F_{t,j} = k_t b h_j + k_{te} b \tag{5}$$

$$F_{n,j} = k_n b h_j + k_{ne} b \tag{6}$$

Where $F_{t,j}$ and $F_{n,j}$ are the tangential and normal force components, k_t, k_n, k_{te}, k_{ne} are the tangential, normal, tangential edge, and normal edge cutting force coefficients. b is the axial depth of cut, and h_j is the instantaneous chip thickness. The cutting force in the x and y directions can be calculated through the projections of $F_{t,j}$ and $F_{n,j}$. The vibration displacement caused by the cutting force can also be calculated through the modal displacement equations.

3. Simulation of the surface prediction

The simulator is coded using MATLAB, the tangential cutting force coefficient is 771.2N/mm^2, and the normal cutting force coefficient is 254.5 N/mm^2. The modal parameters of the machine tool structure are provided in Section 3.1, and the other input parameters are listed in Table 1.

Table 1. Parameters for simulation

number of teeth	tool diameter (mm)	helix angle (°)	spindle speed (r/min)	feed speed (mm/min)	axis depth of cut (mm)	radial depth of cut (mm)
3	12	35	3000	1000	5	0. 1

3.1. *Dynamics of machine tool structure*

The modal parameters of machine tool structure in x and y directions are obtained by impact testing. All of the tests in this paper are conducted in a 3-axis vertical machining center DM4600. A micro-accelerometer sensor (DYTRAN 3224A1) is attached to the tool. Seven accelerometers (PCB 356A16) are attached to the spindle head, the work table, the fixture and the work piece. Two hammers are used in this test: the PCB impulse hammer 086C03 is used to impact the tool point and the HDFC-DFC-1 hammer is used to impact the machine tool structure. Both the impact force of the hammers and the vibration responses of all the measurement points are recorded synchronously and then processed using the acquisition and analysis system LMS Test Lab. To reduce the test errors, the frequency response functions from five repeated tests are averaged.

The frequency response functions (FRF) obtained from the test are shown in Figure 1. Figure 1 (a) shows the FRF of the measurement point on the work table. Figure 1 (b) shows the FRF of the work piece. Figure 1 (c) shows the FRF of the spindle head. Figure 1 (d) shows the FRF of the tool. The dominant natural frequencies of the tool are 907 Hz in the x direction and 844 Hz in the y direction. The identified modal parameters of the machine tool structure are in the range of low frequency as listed in Table 2. The dominant natural frequencies of the machine tool structure in the x direction are 14.3 Hz and 45 Hz, and those in the y direction are 15.3 Hz and 54 Hz.

IOP Conf. Series: Materials Science and Engineering **491** (2019) 012008 doi:10.1088/1757-899X/491/1/012008

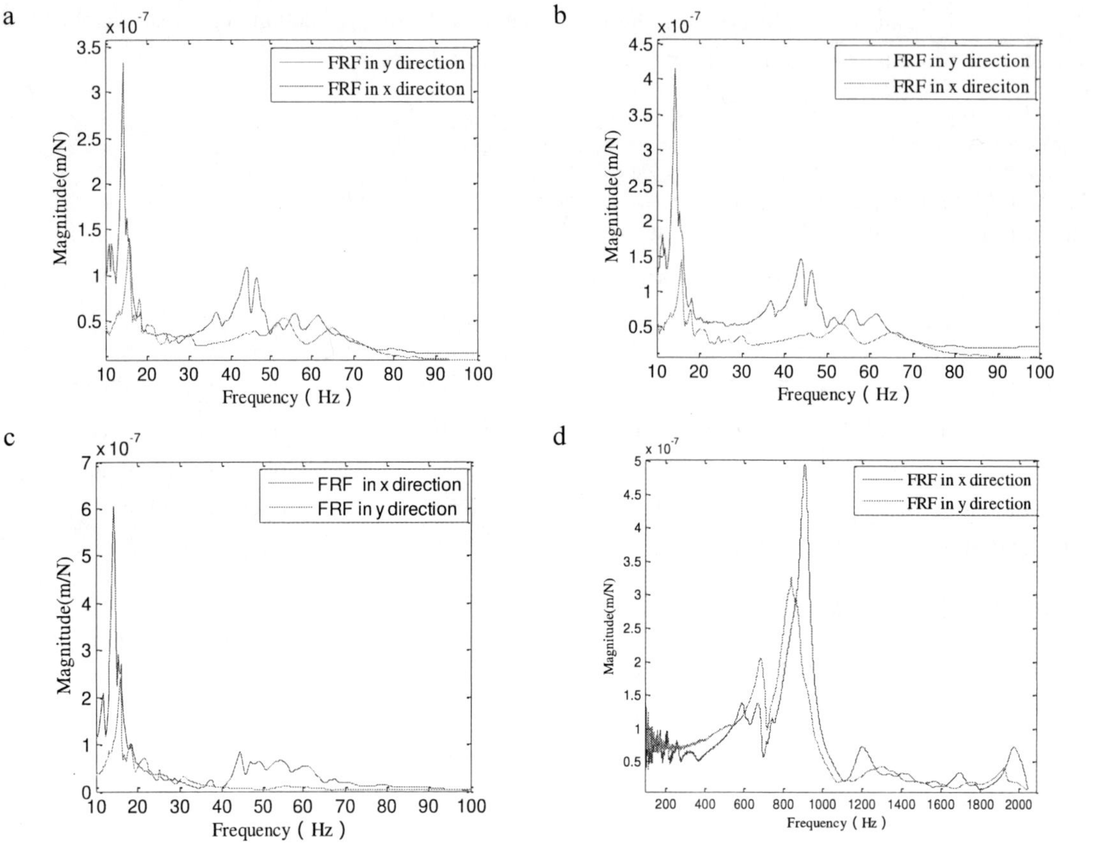

Figure 1. (a) FRF of the measurement point on the work table; (b) FRF of the work piece; (c) FRF of the spindle head; (d) FRF of the tool

Table 2. Modal parameters of machine tool structure

Mode number	Natural frequency (Hz)	Damping ratio (%)
1st mode in x direction	14.3	3.31
2st mode in x direction	36.9	2.35
3st mode in x direction	45	1.17
4st mode in x direction	56	3.1
5st mode in x direction	60.4	4.25
1st mode in y direction	15.3	3.3
2st mode in y direction	30.4	2.1
3st mode in y direction	45	4.62
4st mode in y direction	54	4.95
5st mode in y direction	65	6.96

3.2. Discussion of the simulation result

The surface prediction of a straight-line path machining in the y direction is shown in Figure 2. If the vibration was not considered in the simulation, i.e., the machining system was rigid, then the machined surface waviness was only caused by the feed motion of the tool. The surface shape is the envelope of all of the trochoids produced by the feed motion and the cutter rotation. The distance among the waviness of the surface is equal to the feed per tooth (0.11mm) as shown in Figure 3. Comparing the surface shape considering the dynamics of machine tool as shown in Figure 2 and that without considering the vibration as shown in Figure 3, three types of geometric defects are observed.

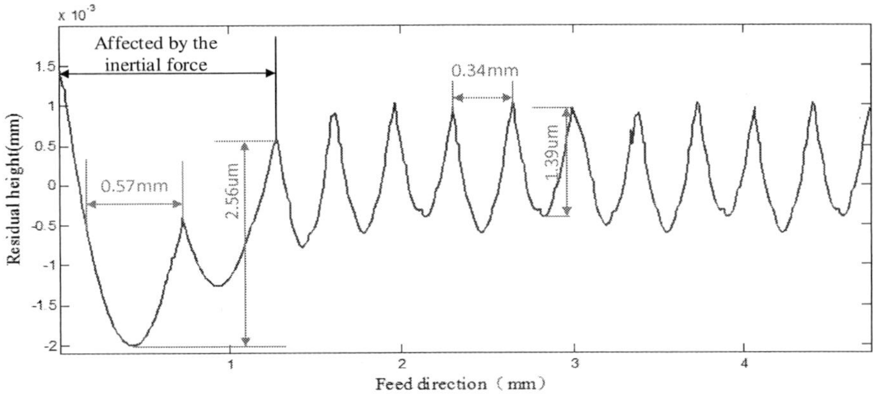

Figure 2. Surface shape of a simulation for a straight-line path

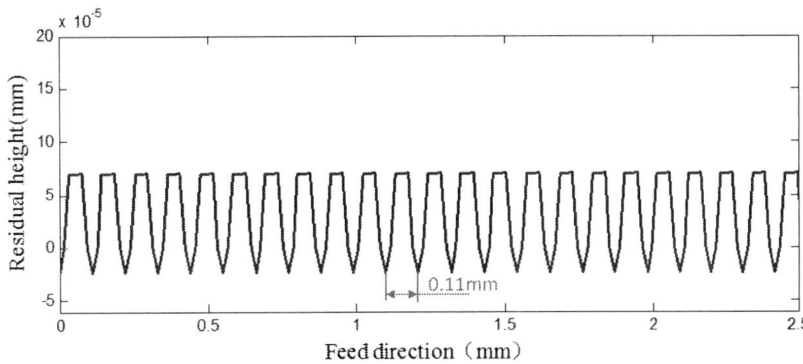

Figure 3. Surface shape from the simulation without vibration

The first geometric defect is the effect of inertia force caused by the change of feed speed. The tool-workpiece is subjected to the inertia force and cutting force; therefore, the amplitude of vibration is larger than that without the excitation of inertia force. As shown in Figure 2, the amplitude of the beginning of the surface waviness is larger and the distance among the waviness is longer.

The second geometric defect is the effect of the dynamics of the structure on the peak-to-peak distance of the surface waviness. In the simulation of Figure 2, the tooth passing frequency is 150 Hz. The dominant natural frequencies in the x direction are 14.3 Hz and 45 Hz. The triple the frequency of 45 Hz is approximately the tooth passing frequency. The triple the feed rate per tooth of 0.11 mm is 0.33 mm, which is close to the average peak-to-peak distance of the surface waviness 0.34 mm as shown in Figure 2. The effect of the dominant natural frequency 14.3 Hz is not very clear in the surface shape shown in Figure 2, whereas in the FFT of the surface waviness as shown in Figure 4, the effect of the dominant natural frequency 14.3Hz can be found.

The surface topography is the coupling of the waviness produced by the feed motion and the vibration caused by the low frequency dynamics. When the vibration amplitude caused by the dominant natural frequency is small, the waviness from the feed motion remains in the coupling waviness as shown in Figure 5. When the vibration amplitude caused by the dominant natural frequency is large, the waviness from the feed motion will be displaced by the vibration as shown in Figure 2.

The third geometric defect is the effect of the dynamics of the structure on the residual height and the roughness. The peak-to-peak surface roughness Rt of the machined profile without considering the vibration can be determined as follows [8]: $Rt = f_{feed/tooth}^2 / 8R$, where $f_{feed/tooth}$ is the feed rate per tooth, and R is the tool radius. The roughness obtained according this formula for the cutting parameter in this case is 0.252 μm, and the roughness considering the vibration obtained through the simulation is 0.453 μm.

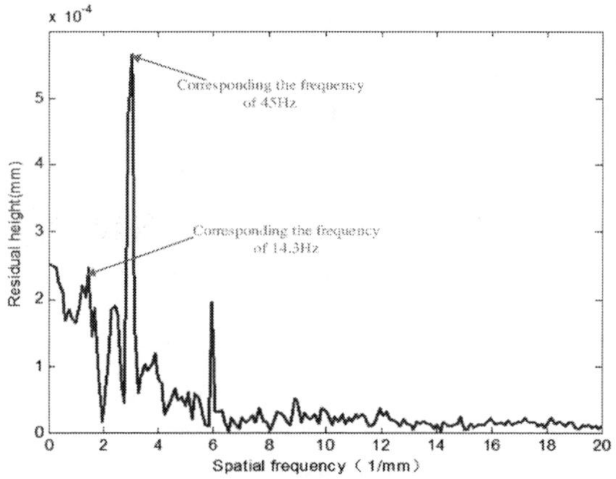

Figure 4. FFT of the surface topography of the simulation

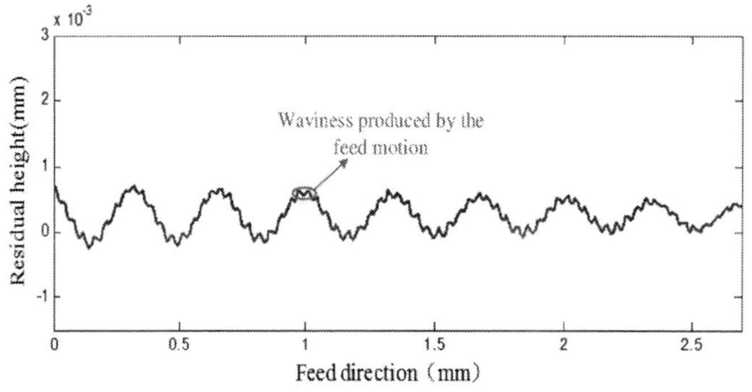

Figure 5. Simulation surface shape with small amplitude of vibration

4. Experimental verification

To verify the accuracy of the model for surface prediction, a set of cutting tests was performed in a 3-axis vertical machining center DM4600. The work piece material is aluminum, and the cross-section of the work piece is square to ensure the same dynamics of the x and y directions. The tool geometry and the cutting parameters are the same as that in the simulation case as shown in Table 1. The PGI-830 comprehensive measurement system for surface profile by Taylor Hopson Limited is used to characterize the machined surfaces. The results of the measurements are shown in Figure 6. A comparison of the simulation and experiment results is presented in Table 3. The peak-to-peak distances of the surface waviness of the experiment are in agreement with the results of simulation, as shown in Table 3. The amplitudes of the surface waviness of the experiment are slightly larger than the prediction, resulting in the smaller values of the prediction of roughness compared to the actual values. The effect of the inertia force is also observed in the surface profile of machined work piece. The model based on the low frequency mode of the machine tool structure provides good predictions of the trend of the machined surface.

Table 3. Comparison of the experiment and simulation result

Average distance of peak to peak		Roughness	
Test(mm)	Simulation(mm)	Test(um)	Simulation(um)
0.34	0.34	0.68	0.45

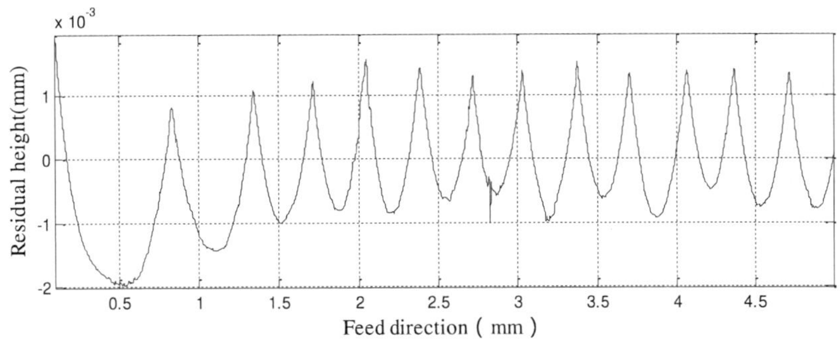

Figure 6. Surface profile of the machined work piece

5. Conclusions

A simulation model based on the low frequency mode of machine tool structure to predict the surface profile was presented in this paper. This model considered the effect of the inertia force caused by the change of feed speed and the dynamics of machine tool structure. This simulation result was verified by a set of cutting tests and was consistent with the result of the tests.

The correlation of the dominant natural frequency of machine tool structure and the surface waviness was studied in this paper. According to the results, the surface shape is the coupling of the waviness produced by the feed motion and the vibration. When the vibration amplitude caused by the dominant natural frequency of machine tool structure is large, the roughness and the distance of peak to peak of the surface waviness are primarily determined by the vibration. When the vibration amplitude caused by the dominant natural frequency is small, the roughness is influenced by both the waviness left by the feed motion and the vibration. The effect of the inertia force on the surface topography was also studied in this paper and in the range of existing the excitation of inertia force, the amplitude of the surface waviness and the peak-to-peak distance is larger.

Acknowledgments

This work was funded by the National Natural Science Foundation of China (NSFC) under grant No. 51775212 and No. 51505084, and by the Key Projects in the National Science and Technology Pillar Program of China under grant no. 2015ZX04000016.

Reference

[1] Montgomery D, Altintaş Y , *Trans. ASME J. Engng Ind.,* 113 (1991), pp. 160-168.
[2] Campomanes M L , Altintas Y, *Trans. ASME J. Manuf. Sci. Eng.,* 125 (2003), pp. 416-422.
[3] Paris H , Peigné G, Mayer R, *Int. J. Mach. Tools Manuf.,* 44 (2004), pp. 1567-1576.
[4] Eksioglu C, Kilic Z M, Altintas Y, *Trans. ASME J. Manuf. Sci. Eng.,* 134 (2012), p. 061006
[5] Yang D,.Liu Z, *Int J Mach Tools Manuf,* 91(2015),pp. 43-53.
[6] Kiran K, Rubeo M, Kayacan M C, Schmitz T, *Precis. Eng.,* 48 (2017), pp. 234-242.
[7] Aguiar M M de , Diniz AE , Pederiva R , *Int J Mach Tools Manuf,* 68 (2013), pp. 1-10.
[8] Peigne G, Paris H, Brissaud D, Gouskov A, *Int J Mach Tools Manuf,* 44 (2004), pp. 1133-1142.
[9] Shimana K, Kondo E, Chifu T, Nakao M, Nishimura Y, *Procedia CIRP,* 46 (2016),pp. 246-249.
[10] Zhang S J , To S, *Int. J. Mach. Tools Manuf.,* 72 (2013), pp. 32-36.
[11] Schmitz T , Ziegert J , Canning J, Zapata R, *Precis. Eng.,* 32 (2) (2008), pp. 126-133.
[12] Law M , Altintas Y, Phani A S , *Int. J. Mach. Tools Manuf.,* 68 (2013), pp. 81-90.
[13] Kono D, Matsubara A, *Procedia CIRP,* 46(2016),pp.529-532.
[14] Wang L, Liu H, Yang L, Zhang J , Zhao W , Lu B, *Int. J. Mach. Tools Manuf.,* 88 (2015), pp. 71-81.
[15] Schmitz T L, Smith K S 2009 *Machining dynamics: frequency response to improved productivity,* Springer, NJ, USA

Trajectory Dependent Structural Vibration and Its Effects on Surface Generation

Mengxiang Yang[1], Xinyong Mao[1,*], Yili Peng[1], Bin Li[1,2], Xuchu Jiang[1]and Liangjie Li[1]

[1]National NC System Engineering Research Centre, Huazhong University of Science and Technology, Wuhan 430074, PR China
[2]State Key Laboratory of Digital Manufacturing Equipment and Technology, Huazhong University of Science and Technology, Wuhan 430074, PR China

*Correspondence E-mail: maoxyhust@hust.edu.cn, maoxyhust@163.com (X. Mao).

Abstract. The vibration of machine tool plays an important role in surface generation. However, the vibration characteristics is different for different machine tool structures and different machining path. This paper proposed a new method of structural sensitivity analysis to identify trajectory dependent structural vibration and its effect on surface generation in milling. The experimental and theoretical results verify that (i) Multimode low-frequency vibration of the sensitive structures (MLFVSS) directly determines the surface topography in milling; (ii) The sensitivity is different for different structures or the same structure at different modes; (iii) Structural vibration characteristics is related to the machining path, resulting in different influence of the structural vibration on surface generation at different position of the machining path. Finally, a surface topography prediction model was established based on the experimental results. This study is helpful to the optimization of machine tool structures and the selection of machining parameters.

1. Introduction

The surface quality of products reflects the machining accuracy, but also directly affects the products' functions. In recent years, researchers have explored a series of factors influencing the surface generation, including machining conditions [1], material swelling and recovery [2, 3], crystal orientation [4], the vibration of machine tool [5, 6], etc. There are two primary sources of affecting the surface generation caused by machine tool vibration: one is the high frequency vibration source, mainly refers to cutting tool vibration; the other is the low frequency vibration source, mainly refers to the structural vibration.

However, the above researches considered the high frequency vibration of the cutting tool, usually higher than 10000Hz. While the structural low-frequency vibration has not been taken into account.

On the other hand, the vibration characteristics is different for different structures which will cause different influences on surface generation. Chen, G., et al. carried out sensitivity analysis in the design of five-axis ultra-precision machine tool to find out the sensitive structures causing the machining error.

This paper proposed a new method to identify the trajectory dependent structural vibration and its effect on surface generation in milling based on structural sensitivity analysis.

2. Theory of sensitivity analysis

Stiffness is different for different structures due to the existence of dynamics joints. In this part, a theoretical derivation of modal mass distribution matrix (MMDM) is performed which can reflects the structural vibration energy. If the local amplitude is higher in the MMDM, it means that the corresponding components are the sensitive structures. The following is the theoretical derivation:

For general machine tool system, the modal mass of ith mode can be derived:

$$m_i = \{\varphi_i^T\}[M]\{\varphi_i\} \tag{1}$$

$$[M] = \begin{bmatrix} M_1 & \cdots & 0 \\ \vdots & \ddots & \vdots \\ 0 & \cdots & M_n \end{bmatrix} \tag{2}$$

Then Eq. (1) can be replaced with:

$$m_i = [\varphi_{1i} \quad \varphi_{2i} \quad \cdots \quad \varphi_{ni}] \begin{bmatrix} M_1 & \cdots & 0 \\ \vdots & \ddots & \vdots \\ 0 & \cdots & M_n \end{bmatrix} \begin{bmatrix} \varphi_{1i} \\ \varphi_{2i} \\ \vdots \\ \varphi_{ni} \end{bmatrix} = \sum_{k=1}^{n} M_k \varphi_{ki}^2 \tag{3}$$

Then all modal mass values can form a matrix in spatial distribution called modal mass distribution vector (MMDV):

$$\{m_i'\} = \left[M_1\varphi_{1i}^2 \quad M_2\varphi_{2i}^2 \quad \cdots \quad M_n\varphi_{ni}^2\right]^T = [M]\{\varphi_i\} * \{\varphi_i\} \tag{4}$$

Where * represents an overlapping multiplication operator. $\{m_i'\}$ represents a vector consisting of a column elements of the modal mass matrix. All MMDV constitute the MMDM:

$$[m'] = \left[\{m_1'\} \quad \{m_2'\} \quad \cdots \quad \{m_n'\}\right] = [M][\varphi] * [\varphi] \tag{5}$$

Where $[\varphi]$ is the scaled mode shape matrix under unit modal mass (UMM) principles.

As shown in Eq. (5), the modal mass expressions is similar to the kinetic expression. If the kinetic energy is mainly concentrated in some local DOFs, it can be deduced that the corresponding structure of those DOFs is the sensitive component.

3. Verification of the MLFVSS method

Before implementation, 0-400Hz was selected as the interest and analysis frequency band. In order to validate the MLFVSS method, the following steps need to be implemented.

Firstly, experimental modal analysis (EMA) was conducted through an impact testing to identify the natural frequencies of the whole machine tool structures between 0Hz and 400Hz.

Secondly, sensitivity analysis was implemented to identify the sensitive structures at these natural frequencies through the modal mass distribution matrix proposed in section 2. Thirdly, a trajectory-related milling experiment was carried out and the surface topography was measured to validate the trajectory dependent structural vibration and the MLFVSS method.

3.1. Impact testing setup

In this part, an impact testing was conducted on a milling machine tool (model number TC500). The experimental machine tool, impact point and measurement points as shown in Figure 1.

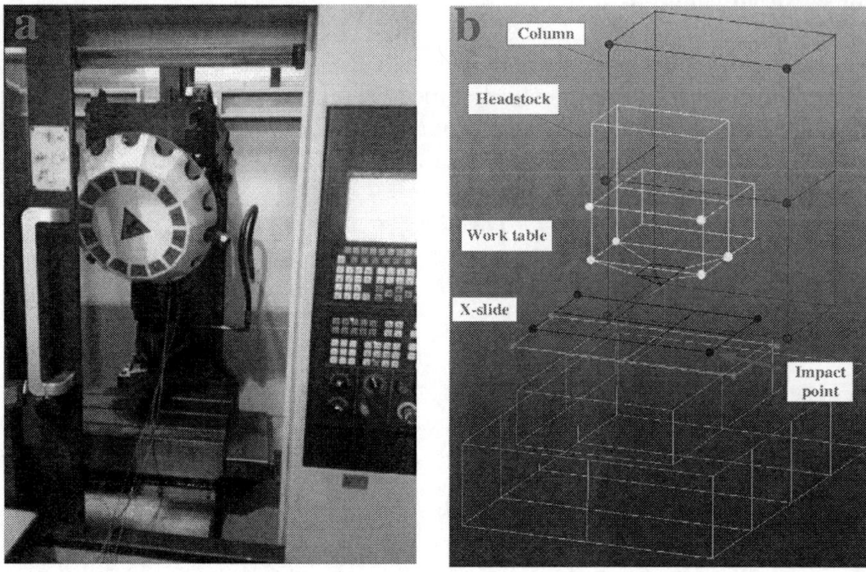

Figure 1. Impact testing system

3.2. EMA and sensitivity analysis

In this part, the natural frequencies of the whole machine tool structures between 0-400Hz were estimated through EMA using the LMS PolyMAX algorithm. Then the sensitive structures were identified at each natural frequency based on MMDM. Table 1 shows the natural frequencies of the TC500 machine (0-400Hz).

Table 1. Natural frequencies of TC500 machine tool structure (0-400Hz)

Mode	1	2	3	4	5	6
Frequency (Hz)	151.6	225.9	289.5	315.7	330.9	381.6

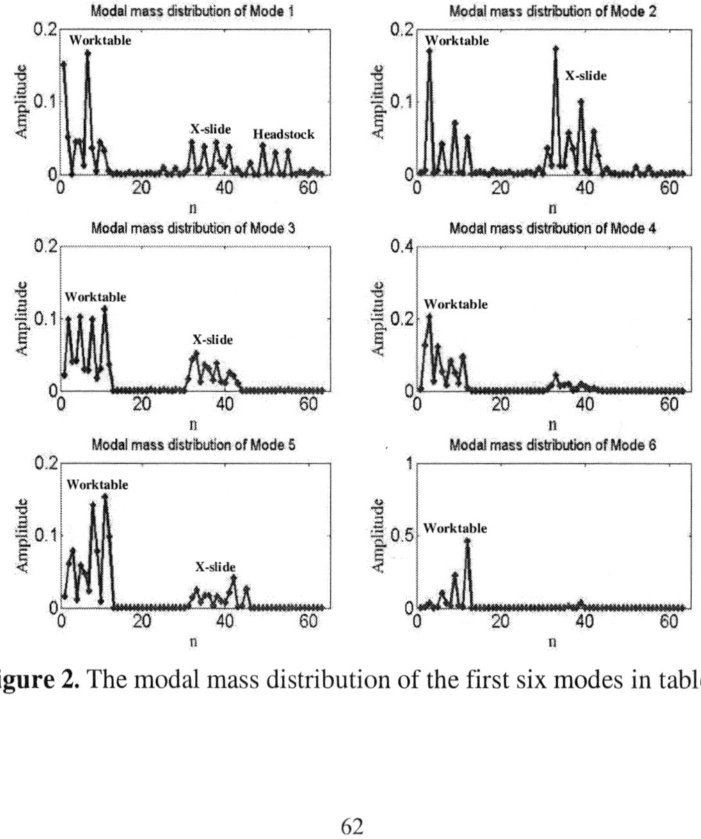

Figure 2. The modal mass distribution of the first six modes in table 1

Figure 2 is the modal mass distribution of the first six modes in Table 1 respectively, where the 'Amplitude' is the value of the modal mass matrix m_i, 'n' represents the corresponding DOFs of measuring points. In the experiment, 21 measurement points were selected and each point is composed of three DOFs (X, Y, Z). The No. 1 to No. 12 DOFs belong to the worktable, No. 13 to No. 30 DOFs belong to the column, No. 31 to No. 45 DOFs belong to the X-slide, No. 46 to No. 63 DOFs belong to the headstock. The sensitive structures of the TC500 machine are worktable and X-slide.

3.3. Verification of MLFVSS method

3.3.1. Trajectory-related milling experiment. Figure 3a shows the cutting experiment device and data acquisition system.In order to better reflect the influence of structural vibration on the surface topography, a curved trajectory and flank milling were adopted (as shown in Figure 3b). The spindle speed and feed rate were set to 2200rpm and 1500mm/min respectively. The transverse cutting depth was set to 0.1mm per cycle.

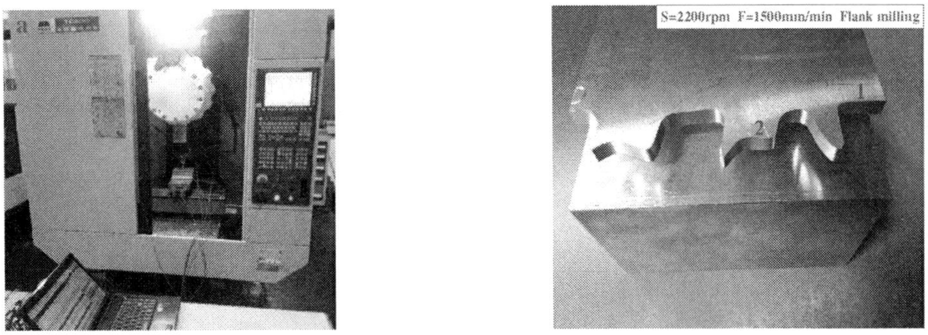

Figure 3. The photograph of milling experiment and work piece

3.3.2. Verification of trajectory dependent structural vibration. The time-varying cutting direction and cutting force at different positions will cause different performance of the structural sensitivity, resulting in different influence of the structure vibration on surface topography. Figure 4 and Figure 5 show the local and filtered surface topography map of the work piece at position 1 and position 2, respectively. From the comparison of these two pictures, it can be seen that there are some regular 'small peaks' in the surface topography of the work piece in Figure 5, separated by the green dashed line (as shown in Figure 5), but not in Figure 4. Through measurement, the width of the 'small peaks' in the surface topography is between 0.07mm-0.09mm. According to the feed rate (F=1500mm/min), the frequency reflected by the 'small peaks' is 277.8Hz to 357.1Hz. Therefore, a conclusion can be inferred that the 'small peaks' are caused by structure vibration, rather than the tool vibration (usually higher than 10000Hz).

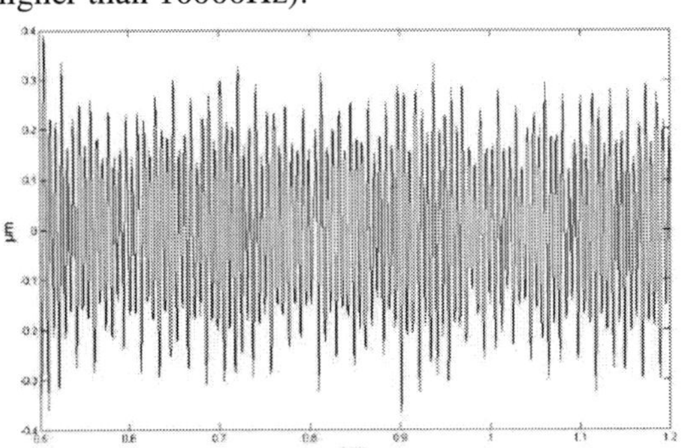

Figure 4. The filtered surface topography map of

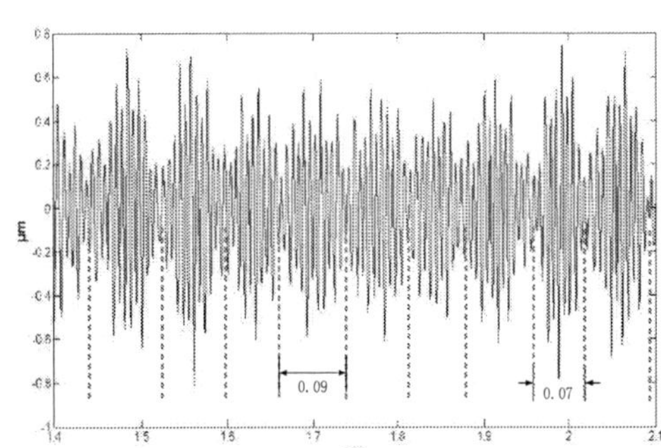

Figure 5. The filtered surface topography map

IOP Conf. Series: Materials Science and Engineering **491** (2019) 012009 doi:10.1088/1757-899X/491/1/012009

the work piece at position 1 of the work piece at position 2

Through the comparison of Figure 4 and Figure 5, it can be deduced that the effect of structural vibration on surface topography is different at different machining positions. Figure 6 shows the single path processing responses of work-piece in X and Y direction. From the comparison of work-piece responses at position 1 and position 2, the amplitude of the response signal at position 2 is higher and more complex than at position 1. The result indicate that the structure sensitivity of the machine tool is related to the machining trajectory, resulting in different performance of the structure vibration on surface topography at different machining position.

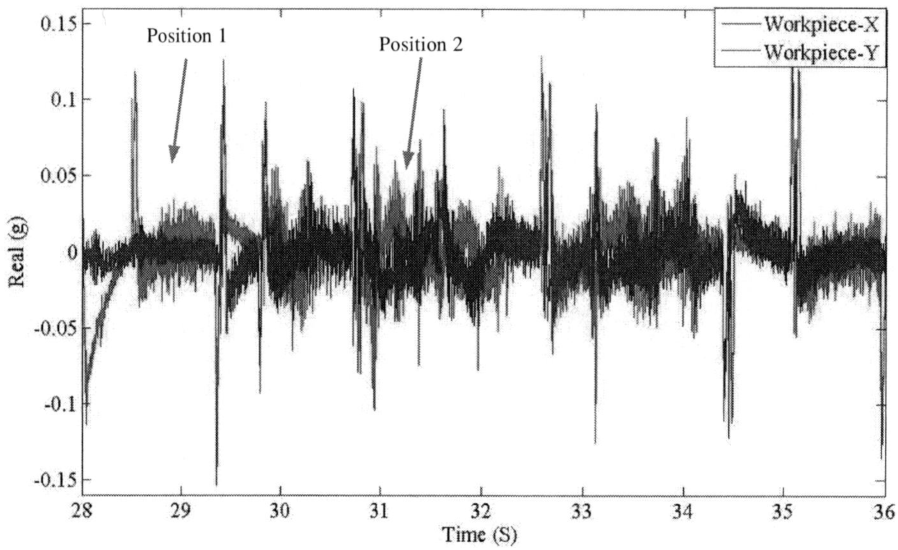

Figure 6. The single path processing responses of workpiece in X and Y direction

3.3.3. Verification of MLFVSS method. In order to validate the proposed method that the multimode low-frequency vibration of the sensitive structures (MLFVSS) has a significant impact on the surface topography. The FRFs signals of different structures of the TC500 machine needed to be compared. Figure 7 shows the comparison of FRFs of different structures. As shown in Figure 7, the sensitive structures (worktable, X-slide) have higher amplitude rather than non-sensitive components (headstock, column) at the natural frequencies in Table 1, indicated that the multimode low-frequency vibration of the sensitive structures (MLFVSS) directly determines the surface topography in actual processing.

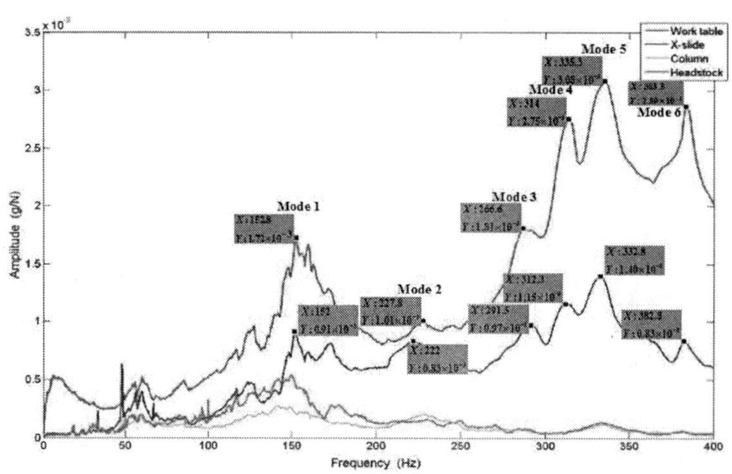

Figure 7. The comparison of FRFs of different structures of TC500 machine.

4. Conclusions

In this paper, a new approach identifying the structure vibration on surface topography in milling machining was proposed combined with the method of structure sensitivity analysis. According to the trajectory-related milling experiments on TC500 machine, this new method was validated and a mathematical model to predict the products' surface topography in milling machining was established.

The following is the conclusions and highlights:

(1) This new method combined structure sensitivity analysis with surface topography prediction.

(2) Multimode low-frequency vibration of the sensitive structures (MLFVSS) directly determines the surface topography in milling machining.

(3) The sensitivity is different for different structures at the same mode or the same structure at different modes.

(4) The structure sensitivity is related to the machining trajectory, resulting in different influence of the structure vibration on surface topography.

(5) The surface topography prediction model in milling was related to the feed rate and the natural frequencies of sensitive structures.

References

[1] Lee, W.B. and C.F. Cheung, A dynamic surface topography model for the prediction of nano-surface generation in ultra-precision machining. International Journal of Mechanical Sciences, 43(4): p. 961-991 (2001).

[2] Liu, K. and S.N. Melkote, Effect of plastic side flow on surface roughness in micro-turning process. International Journal of Machine Tools and Manufacture, 46(14): p. 1778-1785 (2006).

[3] To, S., et al., Influence of material swelling on surface roughness in diamond turning of single crystals. Materials Science and Technology, 17(1): 102-108 (2001).

[4] C.F. Cheung, S. To, W.B. Lee, Anisotropy of surface roughness in diamond turning of brittle single crystals, Materials and Manufacturing Processes, 17(2): 251–267 (2002).

[5] Zhang, S.J., et al., Dynamic characteristics of an aerostatic bearing spindle and its influence on surface topography in ultra-precision diamond turning. International Journal of Machine Tools and Manufacture, 62: p. 1-12 (2012).

[6] Kim DS, Chang IC and Kim SW. Microscopic topographical analysis of tool vibration effects on diamond turned optical surfaces. Precis Eng, 26(2): 168–174 (2006).

IOP Conf. Series: Materials Science and Engineering **491** (2019) 012010 doi:10.1088/1757-899X/491/1/012010

Analysis of Co-Flow Jet Effect on Dynamic Stall Characteristics Applying to Rotor Airfoils

Xu-Dong Yang, Wei-Rong Jiang and Shun-Lei Zhang*

School of Aeronautics, Northwestern Polytechnical University, Xi'an 710072, China

*Corresponding author: zhangslnwpu@163.com

Abstract. The performance of helicopter is limited by the phenomenon of dynamic stall. Dynamic stall is attributed to the time-varying angle of attack (AOA) of blade element resulting from blade flapping cyclic pitch inputs and wake flow. Co-flow jet(CFJ), as a novel flow control technique proposed by the joint research team from University Miami (UM) and AFRL has shown great improvement on lift enhancement, drag reduction due to its strong circulation improvement and turbulent mixing effect. Focusing on dynamic stall characteristics, this paper studies the effect of co-flow jet technique applying to rotor airfoil OA309. Numerical simulations show that CFJ can remove dynamic stall phenomenon occurring on rotor airfoils but the momentum coefficient should be carefully chosen so as to obtain high-efficient control effect.

1. Introduction

The phenomenon of dynamic stall has long been known to be an essential factor that limits helicopter performance [1]. Sources of dynamic stall are very complex. When a helicopter is moving forward, the advancing blade operates at low AOA whereas retreating blade operates at much lower Mach numbers but encounters very high AOA. Besides that, blade elastic deformation, rotor wake and tip vortices also affect the AOA environment of a blade element. The motion of the rotor blade can be simplified as sinusoidal pitching and plunging. The physics of the dynamic stall and related experiments can be seen in references [2-7]. Although Dynamic stall can delay the onset of flow separation, it will bring a rapid aft movement of the center of pressure and dramatic drag increase due to the unstable vortex shedding process. Thus, the dynamic stall of helicopter rotor should be avoided as far as possible.

The novel flow control technique, Co-flow jet, proposed by the joint research team from University of Miami (UM) and AFRL, has shown great improvement on airfoil performance by massive CFD simulations and wind-tunnel experiments [8-12]. Its fundamental configuration is sketched in Figure 1. Near the leading edge on the suction surface of the airfoil, a high energy jet is injected tangentially and the same amount of mass flow is sucked in near the trailing edge. The turbulent shear layer between the main flow and the jet causes a strong turbulence diffusion and mixing which enhance the ability of main flow to overcome the severe adverse pressure gradient and keep flow attached. Moreover, because of the increase of circulation, the lift is also significantly increased. Since Co-flow jet technique poses strong ability of delaying flow separation, it can be predicted that this technique is able to remove the dynamic stall of rotor airfoils.

Focusing on dynamic stall characteristics, the objective of the present work is to validate the viability of Co-flow jet technique using computational fluid dynamic and discuss the control effect of this technique applying to the OA309 airfoil.

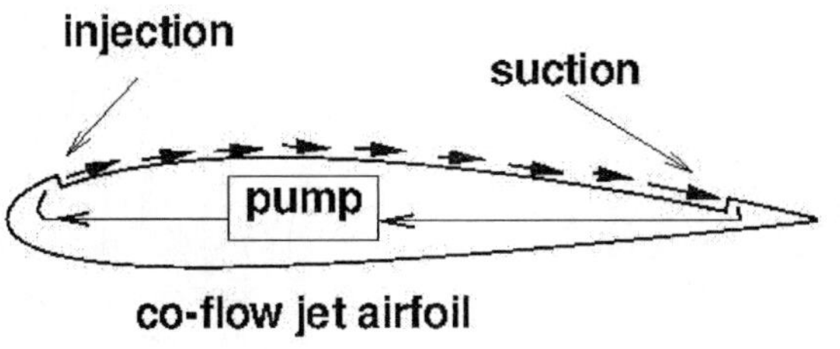

Figure 1. Co-flow jet airfoil configuration.

2. Overview of CFD method

2.1. Geometry and Mesh
The OA309 airfoil is chosen as baseline to construct CFJ-foil naming CFJ309 for short. Its configuration parameters are shown in Table 1.

Table 1. CFJ309 airfoil parameters.

Parameters	Values
Injection location	6%c
Injection size	0.75%c
Suction location	80%c
Suction size	1.35%c

The computational domain used for CFJ airfoils is shown in Figure 2, in which is clear that it is composed of two subdomains (stationary outer subdomain and rotating inner domain). Between the subdomains there is a sliding interface that allows continuous flux of mass and momentum. Rotating subdomain has a diameter of 3 times chord length while the diameter of stationary outer subdomain is about 50 times chord length.

2.2. Computation set-up
The *SST* k-ω turbulence model is used to model the turbulent flow characteristics. The spatial and temporal discretization of all equations are specified with a second order scheme to obtain good accuracy in the results. An adaptive time stepping scheme is used with decrease factor 0.86 and increase factor 1.06. The target inner loop is settled as 20 and the maximum inner loop is 40.

The implemented boundary conditions consist of: the velocity inlet on the left side, the pressure outlet on the right side, the sliding interface condition applied between these two domains and no slip walls used for the airfoil and the duct surfaces.

The y^+ is set to about one near the walls so as to resolve the turbulent boundary layer. Fixed mass flow rate is allowed to go into duct across the injection boundary. Meanwhile, the suction boundary keeps adjusting static pressure in order to matching the injection mass flow rate.

A sinusoidal pitching motion around the quarter chord point is performed to the rotation domain. The motion expression can be written as:

$$\alpha=\alpha_0+\alpha_1\sin(2\pi k\tau) \quad (1)$$

Where, α is the free stream AOA, α_0 is the mean AOA, α_1 is the oscillation amplitude of AOA, the reduced frequency k characterizes the degree of unsteadiness of the aerodynamics and τ is a non-dimensional time variable.

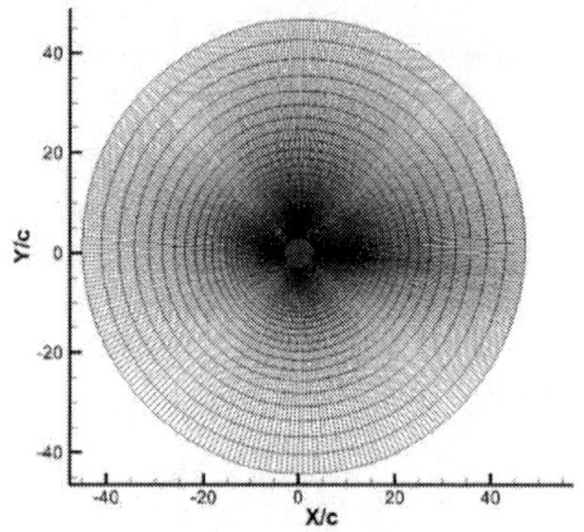

(a) Station(black) and rotation domain(blue)

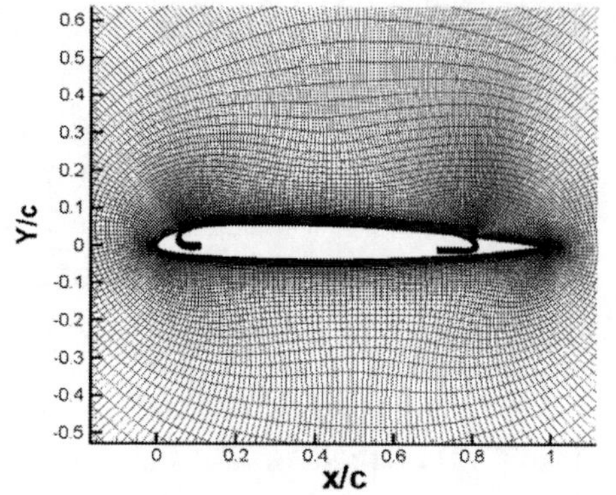

(b) CFJ309 airfoil mesh

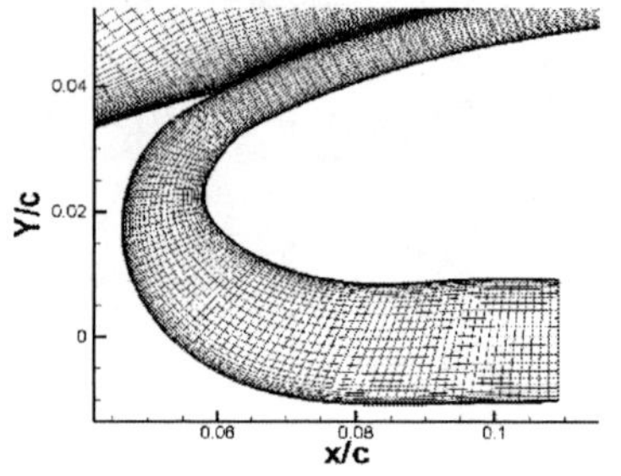

(c) Mesh near the injection slot

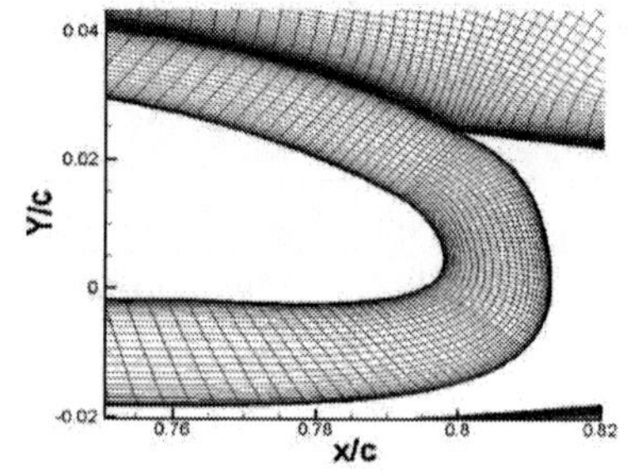

(d) Mesh near the suction slot

Figure 2. Computational mesh

2.3. Validation based on the NACA0012 airfoil

Due to the lacking of experimental data about dynamic stall of OA309 airfoil, the NACA0012 airfoil, whose dynamic stall characteristics has been tested by NASA [2], is selected to validate numerical methods. Table 2 shows the related sinusoidal motion parameters of NACA0012 airfoil. The Mach number is set to 0.301 and the Reynolds number is 3,770,000. The mesh size is about 60,000. Figure 3 presents the numerical and experimental results.

Results show that the numerical lift coefficient agrees well with the experimental lift coefficient, but slightly higher than that when approaching to maximum AOA. Regarding to moment coefficient, as shown in Figure 3, the numerical moment coefficient tends to lower than the experimental moment coefficient when AOA grows bigger, which indicates that numerical simulation over-predicts the dynamic stall effect slightly. The same conclusion can be obtained from the comparison on drag coefficient.

Table 2. Sinusoidal motion parameters of NACA0012 airfoil.

Parameters	Values
α_0	11.48°
α_1	9.84°
k	0.10

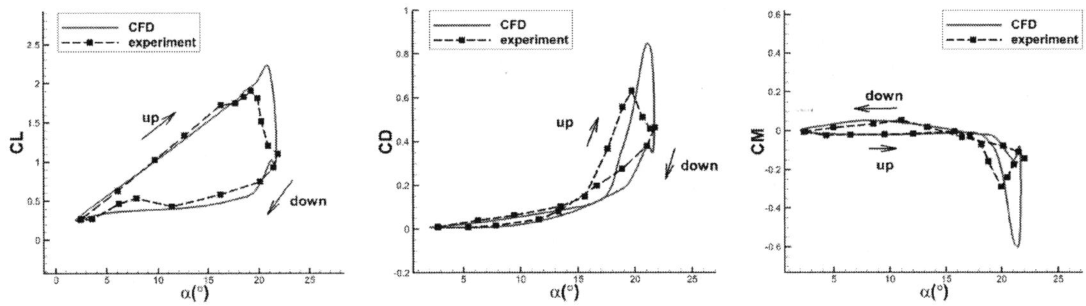

(a) Lift coefficient (b) Drag coefficient (c) Moment coefficient

Figure 3. Lift, drag and moment coefficient of NACA0012 airfoil during dynamic stall

3. Study of CFJ effect on dynamic stall

This section studies the performance of CFJ concept on suppressing dynamic stall. Three sinusoidal motion laws are chosen as shown in Table 3. The Mach number is set to 0.3, Reynolds number is 4,000,000 and the mesh size is 60,000. In order to quantify the strength of injection flow related to free stream, the jet momentum coefficient C_μ is defined in equation (2) (see Reference [12]). When the simulation is implemented, fixed mass flow $\dot{m}$ is set on injection slot and the V_{j1} is undetermined. For this reason, the nominal jet momentum coefficient is introduced based on V_{j1}^* (see equation 2, 3, 4).

$$C\mu = \frac{\dot{m} V_{j1}}{0.5 q_\infty S_{ref}} \quad (2)$$

$$C_\mu^* = \frac{\dot{m} V_{j1}^*}{0.5 q_\infty S_{ref}} \quad (3)$$

$$V_{j1}^* = \frac{\dot{m}}{\rho_\infty S_{j1}} \quad (4)$$

In equation (4), S_{j1} is the area of injection slot size.

Table 3. Sinusoidal motion parameters of OA309 and CFJ309 airfoil.

Items	α_0	α_1	k	C_μ^*
Case1	5°	10°	0.10	0.08
Case2	10°	10°	0.10	0.08
Case3	13°	13°	0.10	0.08

3.1. Case 1

Figure 4 shows the force coefficients vary versus AOA under circumstance of case 1 motion law in Table 3. It should be noted that all the force coefficients are consisted of two parts: surface force, injection and suction reacting forces caused by flow injection and suction. In case 1, since the maximum AOA is slightly higher than the static stall AOA, both airfoils still keep flow attached. The lift coefficients of both half-cycles approach each other. Due to unsteady boundary-layer displacement thickness effects, the moment coefficients during the upstroke phase doesn't correspond with those of downstroke. Besides, the moment coefficients of CFJ309 airfoil tend to descends during both upstroke and downstroke motions.

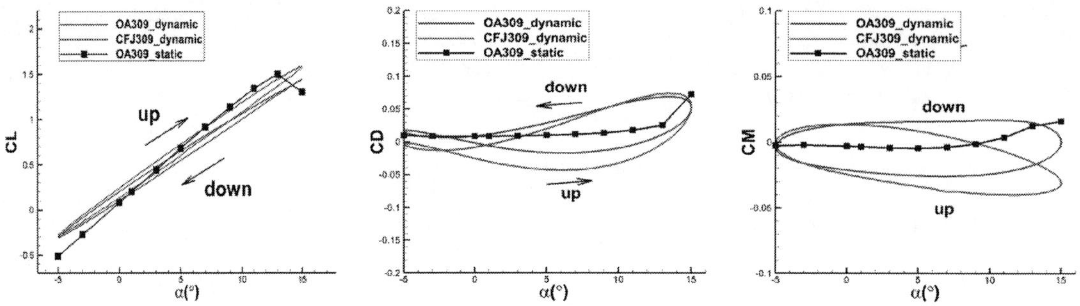

(a) Lift coefficient (b) Drag coefficient (c) Moment coefficient

Figure 4. Lift, drag and moment coefficient comparison between OA309 and CFJ309 airfoil during case 1 motion

3.2. Case 2

As the maximum AOA increases to 20deg in case 2, OA309 airfoil has already been in deep stall status while CFJ309 airfoil remains flow attached, as shown in Figure 5. Vortices contours of upstroke and downstroke motions have shown in Figure 6. Flows around the two airfoils are naturally attached when AOA is 15deg in upstroke motion and no separation appears. When AOA is 20deg, the blowing jet near leading edge forces flow to attach the CFJ309 airfoil surface while separation appears around the OA309 airfoil surface and then propagates downstream quickly. Finally, When AOA is 25deg, flow is fully separated from leading-edge and vortex shedding is observed around the OA309 airfoil.

There exists three layers of vortices near the injection slot of CFJ309 airfoil: (a) clock-wise boundary vortex(blue color) resulted from no-slip wall condition, (b) strong counter-wise CFJ vortex(red color) due to high speed injection effect near leading edge, (c) clock-wise secondary vortex between free stream and injection flow caused by large velocity gradient(blue color). These vortex structures transport the energy of injection flow into free stream so that flow remains attach to the airfoil surface when the maximum AOA is very large, i.e. 20deg.

3.3. Case 3

When the maximum AOA is large enough approaching to 13deg, corresponding to case 3 motion law, both of OA309 and CFJ309 airfoil undergo deep stall as we can see in Figure 7. The shapes of force hysteresis loop of CFJ309 airfoil are similar to those of OA309 airfoil except that the hysteresis loop areas of CFJ309 airfoil are smaller which means Co-flow jet can weaken the dynamic stall.

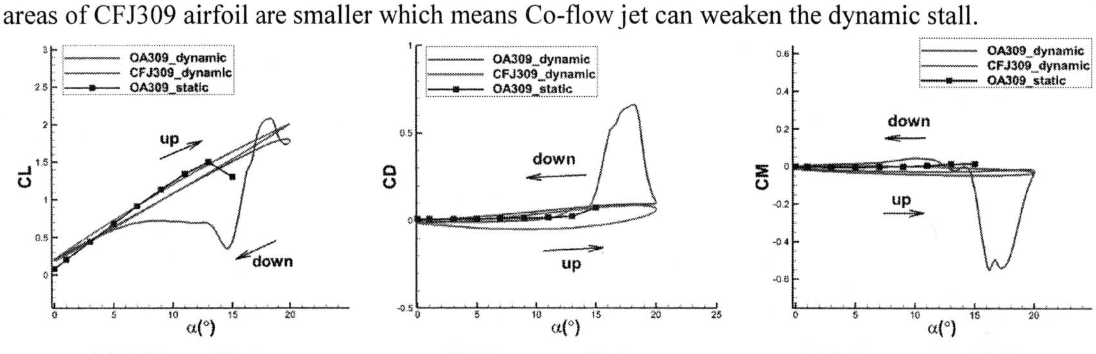

(a) Lift coefficient (b) Drag coefficient (c) Moment coefficient

Figure 5. Lift, drag and moment coefficient comparison between OA309 and CFJ309 airfoil during case 2 motion

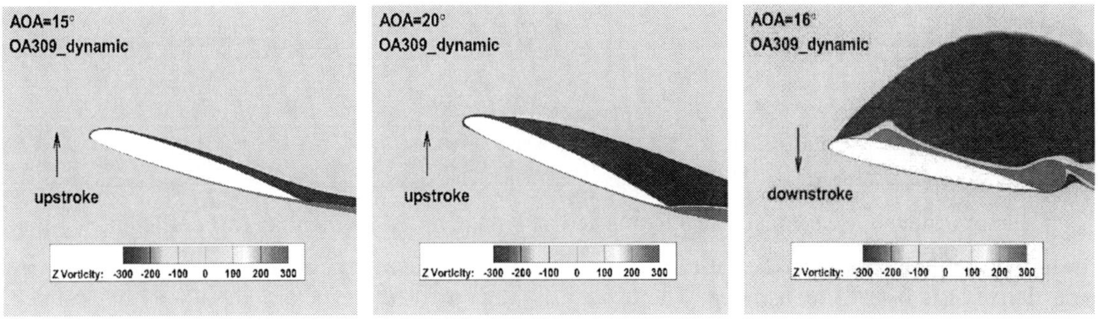

(a) OA309 airfoil (AoA=15° , upstroke)
(b) OA309 airfoil (AoA=20° , upstroke)
(c) OA309 airfoil (AoA=16° , downstroke)

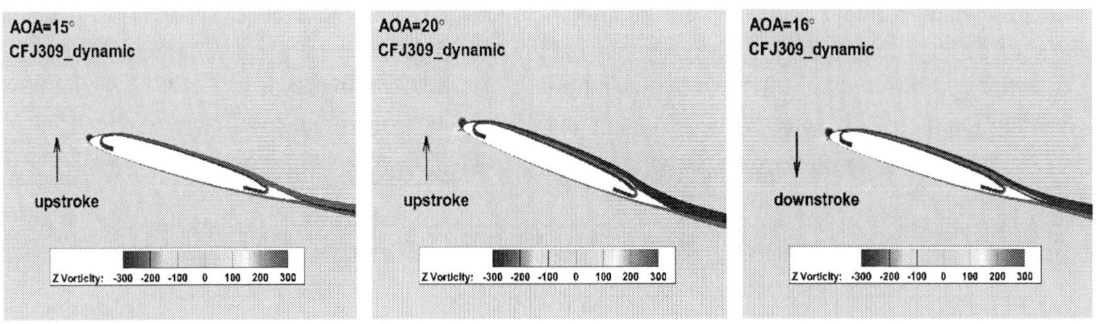

(d) CFJ309 airfoil (AoA=15° , upstroke)
(e) CFJ309 airfoil (AoA=20° , upstroke)
(f) CFJ309 airfoil (AoA=16° , downstroke)

Figure 6. Vortex contours comparison between OA309 and CFJ309 airfoil during case 2 motion

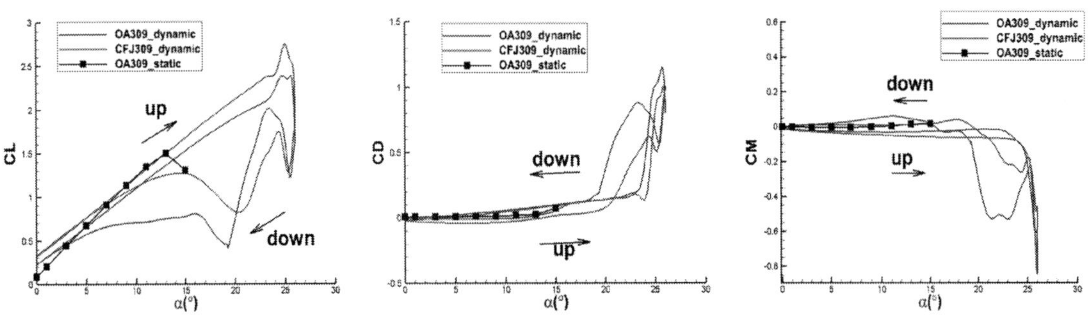

(a) Lift coefficient
(b) Drag coefficient
(c) Moment coefficient

Figure 7. Lift, drag and moment coefficient comparison between OA309 and CFJ309 airfoil during case 3 motion

3.4. Effect of jet momentum coefficient

This part studies how momentum coefficient influences the control of dynamic stall and evaluates efficiency of Co-flow jet technique taking into account the energy consumption of pump (power unit). The pump is used to create high speed jet and suck in equal air at the mean time. Power and efficiency coefficients are defined by equation (5), (6), respectively [12].

IOP Conf. Series: Materials Science and Engineering **491** (2019) 012010 doi:10.1088/1757-899X/491/1/012010

$$P_c = \frac{\dot{m} C p T_{t2}}{\eta}(\Gamma^{\frac{\gamma-1}{\gamma}} - 1) \tag{5}$$

$$(L/D) = \frac{CL}{CD + P_c} \tag{6}$$

The motion law of case 2 in Table 3 is selected and the nominal momentum jet coefficient C_μ^* ranges from 0.06 to 0.10. In Figure 8, the effect of nominal jet momentum coefficients on force and power coefficients has been demonstrated. The hysteresis loops of lift coefficient are almost overlapped at different jet momentum coefficients and so is for the moment coefficient. The drag coefficient decreases a bit when the jet momentum coefficient increases. Although, force coefficients change little, the power consumption rises quickly with the increase of jet momentum coefficient. Table 4 provides the time average quantities about different configurations. $(L/D)_e$ of OA309 airfoil is apparently lower than those of CFJ309 airfoil due to deeply dynamic stall. Although larger jet momentum coefficient gives higher lift and lower drag, the power consumption grows that leading to decrease of $(L/D)_e$. Power consumption should be concerned as selecting an adequate jet momentum coefficient.

Table 4. Time average quantities of CFJ309 airfoil at different C_μ^*.

Items	CL	CD	P_c	$(L/D)_e$
OA309 airfoil	0.9796	0.09210	0.0000	10.6
$C_\mu^* = 0.06$	1.0941	0.01524	0.02102	30.2
$C_\mu^* = 0.08$	1.1224	0.00951	0.03591	24.7
$C_\mu^* = 0.10$	1.1426	0.00330	0.05394	20.0

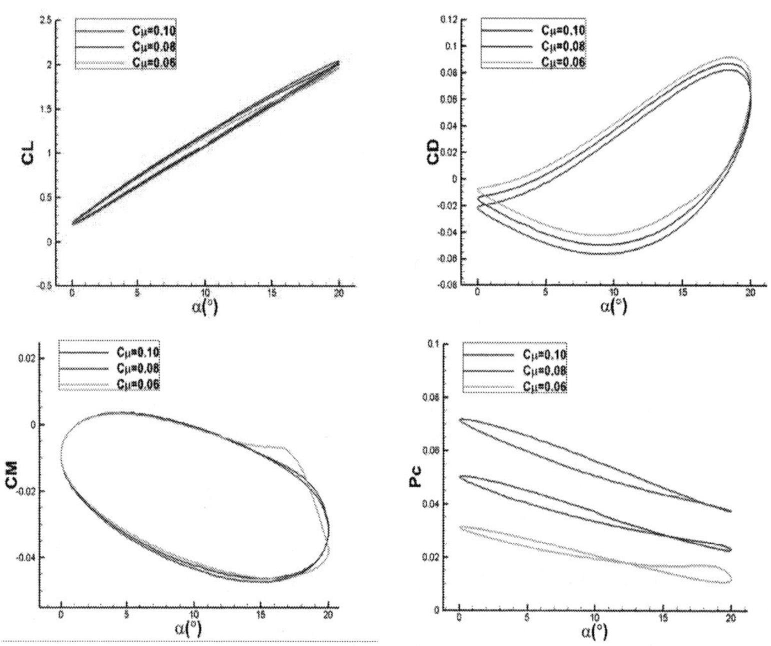

Figure 8. Lift, drag and moment coefficient comparison between OA309 and CFJ309 airfoil at different jet momentum coefficients

4. Conclusions

In this study, the control effect of Co-flow jet flow technique on dynamic stall are researched. The dynamic stall characteristics of OA309 airfoil are studied at three different types of sinusoidal motion law. Numerical simulations prove that co-flow jet can suppress dynamic stall and prevent massive vortex shedding. Vortex contours show that several vortex layers existing on the suction surface of CFJ309 airfoil contribute to energy mixing and CFJ309 airfoil keep flow attached while OA309 airfoil has been in deep stall. When the maximum AOA is big enough, both of OA309 and CFJ309 airfoil undergo deep stall, but the stall of CFJ309 airfoil can been lessened. The study of jet momentum coefficient shows that the jet momentum coefficient should be carefully chosen so as to obtain high-efficient control effect.

Acknowledgments

This research was funded by the National Natural Science Foundation of China (NSFC) under grant No.11672243.

References

[1] Leishman J G 2016 *Principles of helicopter aerodynamics* (London: Cambridge Univ. Press)
[2] Mccroskey W J *et al* 1981 *The Phenomenon of Dynamic Stall (NASA-TM-81264)*
[3] Mccroskey W J et al 1982 *An Experimental Study of Dynamic Stall on Advanced Airfoil Sections (NASA-TM-84245-Vol-3)*
[4] Choudhry A, Leknys R, Arjomandi M and Kelso R 2014 An insight into the dynamic stall lift characteristics *Exp. Therm. & Fluid Sci.* **58** 188-208
[5] Rival D and Tropea C 2010 Characteristics of pitching and plunging airfoils under dynamic-stall conditions *J. of Aircraft* **47** 80-86
[6] Mulleners K and Markus R 2013 Dynamic stall development *Exp. in Fluids* **54** 1–9
[7] Geissler W and Haselmeyer H 2006 Investigation of dynamic stall onset *Aero. Sci. & Tech.* **10** 590-600
[8] Zha G C and Paxton C D 2006 Novel flow control method for airfoil performance enhancement using co-flow jet *Applications of Circulation Control Tech. Progress in Astronautics and Aeronautics* 293-314
[9] Wang B Y, Haddoukessouni B, Levy J and Zha G C 2007 Numerical investigations of injection slot size effect on the performance of co-flow jet airfoil *25th AIAA Applied Aerodynamics Conf., Fluid Dynamics and Co-located Conf.*
[10] Zha G C and Gao.W 2006 Analysis of jet effects on co-Flow jet airfoil performance with integrated propulsion system *44th AIAA Aerospace Sciences Meeting and Exhibit, Aerospace Sciences Meetings* **2** 1273-91
[11] Liu Z X and Zha G C 2016 Transonic airfoil performance enhancement using co-flow jet active flow control *8th AIAA Flow Control Conf. (AIAA 2016-3472)*
[12] Lefebvre A M and Zha G C 2014 Co-flow jet airfoil trade study part I: energy consumption and aerodynamic efficiency *32nd AIAA Applied Aerodynamics Conf. (AIAA 2014-2682)*

IOP Conf. Series: Materials Science and Engineering **491** (2019) 012011 doi:10.1088/1757-899X/491/1/012011

An Experimental Investigation of the Thermohydrodynamic Performance of a Steady Loaded Hole-Entry Hybrid Journal Bearing

Haiyin Cao[1,2], Ruolin Wu[1,2,*], Huiping Chen[1,2] and Chao Lu[1,2]

[1]State Key Lab of Digital Manufacturing Equipment & Technology, Huazhong University of Science & Technology, 430074, Wuhan, China
[2]School of Mechanical Science and Engineering, Huazhong University of Science & Technology, 430074, Wuhan, China

[*]Corresponding author e-mail: wrl515342@163.com

Abstract. The purpose of this paper is to show the results of parametric experiments about the thermohydrodynamic performance of the steady-state loaded hole-entry hybrid bearing. Because there are many available theoretical in that respect have been studied, it become an urgent need to do related experimental researches. In the present study, the internal pressure and temperature distribution, flow rate and shaft center position under different supply pressures were measured. At the same time, a series of tests under different load and speed conditions were conducted. The experimental results show that bearing performances are significantly affected by the temperature rise of oil film, meanwhile, speed is the most important factor affecting the bearing temperature distribution.

1. Introduction

The hole-entry hybrid journal bearings have the merits of precise positioning, no wear, large direct film stiffness and damping, controllability of critical speeds and harmful vibration attenuation, so they are widely used in high-speed turbo-machinery, machine tool spindles, precision grinding spindles, liquid rocket pumps etc. [1]. To utilize both hydrostatic and hydrodynamic actions in a more efficient way, the oil recess of the hydrostatic bearing has been modified to be shallower, even changed to step oil recess or directly cancelled, meanwhile the land area has been enlarged to fit higher speed. Compared with the traditional recessed bearing, the hole-entry hybrid journal bearing is easier to manufacture, but also retains the advantages of hydrostatic bearing. Geometry of the hole-entry hybrid journal bearing studied in this paper is shown in Fig. 1.

Due to the rapid development of technology, bearings often need to work under severer conditions. When the oil film bearing operate under high speed, heat is generated within the fluid-film for oil-shear, and temperature rise of oil and bearing shell occurs. This, in turn, results in a significant reduction in the viscosity of the lubricating oil, and a lower minimum oil film thickness. As a result, the flow field of the lubricant becomes distorted and the bearing performance is affected. In order to accurately and reliably survey the bearing performance, it becomes imperative to measure the temperature distribution of the bearing systems. In the following paragraphs some recent studies into the hole-entry journal bearing systems, both theoretical and experimental, are reviewed.

The theoretical research on the hole-entry hybrid bearings has acquired remarkable achievements in recent decades. Notable among these is the study by Rowe et al. [2] who demonstrated that the hole-entry bearings were particularly effective for better load support and low energy consumption at zero and high-speed operations when compared with other bearing configurations. A comprehensive review about the development of hydrostatic/hybrid journal bearing systems was presented by Rowe et al. [3]. Sharma et al. [1] developed a theoretical model to study the performance of a hole-entry hybrid journal bearing system by considering variation of viscosity due to temperature rise of the lubricant in the analysis, indicated that the variation of viscosity due to temperature rise of the lubricant film had a quite appreciable influence on the static and dynamic performance of a hole-entry hybrid journal bearing system. Garg et al. [4] combined the non-Newtonian behavior of lubricant with temperature rise to study the dynamic performance of a hole-entry hybrid journal bearing system. Very recently, Khatri et al. [5] studied the influence of textured surface on the performance of non-recessed hybrid journal bearing operated with non-Newtonian lubricant.

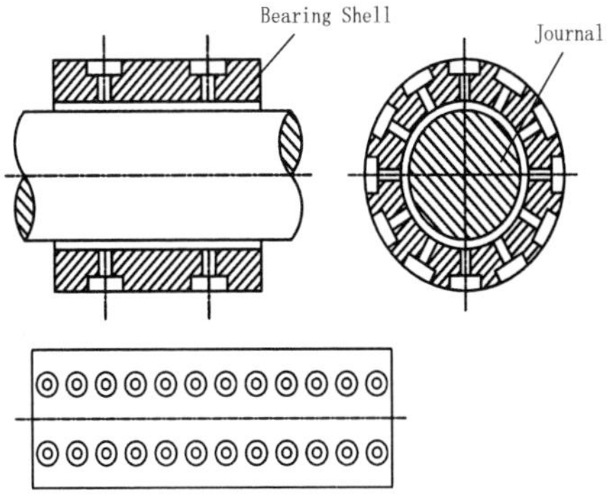

Figure 1. Bearing geometry.

As we know, few experimental researches into the thermohydrodynamic performance of hole-entry hybrid bearings have been performed. Xu et al. [6] designed a bearing test rig to study static and dynamic performance of oil-film journal bearings. The hole-entry bearing studied had 16 uniform distribution holes in double row, with a pair of pressure and displacement sensors orthogonal arranged under the shaft surface to measure the oil film pressure and thickness distribution. But Xu did not consider the effect of oil film temperature rise on the bearing performance. For other types of bearings, a lot of experimental researches appeared in the last few decades. For example, Dowson et al. [7] studied the pressure distribution and the temperature distribution on both bearing shell and shaft of a steady-loaded journal bearing under thermal equilibrium conditions, and concluded that the adiabatic assumption did not agree well with the experimental results and the shaft could be considered isothermal. Ferron et al. [8] studied the theoretical and experimental thermohydrodynamic problem of a finite length journal bearing, concluded that the thermal expansion and deformation of the shaft and the bush must be accounted in the theoretical and experimental processes. Ahmad et al. [9] conducted an experimental study of the effects of groove position on the temperature and pressure distribution of a journal bearing, revealed that when the groove position was in the divergent region, the oil film pressure was lowest. For hydrostatic/hybrid journal bearings, Sawicki et al. [10] experimentally investigated the full linear anisotropic rotordynamic model of a four-pocket, oil-fed, orifice-compensated hydrostatic bearing.

The available theoretical studies clearly indicate that the temperature rise of the lubricant fluid-film affects the performance of fluid-film hydrodynamic journal bearings quite appreciably. So it becomes an urgent need to do related experimental researches. In the present study, extensive experimental work has been conducted to determine the effects of supply pressure, load and speed on the thermohydrodynamic performance of a steady-state loaded hole-entry hybrid bearing.

2. Test Apparatus, Measurements and Bearing Specifications

An original experimental equipment is used in this paper to measure the bearing pressure and temperature profiles, as well as flow rate and steady-state shaft center position. The specific structure is illustrated in Fig. 2, and the photography of the whole test rig is shown in Fig. 3.

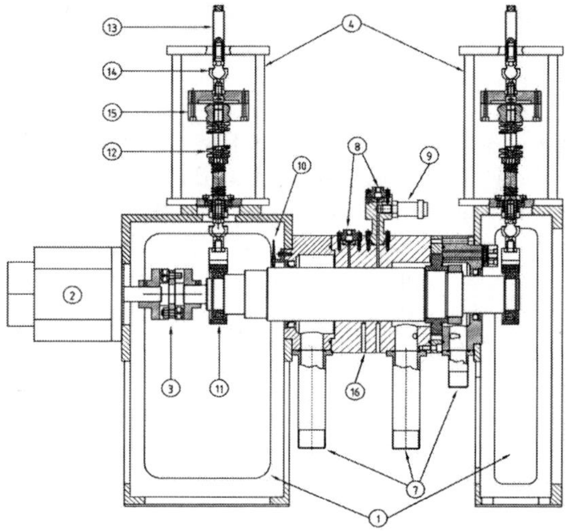

Figure 2. Diagram of the hole-entry hybrid bearing test rig. (1) Support structure, (2) Servo motor, (3) Cross rolling bearing sliding coupling, (4) Screw loading mechanism, (5) Tested bearing, (6) Spindle, (7) Oil return pipe, (8) Orifice restrictor, (9) Pressure transducer, (10) Displacement transducer, (11) Load rolling bearing, (12) Load spring, (13) Adjustment screw, (14) Boll joint universal coupling, (15) Weighing transducer, (16) Temperature transducer.

Figure 3. Photography of the hole-entry hybrid bearing test rig.

An orifice compensated hole-entry hybrid bearing with double row supply holes, each row 12 holes symmetrically, are tested. The bearing geometric and operating parameters are presented in Table 1. The bearing bush is fixed to both ends of the support structure (1), while the shaft is driven by servo motor through a self-aligning cross rolling bearing sliding coupling (3) to eliminate the rotor misalignment caused by the fluctuation of the spindle during the experiment. This structure is obviously different with that of Xu [6], and more close to the actual application of the bearing. As tested, the rotation errors of the spindle are within 1.5μm under different operating conditions, proving the validity of the test rig.

The data collected during the experiment include the flow rate of oil supply, the external load, the pressure of the oil inlet hole after restriction, the temperature distribution of the inner surface of the bearing and the steady shaft center position, successively corresponding to sensors such as flow rate transducer, weighing transducer, oil film pressure transducer, temperature transducer and displacement transducer.

For the loading device, in order to ensure the horizontal alignment of the spindle, the equivalent load at both ends of the spindle is imposed synchronously through hard compression springs. With adjustment screw (13) at the top of the support structure (1), the load can be changed through the compression of load spring (12). The load value can be displayed by the weighing transducer (15). Since the spindle is in a state of high speed rotation, the loading devices are in contact with the spindle through rolling bearings (11). In addition, considering that small oscillation of the spindle will occur in the test process, boll joint universal couplings are adopted in the load transfer structure to avoid restrictions of the shaft center position changes. The measurement range of the weighing transducer is up to 2T.

Table 1. Bearing operating and geometric parameters for the hole-entry hybrid bearing.

Parameter	Value
Journal diameter (D)	100 mm
Bearing length (L)	100 mm
External bearing diameter (D_2)	220mm
Radius clearance (c)	0.05 mm
Bearing land width (a_b)	25 mm
Bush material	Brass
Shaft material	Carbon steel
Supply pressure (P_s)	0~6MPa
Lubricant dynamic viscosity (μ_0)	1.806e-3 Pa·s at 40 ℃, 0.817e-3 Pa·s at 100 ℃
Number of restrictors per row (n)	12
Number of rows of holes in each bearing	2
Speed (N)	0~2200 rpm
Load (W)	0~500 Kg
Type of restrictors	orifice
Diameter of orifice (d_o)	0.5 mm

The measurement of oil flow is through a digital display type turbine flow transducer, which is installed at the outlet of the oil pump. According to the results of numerical calculation, the flow meter range is up to 20 L/min. For the oil film pressure must be measured by the contact-type sensor, the pressure sensor is installed in single row of oil supply holes after the orifice restrictors, so as to avoid the damage of the bearing inner surface and collect enough pressure points at the same time. Then the oil inlet pressure is measured to reflect the distribution of oil film pressure. The temperature distribution is nonuniform as that of the oil film, so it is necessary to collect enough data points to completely reflect the temperature distribution of the oil film. In this paper, contact type platinum resistance thermocouples are used, with temperature range of -30~150 ℃ and combined error of ±0.15 ℃. The distribution of the temperature sensor in expanded view is shown in Fig. 4. Eddy current displacement transducers are adopted here to measure the shaft center position changes, with the range of 0~0.5mm, and accuracy of 0.04um at 100Hz. In this rig, the two displacement transducers are installed orthogonally on the support structure, near the drive end of the shaft. All measurements were carried out in equilibrium state. Essential repeated experiments were carried out to verify the repeatability of the test results.

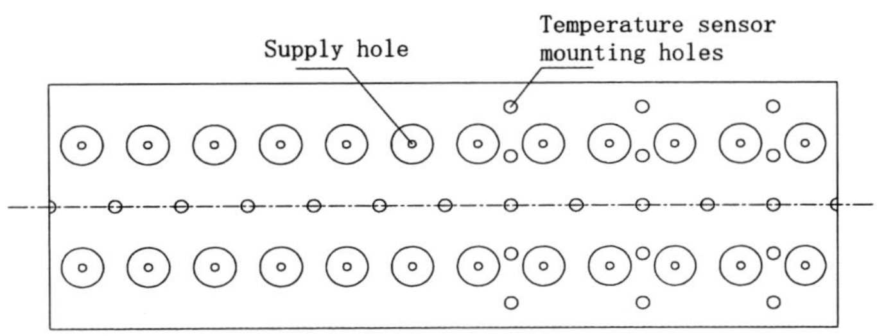

Figure 4. Location of thermocouple and pressure holes.

3. Results and Discussion

The experiment produced a large amount of data, only part of them are presented and analyzed here. Generally, the oil temperature was set at 30 ℃, with several contrast test at 25 and 35 ℃. In order to verify the influence of temperature rise on the bearing performance, transient experiment and steady thermal experiment were carried out in this paper. The difference between them lay in the control of temperature rise during the experiment. For the transient experiment, experimental data was collected before thermal equilibrium, and temperature rise was controlled within 1 ℃. While for the steady thermal experiment, the bearing working time must be long enough to allow the bearing to reach thermal equilibrium. The digital acquisition instrument was used to monitor temperature rise of the bearing. During both experiments, the temperature of supply oil could be set and adjusted automatically through the oil cooling system, and supply oil temperature range was controlled within ±0.5 ℃.

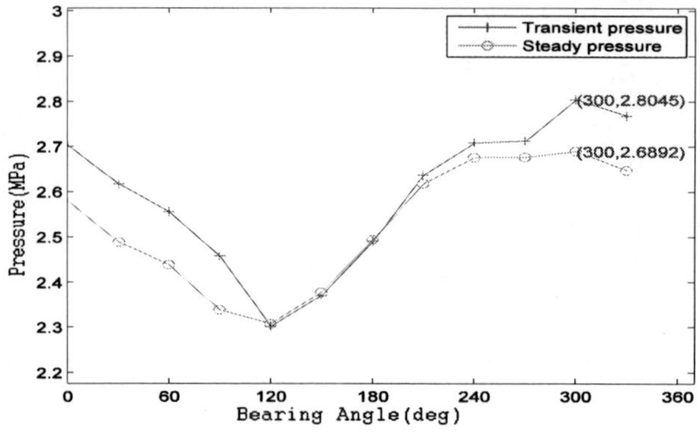

Figure 5. Pressure distribution of transient experiment and steady thermal experiment.

For operating condition of steady load 3KN, supply pressure 4MPa, speed 1000, 1500 and 2000 rpm, the pressure and temperature distribution of transient experiment and steady thermal experiment are shown in Fig. 5. It can be seen that oil film pressure of steady thermal experiment is lower than that of transient experiment. The drop of maximum oil film pressure can reach 4.11%. Data of flow rate and shaft center position were also collected at the same time, flow rate increased from 11.8 L/min to 12.8 L/min, an increase of 8.5%; eccentricity increased from 0.370 to 0.623, an increase of 68.4%. It is verified that the temperature rise of the oil film during the steady operation of the bearing has great influence on the bearing performance. Thus the steady thermal experiment must be carried out to better understand the actual working condition of the bearing.

Under the condition of load 1, 3, 5KN, supply pressure 2, 4, 5MPa, rotating speed 1000, 1500, and 2000 rpm, the pressure measurement results of steady thermal experiment are shown in Fig. 6. Relatively speaking, the effect of rotation speed on the oil film pressure distribution is not obvious. Under high rotational speed (2000 rpm), the oil film pressure is lower than that of low speed, which is caused by the temperature sharp increase at high speed. With the increase of oil supply pressure, the overall trend of the oil film pressure distribution is constant, only increasing in amplitude. The effect of external load on the oil film pressure distribution is obvious, oil film pressure distribution becomes more uneven with the increase of load, the difference between peak and valley values increases.

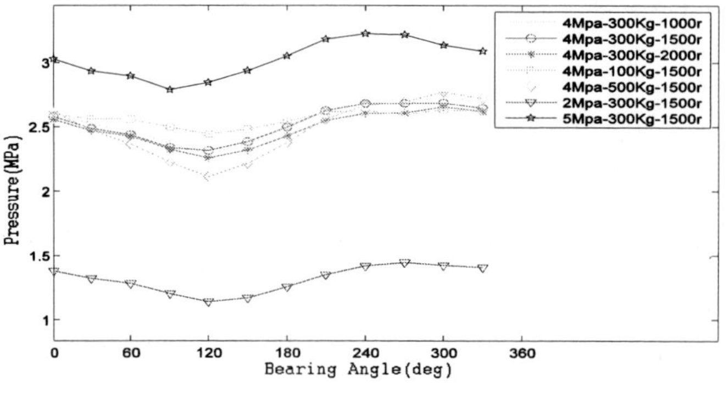

Figure 6. Pressure distribution of steady thermal experiment versus load, supply pressure and speed.

When the steady load is 3 KN, the oil supply pressure is 4MPa, and the speed is 1500 rpm, the temperature and pressure measurement results of the steady thermal experiment are illustrated in Fig. 7 and Fig. 8. The radial temperature distribution compared to the pressure distribution is more complex, overall speaking the loading region has higher temperature; compared to maximum pressure, the maximum temperature moves forward the direction of rotation with 45°. The axial temperature distribution, as illustrated in Fig. 8, is relatively uniform as a whole, showing the trend of high on both sides and low in the middle, which can be explained by the bearing side leakage flow. The basic trend of pressure and temperature distribution is in accordance with the theoretical results of reference [1].

The temperature measurement results of steady thermal experiment are shown in Fig. 9. The steady-state load is 1, 3, 5 KN, the supply pressure is 2, 4, 5 MPa, and the speed is 1000, 1500 and 2000 rpm, with supply oil temperature 30 ℃. The temperature rise of the oil film can reach 6.8 ℃ under the test conditions. It is clear that the effect of speed on temperature rise is the most obvious. At low speed, due to the drop in loading capacity, the temperature rise of the loading region is more pronounced. The shear effect of the oil film increases at high speed, resulting in significant temperature rise in the whole bearing. When the oil pressure is low, the cooling effect of the lubrication oil is relatively weak, which leads to the heat accumulation in the loading region and the temperature rise is remarkable. With the increase of the oil pressure, the temperature rises of the oil film increases overall, but not obviously. With the increase of load, the temperature distribution of oil film is on the rise as a whole, and the oil film temperature rise in the loading region is apparently increased under the condition of heavy load (5 KN).

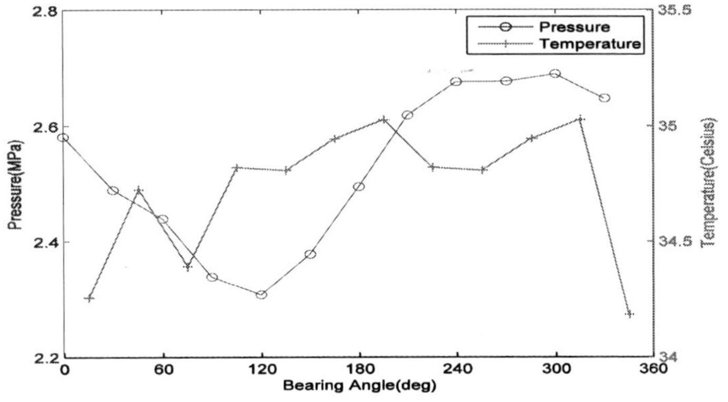

Figure 7. Radial pressure and temperature distribution of steady thermal experiment at load 3 KN, supply pressure 4 MPa and speed 1500 rpm.

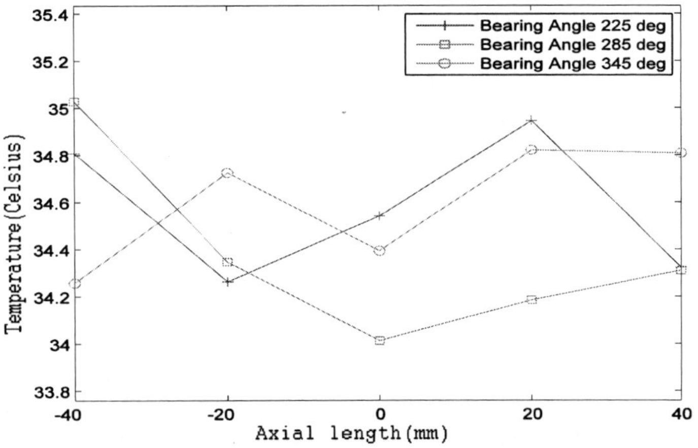

Figure 8. Axial pressure and temperature distribution of steady thermal experiment at load 3 KN, supply pressure 4 MPa and speed 1500 rpm.

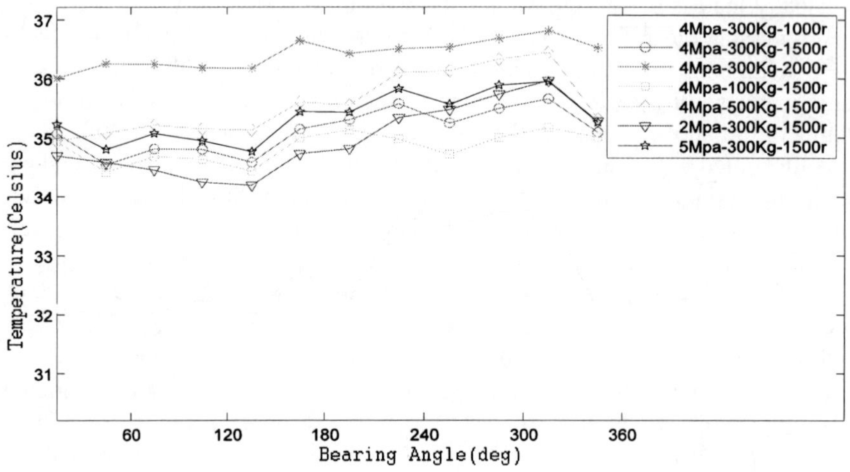

Figure 9. Temperature distribution of steady thermal experiment versus load, supply pressure and speed.

The load capacity of the bearing can be intuitively expressed through the change of the eccentricity ratio, which is illustrated in Fig. 10. The eccentricity ratio increases with the increase of load, and decreases with the increase of speed and oil supply pressure.

It needs to be pointed out that under the experimental conditions, the bearing bush is thick enough and directly contact with the outside space. The effect of heat dissipation is obvious, and the temperature rise is not high. So the influence of the thermal deformation of the bush and shaft is not considered here.

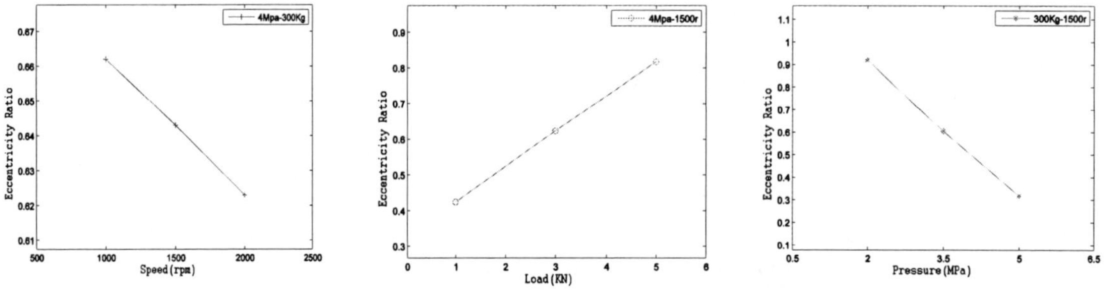

Figure 10. Eccentricity ratio of steady thermal experiment versus load, supply pressure and speed.

4. Conclusions

Experimental measurements of parametric experiments about the thermohydrodynamic performance of a steady-state loaded hole-entry hybrid bearing were performed. Experimental results indicating the effects of temperature rise on bearing performance for different speeds and loads had been presented. For the specific experimental operating conditions employed in this study, the following conclusions can be drawn:

(1) The influence of the oil film temperature rise on the bearing performance cannot be ignored. Bearing internal pressure of steady thermal experiment is lower than that of transient experiment. The drop of maximum oil film pressure can reach 4.11%.

(2) The shaft speed is the main factor affecting the temperature rise of the oil film. With the increase of oil supply pressure, the temperature rises of the oil film changed little, and the lubricating oil acted as the coolant to a certain extent. The oil film temperature rise in the loading region is obviously increased under the condition of heavy load.

(3) The eccentricity ratio, which illustrates load capacity, increases with the increase of load, and decreases with the increase of speed and oil supply pressure.

Acknowledgements
This research is supported by the National Natural Science Foundation of China (NSFC) No. 51875223, Science Challenge Project (TZ2018006-0102-03) and China Postdoctoral Science Foundation (2017M622427).

References

[1] Sharma,Satish C., et al. Study of hole-entry hybrid journal bearing system considering combined influence of thermal and elastic effects[J]. Tribology International,36.12 (2003): 903-920.
[2] Rowe, W. Brian, et al. Hybrid journal bearings with particular reference to hole-entry configurations[J]. Tribology International,15.6 (1982): 339-348.
[3] Rowe,W. B. Advances in hydrostatic and hybrid bearing technology[J]. Proceedings of the Institution of Mechanical Engineers, Part C: Mechanical Engineering Science, 203.4 (1989): 225-242.
[4] Garg, H. C., Vijay Kumar, and H. B. Sharda. Performance of slot-entry hybrid journal bearings considering combined influences of thermal effects and non-Newtonian behavior of lubricant[J]. Tribology International, 43.8 (2010): 1518-1531.
[5] Khatri, Chandra B., and Satish C. Sharma. Influence of textured surface on the performance of non-recessed hybrid journal bearing operating with non-Newtonian lubricant[J]. Tribology International, 95 (2016): 221-235.
[6] Xu, Shangxian. Experimental investigation of hybrid bearings[J]. Tribology transactions, 37.2 (1994): 285-292.
[7] Dowson, D. J. D. B. C. N., et al. Paper 3: An Experimental Investigation of the Thermal Equilibrium of Steadily Loaded Journal Bearings[J]. Proceedings of the Institution of Mechanical Engineers, Conference Proceedings. Vol. 181. No. 2. Sage UK: London, England: SAGE Publications, 1966.
[8] Ferron, J., J. Frene, and R. Boncompain. A study of the thermohydrodynamic performance of a plain journal bearing comparison between theory and experiments[J]. Journal of Lubrication Technology, 105.3 (1983): 422-428.
[9] Ahmad, Mohamad Ali, Salmiah Kasolang, and R. S. Dwyer-Joyce. Experimental study on the effects of oil groove location on temperature and pressure profiles in journal bearing lubrication[J]. Tribology International, 74 (2014): 79-86.
[10] Sawicki, Jerzy T., R. J. Capaldi, and M. L. Adams. Experimental and theoretical rotordynamic characteristics of a hybrid journal bearing[J]. Journal of tribology, 119.1 (1997): 132-141.

Terminal Control Approach Application for Autonomous Airship Docking

Fedorenko Roman[1,*] and Gurenko Boris[2]

[1] Innopolis University, 1, Universitetskaya Str., Innopolis, 420500, Russia
[2] Southern Federal University, 105/42, Bolshaya Sadovaya Str., Rostov-on-Don, 344006, Russia

*r.fedorenko@innopolis.ru

Abstract. The article considers the position-trajectory approach to the synthesis of the terminal regulator for controlling the airship for the docking task. The control system constructed solves the problem of guiding the airship to the docking position at given moment of time in autonomous mode. The airship control algorithm is based on mathematical model of an airship as rigid body, represented by a system of nonlinear differential equations, without linearization. The procedure for the control law synthesizing and the results of modelling by means of MATLAB are given. The task of docking of two airships, as proposed in EU MAAT project, was considered for simulations. The proposals for the further development of this study are formulated.

1. Introduction

Currently, one of the most promising applications of airships is the transportation of goods and people [1]. Various projects related to the achieving of this goal assume, that system is functioning as follows: an aeronautical platform (cruiser) is operating at a cruising altitude and the airship-shuttle preforms docking to this platform for several needs [2, 3].

It should be noted that a similar task of docking control in near-earth space and for underwater vehicles is carried out using classical approaches [4 – 10], for example, by program control [4, 5] by applying a linear-quadratic controller [6], the method of potential fields [7]. The main emphasis in mentioned works is made on the features of the technical implementation of control systems with high reliability and fault tolerance. Linear regulators with the separation of motion into longitudinal and transverse components are also most often used in the synthesis of the control laws of the airships motion [9, 10]. However, the control systems for the docking of airships, built on the basis of such approaches, are operable in a limited range of velocities and disturbances acting on the airship. It is also difficult to use this approaches when solving the problem of docking in three-dimensional space. Multiply connected nonlinear models are used much less often in the synthesis of control systems for airships. This fact does not allow to effectively control airships or expand their functional capabilities.

Thus, the task of terminal control of the docking of the airship, i.e. the task of bringing this aircraft from the initial position to some desired final position in the space at a given time and taking into account the its nonlinear mathematical model is actual.

The most commonly used approach for the problem of terminal control involves two stages:

- selection and calculation of the program trajectory as a certain function of time and

- the solution of the stabilization problem, which ensures the motion of the mobile object along a given program path [11].

However, the complexity of synthesizing the law of terminal control increases substantially if a nonlinear model of motion of a mobile object is used. Therefore, this work considers the problem of terminal control of airships, the model of which is represented by a system of nonlinear differential equations.

2. Airship mathematical model and statement of the task of terminal control of airship docking

To represent the mathematical model of an airship, we will use two coordinate systems, movable and fixed, shown in Fig. 1.

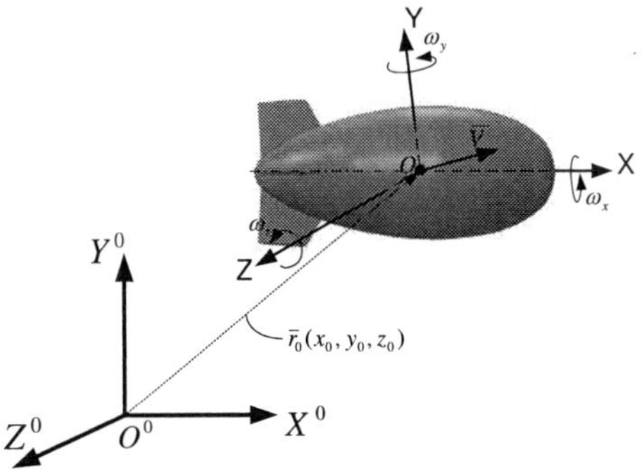

Figure 1. Coordinate systems K ($OX^0Y^0Z^0$) and K (OXYZ)

Let the mathematical model of an airship, constructed on the basis of the known equations of a rigid body, be given by a system of nonlinear differential equations and be represented in the following vector-matrix form [12]:

$$\dot{Y} = \begin{bmatrix} R_v & 0_{3x3} \\ 0_{3x3} & R_\omega \end{bmatrix} X \quad (1)$$

$$M\dot{X} = F_u(X, Y, \delta, l, t) + F_d(P, V, W) + F_v(G, A, R) \quad (2)$$

$$T_{иy}\dot{\delta} + \delta = \Psi_{иy}(U, \delta), \quad (3)$$

where $T_{иy}$ – diagonal (m x m) - matrix of time constants of actuators; $\Psi_{иy}(U, \delta)$ – m-vector of non-linear functions of the right sides of actuators equations; δ – m-vector of controlled coordinates (angles of deviation, thrust); U – m-vector of controls generated by the airship control system depending on the arrangement of actuators, $X = \begin{bmatrix} R_v \\ R_\omega \end{bmatrix}$ – m-vector of internal coordinates (state coordinates); M – $(m\times m)$-matrix of mass-inertial parameters, the elements of which are mass, moments of inertia, added masses of the airship; $F_u(X, Y, \delta, l, t)$– m- vector of control forces and moments, l – vector of design parameters; $F_d(P, V, W)$ – m-vector of nonlinear elements of airship dynamics; $F_v(G, A, R)$ – m-vector of measured and non-measurable external disturbances; $Y = \begin{bmatrix} P \\ \Theta \end{bmatrix}$ – n-vector of position P and orientation Θ of the body coordinate system *K(OX Y Z)* relative to the fixed coordinate system *K($OX^0Y^0Z^0$)*. Conversion procedure frome $F_u(X, Y, \delta, l, t)$ to U is described in [13]. It is required to synthesize such a control algorithm $F_u(X, Y, \delta, l, t)$, which would ensure the movement of the airship given by the model (1) - (3) from the initial position $Y_0 = (x_0, y_0, z_0, \varphi_0, \delta_0, \gamma_0)$ to the final position $Y_k = (x_k, y_k, z_k, \varphi_k, \delta_k, \gamma_k)$ at the given time $t = T_k$.

3. Method of terminal control of airship docking

To solve the problem of synthesis of terminal control, we use the procedure described in [11]. We define the program trajectory of the motion of the airship in a polynomial form and represent it in a matrix form:

$$Y^*(t) = CL(t), \quad (4)$$

$$\text{where } C = \begin{bmatrix} x_0 & 0 & \frac{3}{T_k^2}(x_k - x_0) & \frac{3}{T_k^3}(x_k - x_0) \\ y_0 & 0 & \frac{3}{T_k^2}(y_k - y_0) & \frac{3}{T_k^3}(y_k - y_0) \\ \dots & & \dots & \dots \\ \gamma_0 & 0 & \frac{3}{T_k^2}(\gamma_k - \gamma_0) & \frac{3}{T_k^3}(\gamma_k - \gamma_0) \end{bmatrix}, \quad L(t) = \begin{bmatrix} 1 \\ t \\ t^2 \\ t^3 \end{bmatrix}. \tag{5}$$

Thus, C – (n x 4)-matrix of constant coefficients that depend on the initial and final position of the airship, as well as on the given time point of arrival to the positioning point T_k . Moreover, it can be shown that

$$\dot{Y}^*(t) = CD_1L(t), \tag{6}$$

$$\ddot{Y}^*(t) = CD_2L(t), \tag{7}$$

$$\text{where } D_1 = \begin{bmatrix} 0 & 0 & 0 & 0 \\ 1 & 0 & 0 & 0 \\ 0 & 2 & 0 & 0 \\ 0 & 0 & 3 & 0 \end{bmatrix}, \quad D_2 = \begin{bmatrix} 0 & 0 & 0 & 0 \\ 0 & 0 & 0 & 0 \\ 2 & 0 & 0 & 0 \\ 0 & 6 & 0 & 0 \end{bmatrix}. \tag{8}$$

Formation of the law of terminal control is implemented on the basis of position-trajectory algorithms [14]. To do this, we set the error of the deviation of the real trajectory of motion from the specified program path (4) in the form:

$$\Psi_{tr} = Y - Y^* = Y - CL \tag{9}$$

We introduce an additional variable $\Psi = \dot{\Psi}_{tr} + T_1 * \Psi_{tr}$, where T_1- diagonal (n x n) -matrix of time constants, $\dot{\Psi}_{tr} = \dot{Y} - CD_1L$. And we define the desired behavior of a closed system in the form:

$$\dot{\Psi} + T_2 * \Psi = 0 \tag{10}$$

where T_2- diagonal (n x n) -matrix of time constants. Thus, the desired behavior of the system, expressed in terms of Ψ_{tr} is as follows:

$$\ddot{\Psi}_{tr} + T_1 * \dot{\Psi}_{tr} + T_2 * (\dot{\Psi}_{tr} + T_1 * \Psi_{tr}) = 0 \tag{11}$$

where $\dot{\Psi}_{tr} = \dot{Y} - CD_1L, \ddot{\Psi}_{tr} = \ddot{Y} - CD_2L$.

Substituting in (11) the equation of the mathematical model of the airship (1), (2) and expressing the corresponding vector of control forces and moments $F_u(t)$, we obtain the following control law:

$$F_u(t) = -MR^{-1}[\dot{R}X - CD_2L + T_1\dot{\Psi}_{tr} + T_2 * (\dot{\Psi}_{tr} + T_1\Psi_{tr})] - F_d - F_v \tag{12}$$

Structure of closed-loop control system, corresponding to the resulting algorithm is shown in Fig. 2 and includes the following elements: D_X-sensors of internal coordinates X, D_Y-sensors of external coordinates Y, D_V-external environment sensor, planner and controller $F_u(t)$.

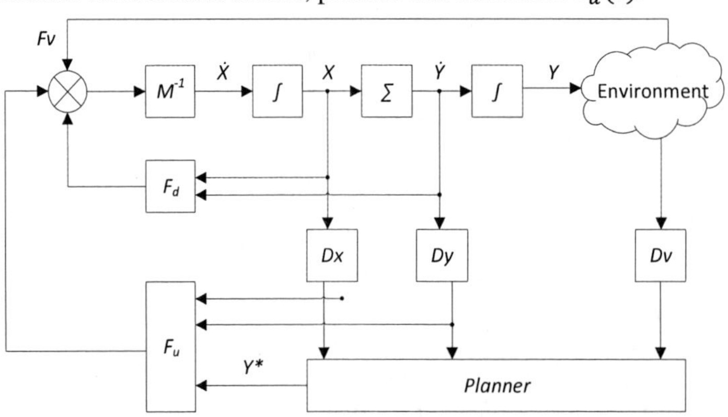

Figure 2. Structure of a closed system with model (1), (2) and control (12)

The system functions as follows: at the outputs of the sensors of internal and external coordinates D_X, D_Y current values of coordinates X and Y are formed. The planner based on the readings of the sensors of the external environment D_V forms the coefficients of the program trajectory $Y^*(t) = CL(t)$. According to the current values of the coordinates X, Y and the specified trajectory $Y^*(t)$, the control

IOP Conf. Series: Materials Science and Engineering **491** (2019) 012012 doi:10.1088/1757-899X/491/1/012012

calculation module F_u generates control actions in accordance with the expression (6). The value of forces F_d, F_v is formed either by the corresponding sensory system, or by an observer of external disturbances.

4. The terminal control of the airship docking simulation results

Simulation of docking of feeder to cruise was performed for two strategies:

- both airships (cruiser and feeder) are controlled by the described terminal regulator in order to simultaneously arrive at the point of docking;
- the cruiser does not change its mission and does not stop at docking point, the cruiser chases it up at a given time and moves with same speed at the final moment of the terminal control operation.

The results of the simulation of the first docking strategy are presented in Figures 3 – 5. The cruiser airship starts from the point with coordinates (0; 1100; 0) m, the feeder from the point (200; 900; 100) m. The specified docking point is (100; 1000; 200) m, the given terminal task duration is 100 seconds.

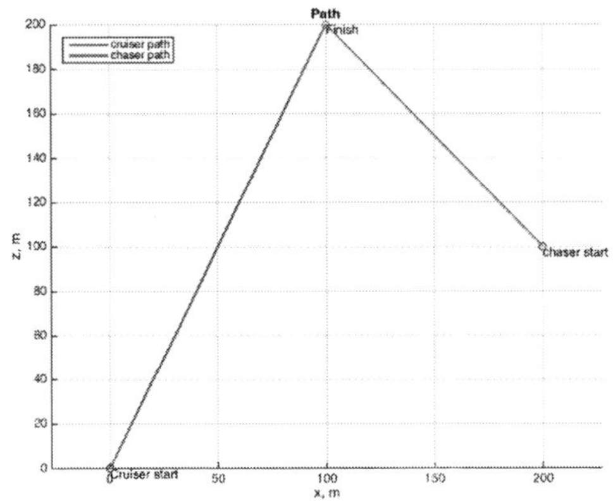

Figure 3. Trajectories of the cruiser and feeder for the first docking strategy

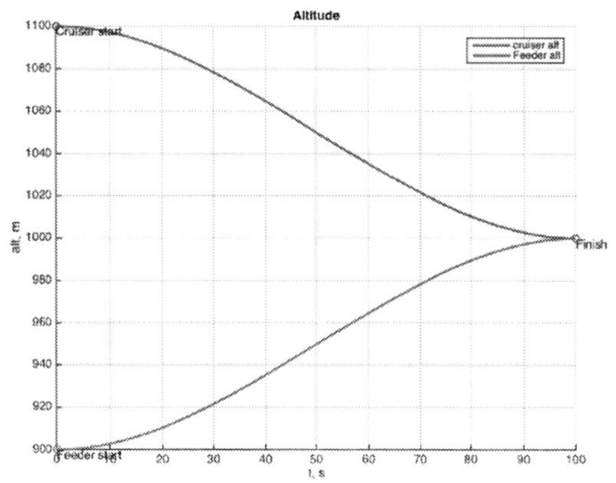

Figure 4. Cruiser and feeder altitudes for the first docking strategy

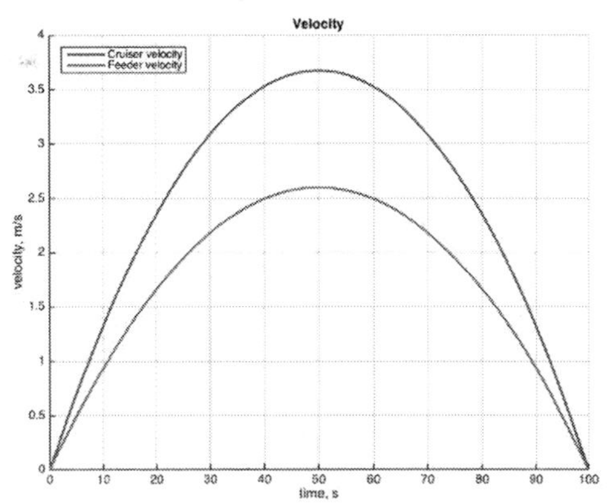

Figure 5. Cruiser and feeder velocities for the first docking strategy

Proposed approach efficiency is confirmed the above simulation results.

Comparing the plots of the cruiser and feeder speed changes in the first case, it can be seen that the maximum cruiser speed is 3.7 m/s and is higher than the maximum feeder speed of 2.6 m/s (Fig. 5), because the trajectory of the cruiser is longer, and in order to reach the final point at a given time Tk, the speed of its movement must be higher. Note, that there are possible situations when the energy capabilities of the airship will not allow it to reach a given point at a given moment in time, with a certain increase in the length of the trajectory.

The results of the second variant of docking simulation (when the cruiser airship does not change its mission and the feeder catch it up with it at given time) are presented in Figures 6-8. The cruiser airship moves from the point with coordinates (0; 1000; 100) m, feeder - from the point (0; 900; 0) m. The docking point is not defined (but determined by the position of the cruiser), the time of solving the terminal task is 100 seconds. The cruiser airship is moving along a given mission in a straight line at a constant speed at a constant altitude.

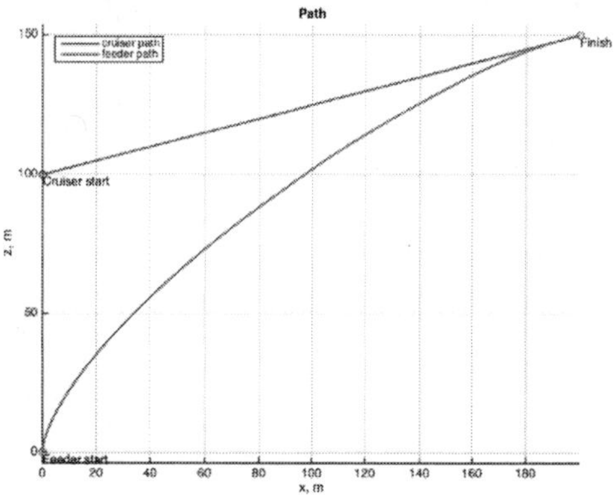

Figure 6. Trajectories of the cruiser and feeder for the second docking strategy

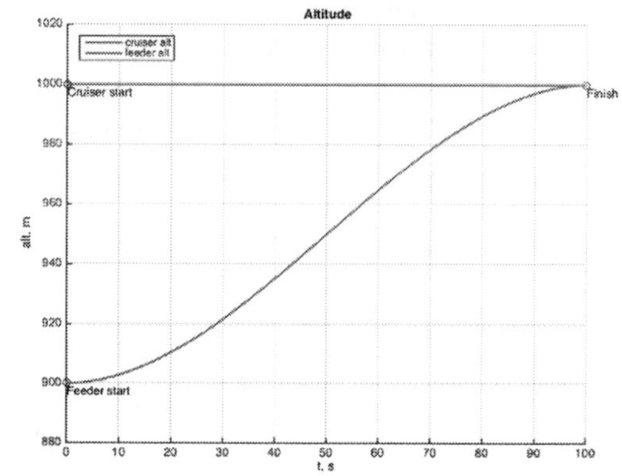

Figure 7. Cruiser and feeder altitudes for the second docking strategy

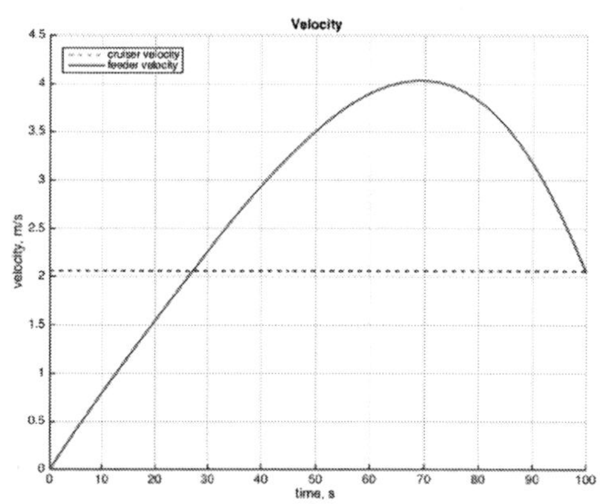

Figure 8. Cruiser and feeder velocities for the second docking strategy

Figure 8 shows that at a given point in time (after 100 seconds), the feeder airship chases the cruiser and moves with the same speed.

5. Conclusion

In the article the problem of terminal control of an airship, the model of which is represented by a system of nonlinear differential equations, is solved.

It is worth noting that to solve the terminal problem it is necessary to form a terminal point in time T_k taking into account the energy capabilities of the airship, i.e. it is necessary to take into account the limitations on the control actions and the power of the actuators. The decision of the task of forming the program trajectory of the airship movement, taking into account the limitations on control actions and the power of the actuators, is planned in the future work. It is also necessary to evaluate the influence of external and internal disturbances. In real conditions, in the process of docking of an airship, stationary or non-stationary obstacles may exist in the area of its operation. To avoid obstacles, there are many methods, in the future work it is proposed to use the obstacle avoidance method based on the use of unstable regimes.

Acknowledgments

The work was supported by grants of the President of the Russian Federation for the state support of young Russian scientists MK-142.2017.8 "Development of methods for airships control systems design unified to their technical characteristics and actuators", MK-3089.2017.8 "Methods for developing of an intelligent group control system of autonomous marine robotics", internal grant of the Southern Federal University "Development of theoretical foundations and intellectual methods of controlling the terminal state of mobile vehicles under conditions of uncertainty" VnGr-07 / 2017-19.

The authors are grateful to the Council on Grants of the President of the Russian Federation and the Southern Federal University for this support.

References

[1] Control of Feeder of Novel Cruiser/Feeder MAAT System," SAE Technical Paper 2012-01-2099, 2012, https://doi.org/10.4271/2012-01-2099.

[2] Vucinic, D., Sunol, A., Trancossi, M., Dumas, A. et al., "MAAT Cruiser/Feeder Airship: Connection and Passenger Exchange Modes," SAE Technical Paper 2013-01-2113, 2013, https://doi.org/10.4271/2013-01-2113.

[3] Pshikhopov, V., Medvedev, M., Krukhmalev, V., Fedorenko, R. et al., "Method of Docking for Stratospheric Airships of Multibody Transportation System," SAE Technical Paper 2014-01-2162, 2014, https://doi.org/10.4271/2014-01-2162.

[4] Hablani H.B., Tapper M.L., Dana-Bashian D.J. Guidance and relative navigation for autonomous rendezvous in a circular orbit //Journal of Guidance, Control, and Dynamics. – 2002. – T. 25. – No. 3. – C. 553-562.

[5] Ichikawa A., Ichimura Y. Optimal impulsive relative orbit transfer along a circular orbit // Journal of guidance, control, and dynamics. – 2008. – T. 31. – No. 4. – C. 1014- 1027.

[6] Bevilacqua R., Lehmann T., Romano M. Development and experimentation of LQR/APF guidance and control for autonomous proximity maneuvers of multiple spacecraft //Acta Astronautica. – 2011. – T. 68. – No. 7. – C. 1260-1275.

[7] LopezI., McInnes C.R. Autonomous rendezvous using artificial potential function guidance // Journalof Guidance, Control, and Dynamics. – 1995. – T. 18. – No. 2. – C. 237-241.52. Oh M. H. et al. Homing and docking control of AUV using model predictive control //The Fifth ISOPE Pacific/Asia Offshore Mechanics Symposium. – International Society of Offshore and Polar Engineers, 2002.

[8] Smith S. M. et al. Fuzzy logic control of an autonomous underwater vehicle //Control Engineering Practice. – 1994. – T. 2. – No. 2. – C. 321-331.

[9] Batista P., Silvestre C., Oliveira P. A sensor-based controller for homing of underactuated AUVs //Robotics, IEEE Transactions on. – 2009. – T. 25. – No. 3. – C. 701-716.

[10] Jantapremjit P., Wilson P.A .Control and guidance for homing and docking tasks using an autonomous underwater vehicle //Intelligent Robots and Systems, 2007. IROS 2007. IEEE/RSJ International Conference on. – IEEE, 2007. – C. 3672-3677.

[11] Pshikhopov V.Kh. Analytical construction of nonlinear systems of terminal control // Sb. RAS "New concepts of general management theory". M.-Taganrog. 1995. P. 125-141

[12] V.Kh. Pshikhopov, M.Yu. Medvedev, R.V. Fedorenko et al., "Mathematical model of robot on base of airship," 52nd IEEE Conference on Decision and Control, Firenze, 2013, pp. 959-964.

[13] V. Kh. Pshikhopov, M.Yu. Medvedev, A.R. Gaiduk, R.V. Fedorenko, V.A. Krukhmalev, B.V. Gurenko, Position-Trajectory Control System for Unmanned Robotic Airship, IFAC Proceedings Volumes, Volume 47, Issue 3, 2014, Pages 8953-8958, ISSN 1474-6670

[14] Pshikhopov, V., Medvedev, M., Kostjukov, V., Fedorenko, R. et al., "Airship Autopilot Design," SAE Technical Paper 2011-01-2736, 2011, https://doi.org/10.4271/2011-01-2736.

Performances of Straight Head and T-head Groynes as River Training Structures

D Dutta and H M Kalita*

National Institute of Technology Meghalaya, Shillong, 793003, India

*Email: hridaymanikalita@gmail.com

Abstract. Any structure which is constructed within a river, for managing the flow of the river to make it beneficial for mankind is termed as river training structure. Groyne is one such type of hydraulic structures, which is generally constructed for various purposes including bank erosion prevention, facilitating smooth navigation, etc. These structures are generally constructed from the river bank and extend towards the main channel flow. Depending upon the head shape, groynes may be further classified into straight and T-head types. Literature survey has revealed that comparative analysis of performances of straight head and T-head groynes is still not sufficient enough. Therefore, in the present work numerical models have been developed for both the types of groynes specifically in channel bend and their performances are compared. In order to numerically simulate the flow processes, MIKE21C modeling tool is used here. From the comparison it appears that in case of T-head groyne, the shifting of the thalweg line from the outer is more. Moreover, higher value of sediment deposition is also observed in the T-head groyne field. All these results finally establish that T-head groynes are better than straight head groyne in river bend.

1. Introduction

Any kind of structures those are constructed in a river for guiding/managing the river water for beneficial purpose of mankind are termed as river training structures. Groyne or spur dike is one of the most commonly used river training structures generally constructed from the river bank and extend towards the main channel flow. Depending upon the head shape, they may be classified into straight head and T-head groynes. They are constructed for various purposes including, deflecting/repelling the flow away from bank for safety, increasing flow depth for smooth navigation, etc. Construction of these structures leads to blockage in the flow and always leads to changes in the local hydrodynamics as well as the morphology of the river. At the same time, the flowing water also imparts some effects on the groyne. These effects may be in the form of increment in the flow speed near the tip of the structure, development of local scouring around the structure due to pressure difference between the top and bottom water pressures near the structure. These local scouring many a time even may leads to failure of the structure itself. Moreover, the opposite river bank also faces high speed and shear stress upon construction of these structures. Therefore, before implementing a project deals with construction of groyne, it is always essential to conduct a model study to evaluate the probable effect of the groyne on the river as well as the reverse effect of the river on the structure itself. These model studies however, may be based on either experimental approaches or numerical approaches. However, experimental approaches are not generally preferred due to their large investment cost, repetitive approach etc. Numerical modeling however has become very popular techniques in this regard.

Ahmad [1] conducted one of the first experimental studies for simulating scouring process around straight head groynes. After that several numerical and experimental studies [2, 3, 4] are reported in the literatures for flow simulation around straight head groynes. Application of T-head groynes in rivers is comparatively a recent topic in computational hydraulics. A few literatures are also available [5, 6] for flow simulation around T-head groyne. Mansoori et al. [7] first compared the performances of straight head and T-head groynes in straight channels. Results have shown that the magnitudes of velocity vectors inside the T-head groyne's fields are much smaller than the similar areas in straight groynes, leading to more safety of the near bank, using T-head groynes. So far the authors' knowledge, no study has been taken up yet for comparing the performance of straight head and T-head groynes in channel bend. Therefore, the present work tries to simulate series of straight head and T-head groynes in channel bend to appraise their performances for bank stability.

2. Numerical model for morphological flow simulation

2.1. MIKE21 governing equations and solution techniques

MIKE 21C is an integrated river morphology modelling tool based on a curvilinear version of the 2D surface water model MIKE 21. MIKE 21C simulates the changes in the river bed and plan form, including bank erosion, scouring associated with construction work and changes in the hydraulic regime. The hydrodynamic equations solved in MIKE21C are;

$$\frac{\partial H}{\partial t}+\frac{\partial p}{\partial s}+\frac{\partial q}{\partial n}-\frac{q}{R_s}+\frac{p}{R_n}=0 \tag{1}$$

$$\frac{\partial p}{\partial t}+\frac{\partial}{\partial s}\left(\frac{p^2}{h}\right)+\frac{\partial}{\partial n}\left(\frac{pq}{h}\right)+2\frac{pq}{hR_n}+\frac{p^2-q^2}{hR_s}+gh\frac{\partial H}{\partial s}+\frac{g}{C^2}\frac{p\sqrt{p^2+q^2}}{h^2}=RHS \tag{2}$$

$$\frac{\partial q}{\partial t}+\frac{\partial}{\partial s}\left(\frac{pq}{h}\right)+\frac{\partial}{\partial n}\left(\frac{q^2}{h}\right)+2\frac{pq}{hR_s}+\frac{p^2-q^2}{hR_n}+gh\frac{\partial H}{\partial n}+\frac{g}{C^2}\frac{q\sqrt{p^2+q^2}}{h^2}=RHS \tag{3}$$

where, s, n are co-ordinates in the curvilinear co-ordinate system, p, q are mass fluxes in the s-and n-direction, respectively, H is water level, h is water depth, g is gravitational acceleration, C is Chezy roughness coefficient, Rs, Rn are radius of curvature of s and n line, respectively and RHS is the right hand side in the force balance contains Reynolds stresses, Coriolis force and atmospheric pressure. The governing hydrodynamic equations are solved here using Alternate Direction Implicit (ADI) scheme.

The sediment continuity equation solved in MIKE21C is;

$$(1-n)\frac{\partial z}{\partial t}+\frac{\partial S_x}{\partial x}+\frac{\partial S_y}{\partial y}=\Delta Se \tag{4}$$

where, Sx is total sediment transport in x-direction, Sy is total sediment transport in y-direction, n is bed porosity, z is bed level, t is time, (x, y) are Cartesian co-ordinate system and ΔSe is the lateral sediment supply from bank erosion.

From equation (4) the calculation of sediment transport of bed material (bed load and suspend load), the bed level change can be computed. The total sediment transport is the sum of the bed load and suspended load. A space centred-time forwarded difference scheme is applied here for solving the equation. The time step is limited by the Courant criterion i.e. that the Courant number should be less than 1.

2.2. Validation of the Numerical Model

A numerical model has been developed to validate the model with available analytical data presented by Mohapatra and Bhallamudi [8]. Figure 1 show the channel which is gradually expanding in nature.

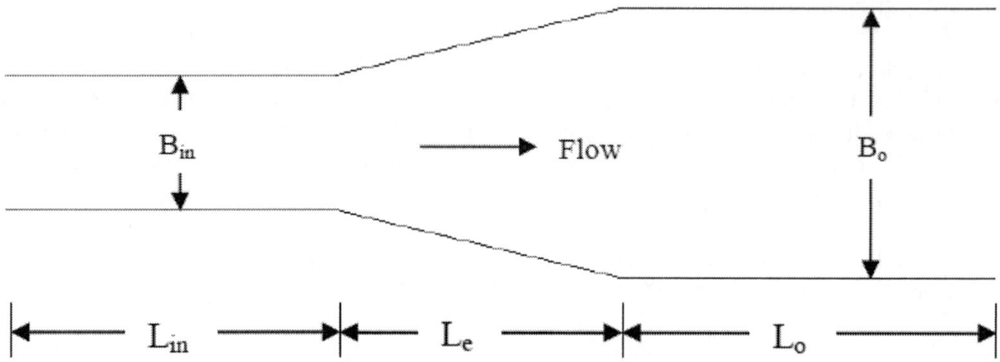

Figure 1. Definition Sketch (Plan) (Mohapatra et al.1994) ; Lin = 5.6 m, Le = 4.0 m, Lo = 6.4 m, Bo= 0.9 m and Bin = 0.3 m.

Grids for the considered channel have been developed in MIKE21C Grid Generator. Along the channel no of grid points are 161 and across the channel no of grid points are 61. Initial surface elevation has been considered as 0.107 meter throughout the channel. For upstream boundary, a discharge of value 0.019 m3/s and for downstream boundary water depth of 0.107 meter has been considered. Time step interval has been considered as 0.06 sec, with this maximum Courant no has been found out as 9. As mentioned in literature by Mohapatra and Bhallamudi [8], sediment is considered as Non cohesive in nature, with a porosity of 0.4. The mean diameter of grains is reported as 0.28 mm. Englund-Hansen sediment transport formula is chosen to find morphological changes in channel. The model ran for total 6hr 40min. Figure 2 shows the comparison of steady state bed level for analytical and numerical cases. This comparison shows the capability of the present numerical model for morphological flow simulation.

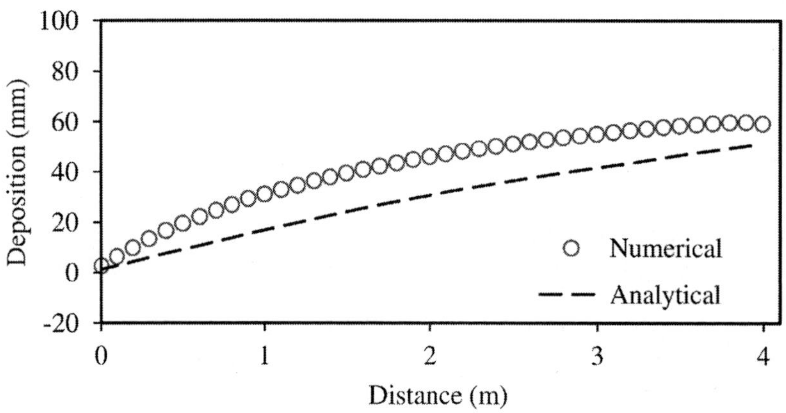

Figure 2. Comparison between simulated and analytical bed level along the centre line.

3. Results and discussion

A hypothetical channel bend is considered as shown in figure 3. It consists of two straight reaches of length 1000 m each, joined by a 600 bend. Width of the channel is considered as 200 m, with a bed slop of 0.0002.

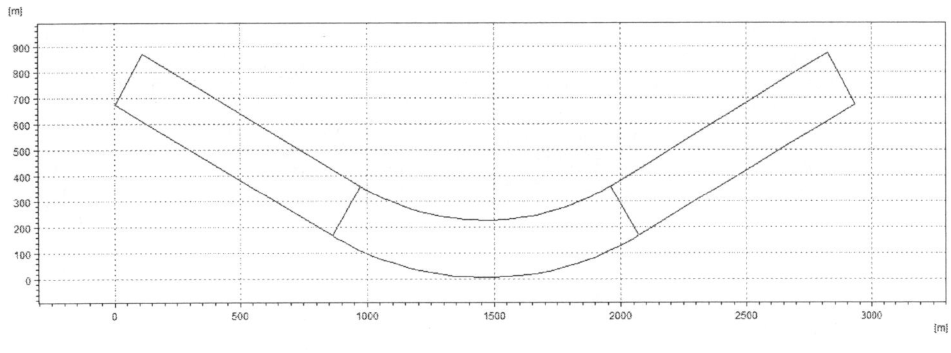

Figure 3. Hypothetical channel bend

An upstream discharge of 2000 cumec is considered for simulation; along with a downstream flow depth of 7.05 m. Amount of sediment transport is calculated using Englund-Hansen formula. Uniform sediment of D50=1.5 mm is considered in the channel. The specific gravity (s) and the porosity (λ) of the considered sediment particles are, 2.65 and 0.4, respectively. For numerical simulation, the whole domain is discretized into 6480 grid. First of all, the model is run without any groynes. Figure 4 shows the bed level variation specifically on the bend. The bed level is found to be scoured near the outer bank and the sediments are deposited near the inner bank, as intuitively expected.

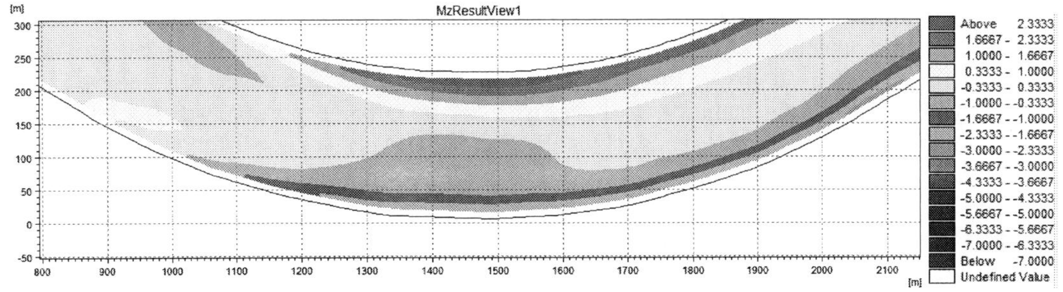

Figure 4. Bed level variation in the hypothetical bend withpout any groyne

The developed model is again run with 7 straight head groynes and the steady state result is shown in figure 5. Figure 6 shows the result with 7 T head groynes. In both the cases, the thalweg line is found to be shifted from the vulnerable outer bank.

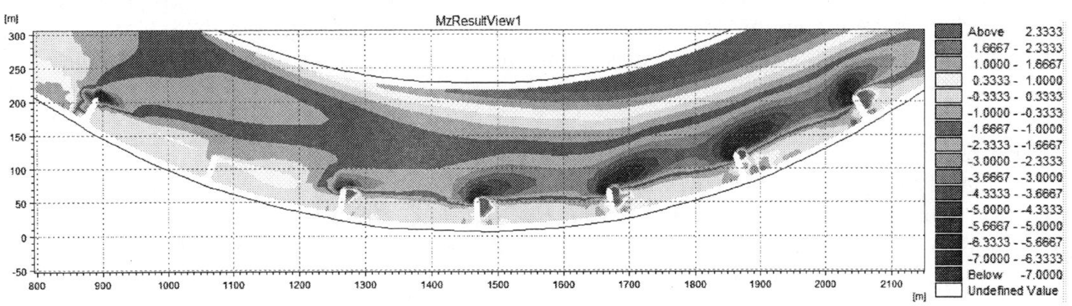

Figure 5. Bed level variations in the hypothetical bend with 7 straight head groynes

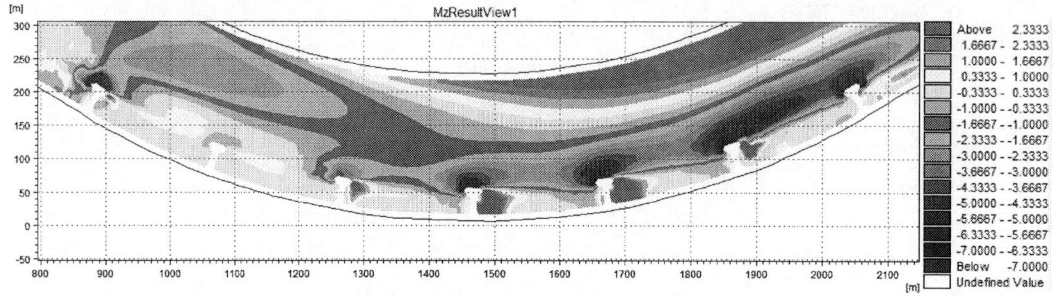

Figure 6. Bed level variation in the hypothetical bend with 7 T-head groynes

In order to compare the performances, three parameters such as, average distance of thalweg line from the groynes, maximum scour depth near the groyne tip and maximum deposition depth in the groyne field, are chosen. Table 1 shows the corresponding values for both the cases.

Table 1. Performance parameters for straight and T head groynes

Groyne system	Average distance of thalweg line from the groynes (m)	Maximum scour depth near the groyne tip (m)	Maximum deposition depth in the groyne field (m)
Straight head	21.43	5.88	0.46
T-head	25.71	6.25	0.87

It may be observed from the table 1 that T-head groynes shift the thalweg line more, leading to more safety of the outer bank. Deposition in the groyne field is also more for T-head groynes. The scour depth near the T-head groyne tips is found to be slightly more. However, it is believed that this scouring will not affect the stability of the groynes, as the flange portion will impart additional stability. From these results it may be conclude that T-head groynes are better for stability of the outer bank, in channel bend.

4. Conclusion

In this study, the performances of straight head and T-head groynes are compared for vulnerable channel bend. For this purpose, numerical models have been developed using the user friendly MIKE21C modeling tool. A hypothetical channel bend is considered first and the morphological modeling has been done. As intuitively expected, the thalweg line is found to be near the outer bank leading to vulnerability for the bank. After that, series of straight head and T-head groynes are installed on the outer banks and the numerical models are run separately. In both of the case, the thalweg lines are found to be shifted from the outer bank. In order to compare the performances, three parameters are considered, such as, average distance of thalweg line from the groynes, maximum scour depth near the groyne tip and maximum deposition depth in the groyne field. Comparing these parameters for both the groyne cases, it is found that T-head groynes are better for stability of the outer bank.

References

[1] Ahmad M 1953 *Proc. Int. Hydraul. Convention* 145-159
[2] Rajaratnam N and Nwachukwu B 1983 *J. Hydraul. Eng.* **109(3)** 463-480
[3] Molinas A, Kheireldin K and Baosheng W 1998 *J. Hydraul. Eng.* **124(8)** 822-830
[4] Sharma K and Mohapatra P 2012 *J. Hydraul. Eng.* **138(10)** 897-901
[5] Safarzadeh A, Salehi Neyshabouri S A A, Ghodsian M, 2010 Proc. of 5th Int. Conf. Fluvial Hydraul. 651-657
[6] Saberi O and Galoie M 2016 *American J. Water Sci. Eng.* **2(6)** 43-52
[7] Mansoori A R, Nakagawa H, Kawaike K, Zhang H and Safarzade A 2012 *Annual Disaster Prevention Res. Inst. Kyoto Univ.* **55(B)** 453-458
[8] Mohapatra P K and Bhallamudi S M 1994 *J. Irrig. Drain. Eng.* **120(6)** 1114-1121

IOP Conf. Series: Materials Science and Engineering **491** (2019) 012014 doi:10.1088/1757-899X/491/1/012014

Mechanical Strength of Galvanized Steel Wire Mesh (GSWM) as a Strengthening Material of Short RC Column

C Marthong[1, *], D Sutnga[1], O Kharshandi[1], I J Khryiemmujat[1] and A Shangpliang[1]

[1]National Institute of Technology Meghalaya, Shillong, 793003, India

[*]Email: commarthong@nitm.ac.in, Phone: +919485177027, Fax: +91 364 2501113

Abstract. Columns of varying shapes have been adopted in construction of various Reinforced Concrete (RC) building. The revision of the design codes or change in usage, strengthening is required to enhance the performance of the column whilst extending its design life. In this paper, three grades of concrete, i.e. M15, M20 and M25 are adopted for casting five different column shapes ranging from circular, square, rectangular, L-Shape and polygon and tested under axial compression. All the columns were wrapped with two layers of locally available Galvanized Steel Wire Mesh (GSWM) jacketed with cement mortar. A total of 30 RC columns were casted in which half of them were treated as control specimens. The columns were design as per IS 456:2000. The test results of the experimental investigation shows that the load carrying capacity of the columns can be increase to a considerable extent by using GSWM. The load carrying capacity decreases as the column shape change from circular to polygon. This shows that the column shapes plays an important role in determining its structural behaviour. In addition to this, the confined column shows a higher deformation, displacement ductility, stiffness degradation and energy dissipation as compared to the control columns. This improvement in the column behaviour may be attributed to the effectiveness of the GSWM jacket as a strengthening material and significant for low grade concrete as compared to higher grade.

1. Introduction

Columns of varying shapes are used in residential as well as industrial building due to the architectural design requirement and the configuration of the structures. This paper focussed on the behaviour of the five different shapes of column ranging from circular to polygon, strengthened with a galvanized steel wire mesh (GSWM) jacketed with mortar. Due to change in usage of a building or revision of design codes strengthening of column is necessary to enhance its performance. As per Indian standard coded of practice [1, 2], the minimum grade of concrete for construction of RC structures is M15 and M20 respectively. However the latest Indian Standard Codes for design of RC structures [3], which the minimum grade should be M20 for any building and M25 for building with a height greater than 15 metres for building located in Seismic Zones III, IV and V. Thus, those structures constructed prior to its existence becomes deficient and making them unfit for further use. As a result strengthening is required to improve its serviceability and existing design level which would satisfy the criteria described in IS 13920 [3].

During the last few decades, several method for strengthening of concrete structural members have been reported [4,5,6,7]. Many techniques possess its own practical limitations. To overcome the difficulties recent research efforts have focused on the use of epoxy-bonded fiber-reinforced polymer (FRP). This technique eliminates many limitations. Although the use of FRP composites jackets enhanced the compressive strength and ductility of concrete columns the anchoring of FRP materials however, has evolved as a difficult problem for the effectiveness of this technique [8]. In addition to FRP, composite grid [9] and textile [10] has also been used as strengthening method to improve strength and ductility of concrete columns. However, owing to the high manufacturing and application costs of these materials, the need to investigate other possible wrapping materials has arisen. GSWM is another promising material for strengthening solution. Due to its advantages such as easy availability, no skill labour require, tailorability, GSWM is adopted for this study for strengthening the column of different grade and shapes, taking into consideration the stiffness and ductility criteria. There are varieties of GSWM locally available depending on the thickness of wires and grid opening in which the costs vary from 120 per m^2 to 220 per m^2. The present study used a GSWM which has a grid size of 25 x 25 mm and 1.2 mm thick wire.

2. Experimental program

Different sizes of column corresponding to the shape are made. The dimension of column is shown in Figure 1. The specimens were casted with a concrete mix of grades M15, M20 and M25 which has a compressive strength of not less than the values specified in IS: 456-2000 [2]. The specimens were then cured for 28 days. After curing, half of the specimens were tested under compression load and the rest of the specimens were wrapped with two layers of GSWM jacketed with mortar and cure for another 28 days. Strengthened specimens are presented in Figure 2. All specimens were tested in a compression testing machine as shown in Figure 3.

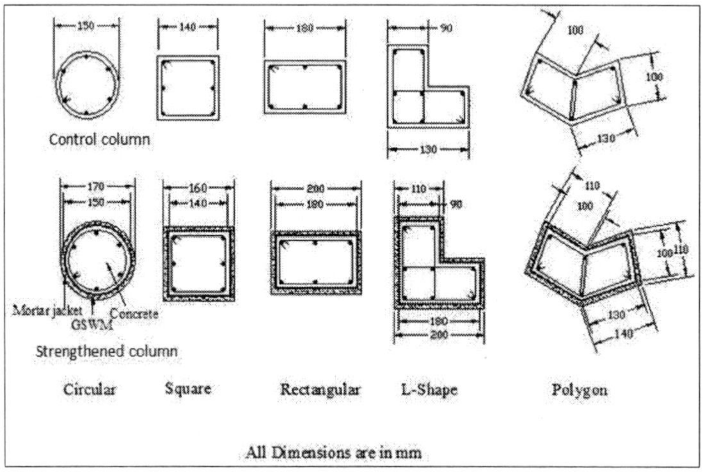

Figure 1. Cross-section of Column

3. Results and discussion

3.1. Variation with respect to Shap

The load carrying capacity of column of different shape for M15, M20 and M25 grade are given in Table 1 and the typical variation of M20 is shown in Figure 4. It is obvious that for the same column shape, RC columns of M25 have a higher load carrying capacity for the same grade and circular columns have a higher load carrying capacity. This increase in the load carrying capacity of circular is due to the uniform distribution of the stress when the axial load is applied on the column. Polygon columns have numerous corners and due to the non- uniformity of the corner, the load at the corners is the maximum causing its corner to crush and fails when loading is applied. Further, from Figure 4 it was observed that the trend of the increase in the load carrying capacity of the GSWM confined columns is similar to that of the unconfined control columns with respect to both shapes and concrete grades. However it was observed that the increase in the load carrying capacity of the GSWM confined columns is higher than those of the unconfined confined columns. This increase in the load carrying capacity and compressive strength may be attributed to the effectiveness of the GSWM as a confining material by providing confinement to resist the additional load.

Figure 2. Strengthening of column

Figure 3. Test setup

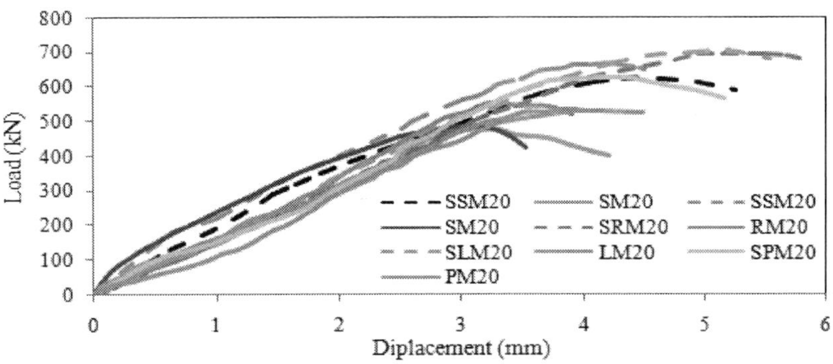

Figure 4. Variation of load carrying capacity of square columns with shapes

3.2. Increase in ultimate load carrying capacity of columns

The effectiveness of the GSWM confinement can be explained by the increase in the load carrying capacity. The increase in the ultimate load is given in Table 1 and the effectiveness of the confinement is higher in case of the Circular columns followed by rectangular, square, L-shape and the polygon shape columns shows the lowest increase in the load carrying capacity. This indicates that the strength of the column is influence by its shape. Also the increase in load carrying capacity for all the column shapes is found to be higher in case of the lower grade of concrete. This implies that the grade of concrete plays an important role in the applicability and effectiveness of GSWM jacket as a strengthening material. Since like other strengthening material, GSWM is adopted mostly for strengthening of structural members of existing old buildings that had been considered outdated by the existing design codes and also a change in usage that are mostly casted with a low strength concrete.

The data obtained shows that GSWM proves to be a viable strengthening material for strengthening the columns casted with M15, M20 and M25 grade concrete.

Table 1. Load carrying capacity of control columns

Concrete grade	Shape	Designation	Load of control column (kN)	Load of strengthen column (kN)	% increase
	Circular	CM15	415.80	643.60	35.39
	Square	SM15	415.45	604.70	31.30
M15	Rectangular	RM15	451.50	571.40	20.98
	L-Shape	LM15	426.99	535.10	20.20
	Polygon	PM15	435.50	525.70	17.47
	Circular	CM20	435.50	621.90	29.97
	Square	SM20	489.90	666.10	26.45
M20	Rectangular	RM20	552.80	705.40	21.63
	L-Shape	LM20	529.10	696.60	24.05
	Polygon	PM20	535.10	626.40	14.58
	Circular	CM20	597.50	770.36	22.44
	Square	SM20	558.40	685.90	18.59
M25	Rectangular	RM20	622.80	726.60	14.29
	L-Shape	LM20	565.30	639.60	11.62
	Polygon	PM20	632.10	672.30	5.98

3.3. Displacement ductility

The ductility of the columns is given by its displacement ductility which is defined as the ratio of its ultimate displacement to its yield displacement. The displacement ductility of the columns of different shapes and grades are shown in Table 2 and is calculated as per the procedure of Shannag *et al.* [11]. From the table it was observed that the displacement ductility of the confined columns is higher than those of the unconfined columns indicating that the confined columns fails in a ductile manner as compared to the unconfined ones which fails directly through crushing of the concrete as shown in the failure patterns of the columns. The GSWM being made of steel provide extra ductility to the confined columns which is a desirable property for a column.

3.4. Energy absorption capacity

Energy absorption gives the capacity of the structure to resist dynamic or impact loads. The energy absorption (kN-mm) is calculated as the area under the load displacement curve as described by the procedure given by Shannag et al [11] and given in Table 2. It's shown that all confined column absorbed higher energy than the unconfined. The energy dissipated of the column with M15 grade is the lowest as column shapes changes. Circular columns in both confined and unconfined absorbed the maximum energy. However, the energy dissipation increases considerably for other confined column as well. Thus, using GSWM improve the energy dissipation of structural members.

3.5. Stiffness degradation

Stiffness indicates the extents to which the columns can resist deformation. It gives the degradation in the strength of the column. The stiffness degradation is calculated as per the formula given by Naeim et al. [12]. The variation in the stiffness is given in Figure 5 and irrespective of the column shapes, the axial stiffness of the GSWM confined columns is higher than those of the unconfined ones indicating that the confined columns can resist higher loads. Also the stiffness of the GSWM confined columns first decrease to some extent and then it rise again before falling permanently. This rise in the stiffness even after it has fall is due to the additional stiffness provided by the GSWM confinement as compared to the continuous dropping of stiffness for the unconfined columns.

Table 2. Displacement ductility and Energy absorption of column

Shape	Parameters	Designation					
		CM15	CM20	CM25	SCM15	SCM20	SCM25
Circular	Displacement ductility	1.64	2.12	2.16	2.15	2.24	2.48
	Energy Absorption	879.31	1238.9	1249.2	1969	2144.4	2705
		SM15	SM20	SM25	SSM15	SSM20	SSM25
Square	Displacement ductility	1.65	1.69	2.21	2.56	2.43	2.02
	Energy Absorption	854.1	1156.8	1509.2	1969	1870.1	2378.9
		RM15	RM20	RM25	SRM15	SRM20	SRM25
Rectangular	Displacement ductility	1.48	1.62	1.63	2.4	1.85	2.04
	Energy Absorption	1328.8	1255.4	1240.4	2310.8	2456.5	1728.2
		LM15	LM20	LM25	SLM15	SLM20	SLM25
L	Displacement ductility	2.44	1.89	2.2	2.72	2.03	2.3
	Energy Absorption	988.74	1466.5	1589.8	1856	2387.7	2097.8
		PM15	PM20	PM25	SPM15	SPM20	SPM25
Polygon	Displacement ductility	1.78	1.59	1.5	2.57	2.2	1.8
	Energy Absorption	1234.5	1153.4	1249.2	1761.2	2003.7	2055.4

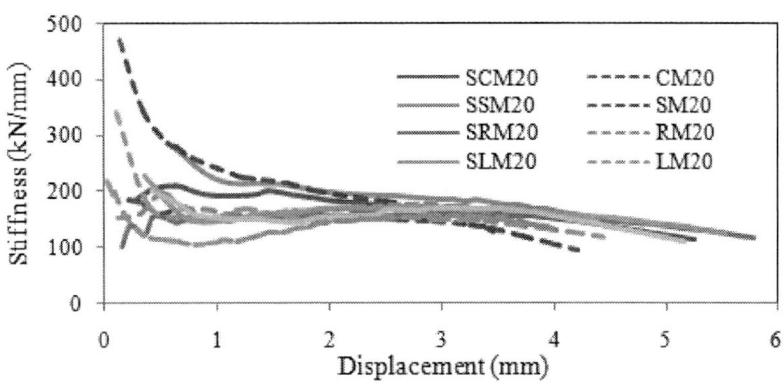

Figure 5. Variation of stiffness for M20 grade columns

3.6. Effect of shape and grade

The variation of the columns behaviour with respect to shape is determine by its confinement ratio which is the ratio of the compressive strength of the confined columns (f'_{cc}) to that of the unconfined columns (f_{cc}). The variation of the confinement ratios with respect to grade and columns shape are given in Table 3 and the confinement ratio decreases from M15 to M25 indicating that the columns casted with the lower grade concrete gives a better response. A similar trend is observed as the shapes of the columns vary from circular to Polygon shape columns. The higher confinement ratio of the column indicates the effectiveness of the GSWM as a confining material for which the circular column offer a full confinement. The polygon column shows a lesser response as the stress concentrations of the columns are mostly concentrated at the corners as a result of which it has a less confinement ratio.

Table 3. Confinement ratio of columns

Column Shape	Confinement ratio (f'_{cc}/f_{cc})		
	M15	M20	M25
Circular	1.54	1.43	1.29
Square	1.45	1.36	1.23
Rectangular	1.27	1.28	1.17
L-Shape	1.25	1.21	1.13
Polygon	1.21	1.17	1.06

4. Conclusion

Based on the experimental investigation and results obtained, the following conclusions can be drawn:

1) The load carrying capacity of the circular columns is higher followed by square, rectangular, L-shape and polygon.
2) Using GSWM, the load carrying capacity of the columns increases significantly for column of M15 and M20 grade which shows that GSWM is very efficient for strengthening columns of existing old building which has been casted with lower grade of concrete.
3) The displacement ductility of confined columns is greater than those of the unconfined columns. This shows that the confined column fails in a ductile manner while the unconfined column fails in a brittle manner.
4) The energy dissipation and stiffness degradation of the GSWM confined columns is higher than those of the unconfined ones. This proves that GSWM is a viable strengthening material.

References

[1] IS 456-1978. *Bureau of Indian Standard New Delhi.*

[2] IS 456-2000. *Bureau of Indian Standard New Delhi.*

[3] IS 13920- 2016. *Bureau of Indian Standard New Delhi.*

[4] He A, Cai J, Chen Q, Li, Z and Xu J. 2016. *Thin-Walled Structures,* 109 25–39.

[5] Tarabia A M, and Albakry H F 2014 *Alexandria Engineering Journal*, 53 615–626.

[6] Zhang B, Yang Y, Weil Y, Liu R, Ding C and Zhang K. 2015 *Str. Engg and Mech* 55(6), *1139-1155.*

[7] Campione G. 2008 *Str. Engg and Mech*, 28(3) 295-316.

[8] Ghobarah A, and Said A. 2002 *Engineering Structures*, 24(7) 881-888.

[9] Bentayeb F. Tahar K. A, and Chateauneuf A. 2008 *Const. Bldg. Mat.*, 22 1624–1633.

[10] Ortlepp R, and Ortlepp S. 2017 *Const. and Bldg. Mat.* 132 150–160.

[11] Shannag M J, and Alhassan M A. 2005 *ACI Structural Journal*, 102(1) 131-138.

[12] Naeim F. Kelly J. M. 1999 *John Wiley & Sons, Inc.*

IOP Conf. Series: Materials Science and Engineering **491** (2019) 012015 doi:10.1088/1757-899X/491/1/012015

An Experimental Investigation on the Analogous Transpiration Cooling Utilizing Hydrogels as Coolants

Kang Qian[1], Yadong Wu[2], Zihe Zhou[1], Nan Wu[1], Fei He[1,*], Jianhua Wang[1] and Meng Wang[1]

[1] Department of Thermal Science and Energy Engineering, University of Science and Technology of China, Jinzhai Road 96, Hefei 230027, PR China
[2] Beijing Institute of Astronautical System Engineering, Beijing 100076, PR China

*hefeihe@ustc.edu.cn

Abstract. In the applications of transpiration cooling, water acting as a liquid coolant shows excellent cooling performances due to its huge heat capacity and latent enthalpy. However, the fluidity of liquid water could cause extra difficulties on occasions where unexpected interferences to the system balance should be avoided. To overcome the above issue, we propose a new transpiration cooling pattern, in which the liquid water is solidified and stored as hydrogels. In this work, the hydrogel was prepared from the superabsorbent polymer (SAP) and thermogravimetric analysis was conducted to test the water release property and upper critical temperature. Respectively using water and hydrogels as coolant, transpiration cooling experiments on a homemade porous plate were carried out to explore the feasibility and cooling performance of the new transpiration cooling pattern. The results implied that the transpiration cooling with hydrogels and water can both slow down the temperature rising of the structure. The hydrogel can provide considerable cooling effect and is superior for prolonging service time.

1. Introduction

In the development of transpiration cooling, the transition of coolant from gases to liquids improves the cooling effect on high heat flux surface greatly [1]. As a common liquid coolant which has huge heat capacity and latent enthalpy, water plays an important role in researches on transpiration cooling [2,3]. The transpiration cooling based on water has been widely investigated and its high efficiency has been validated in protecting the leading edge of hypersonic vehicles and combustion chamber [4–6]. However, there are still many issues to solve to accelerate its applications. Apart from the complex system for water injection control, the movement of liquid water causes extra problems, such as a higher requirement for sealing and the variation of barycenter position, which may even destroy the balance of the whole system at last. Then, it's meaningful to make some improvements on the transpiration cooling with liquid water, and the analogous transpiration cooling (ATC) utilizing solidified water could be a new direction.

In the process of ATC, as shown in Figure 1, solidified water acting as a cold source is pre-sealed in the coolant chamber, and its state decides the good sealing performance that it cannot be pushed into the porous matrix at the initial moment. When the chamber temperature is above the phase-change point, the corresponding vapor will flow out through the porous matrix and forms a protection film layer on the surface. It should be noted that phase change may occur anywhere in the traditional transpiration cooling with liquid water while can only happens in the coolant chamber in ATC.

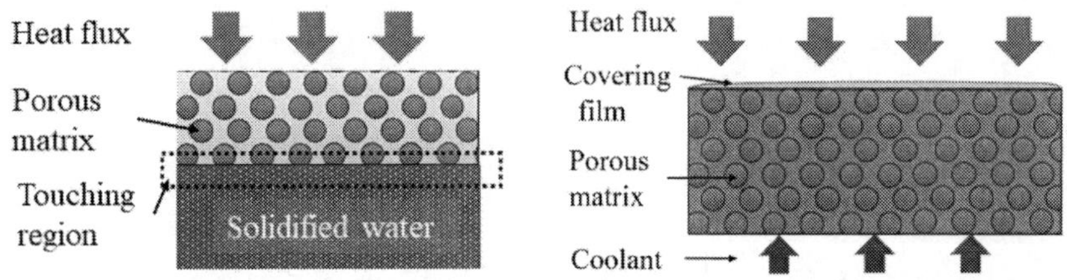

Figure 1. Schematics of analogous transpiration cooling and traditional transpiration cooling

Hydrogels own an integrated ability of water absorption, water retention and water release. And due to the good biocompatibility, hydrogels have been widely studied and applied in the field of agriculture and biomedicine. However, most of the existing researches were focused on the mechanical behaviors, improved synthesis methods and drug sustained release [7–9]. There are little investigations on utilizing hydrogel in the heat dissipation, especially with the application of transpiration cooling. In the work conducted by Huang Z. etc. [10], the bio-inspired nano-skin technique based on the temperature sensitive hydrogels was studied with the aim to enhance the heat dissipation of electron devices.

In this paper, we proposed a new cooling mechanism, i.e. analogous transpiration cooling; then we conducted the experimental investigation on the corresponding cooling performances using hydrogels as coolant, and compared with that of liquid water. The aim is to provide a reference for the designers of heat dissipation.

2. Experimental arrangement

2.1. Experimental system

The experiment was conducted in the open low-speed electrical heated wind tunnel at the University of Science and Technology of China. Compressed air with a controlled flow rate is electrically heated to a given temperature and employed as the mainstream. The measurement instruments include the infrared thermal imaging system (NEC TH504) which captures the temperature distribution on the surface of plate, and several K style thermocouples for the temperature monitor in the coolant chamber at different positions. The accuracy of them is 1% and 0.75%, respectively.

2.2. Test apparatus

The specimen is separated into two main parts, the porous plate and the supporting structure with a chamber inside, as shown in figure 2. The porous plate has an average diameter of 100 μm and a thickness of 4mm, and the size of its upper surface is 42mm*72mm. Three thread holes are drilled in the coolant chamber and the thermocouples are installed to monitor the coolant temperatures at different positions, with a distance of 9mm, 11mm and 6mm to the bottom of the plate, respectively.

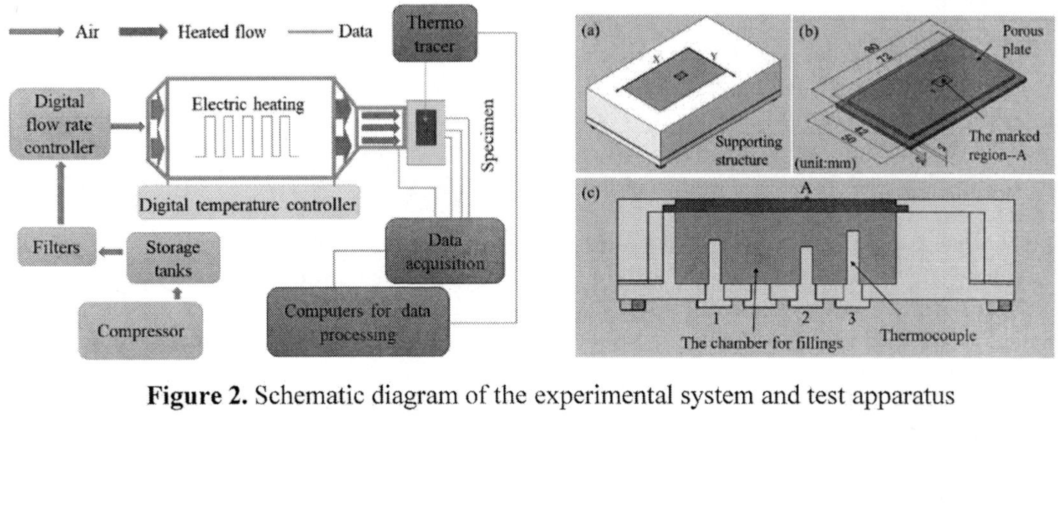

Figure 2. Schematic diagram of the experimental system and test apparatus

2.3. Hydrogels

In this exploratory experiment, an ordinary superabsorbent polymer (SAP) was used to prepare hydrogels. The SAP is mainly based on the low crosslinked polyacrylic resin and its strong water-absorption could meet our primary requirements on water retention and solidification. By means of natural filtration, it is measured that the SAP has a water absorption ratio up to 167 times of its own weight. In the test, we chose the hydrogels with a ratio of 100:1, and it shows good performance in keeping immobilized.

To learn more about its property of water loss at high temperature, a Thermogravimetric Analysis (TGA) test was conducted and the result is shown in figure 3. From figure 3, we can find that there is an obvious weight drop once the heating begins. When the temperature increases to 173.61°C, the ratio of weight loss reaches 98.85%, which means that the water absorbed almost completely releases. Besides as the temperature keeps rising, no more water releases and the SAP decomposes at a temperature about 400.86°C.

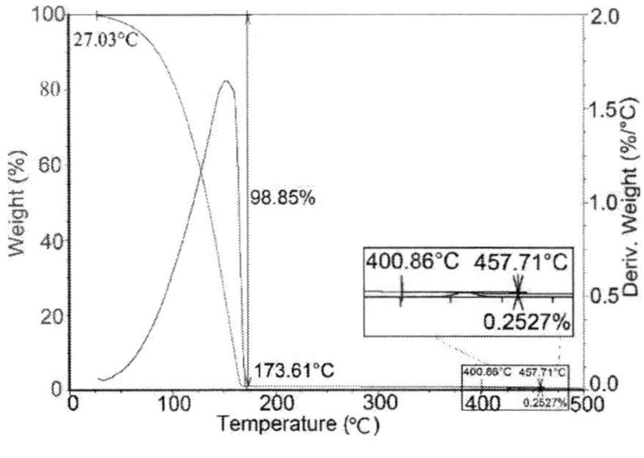

Figure 3. TGA curve of the hydrogel

3. Analysis of experimental phenomena

Under the same mainstream conditions, $T_\infty = 500°C$ and $V_\infty = 1000\mathrm{L/min}$, three groups of comparative experiments were conducted at different chamber conditions, i.e. empty, filled with SAP hydrogel and pure water.

3.1. Temperature distribution on the whole surface

Figure 4 shows the temperature distributions on the surface with an empty chamber at different time intervals, in which the mainstream flows from top to bottom. It should be noted that when the mainstream achieves to a relatively stable state, the moment as the plate is exposed to the hot temperature mainstream is chosen as the initial state, and is marked as Δt=0. As shown in figure 4, the surface temperature of the porous plate increased with time, and the high-temperature region moved towards the outlet of wind tunnel. The highest temperature on the surface monitored is about 322°C.

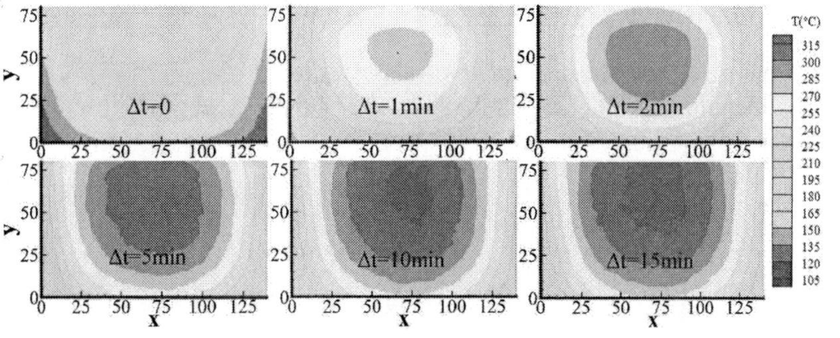

Figure 4. Temperature distributions of the porous plate with an empty chamber

To study the cooling performance of analogous transpiration, the temperature drops of the porous plate using SAP hydrogels and water as coolants from that with the empty chamber are calculated at the same time intervals and exhibited in figure 5 and figure 6.

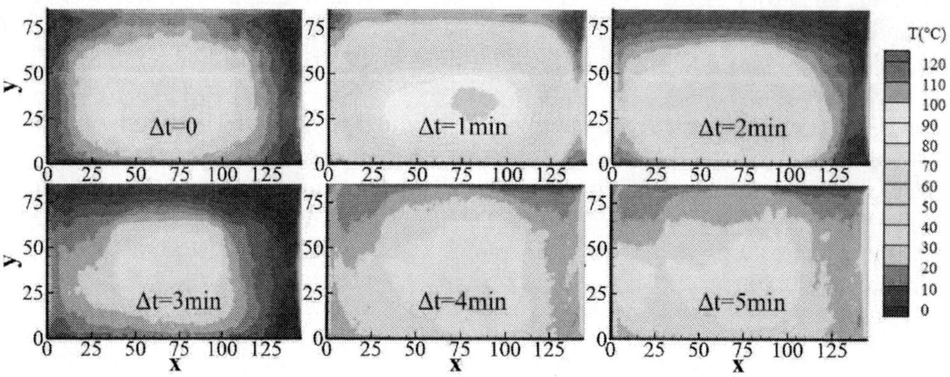

Figure 5. Temperature drop on the surface of porous plate with SAP hydrogels and empty chamber

From the figures, we can find that the existence of coolant slows down the temperature rising rate in both cases, and the temperature drop varies with time. For SAP hydrogels, the temperature drop is 100°C at Δt=1min, and decreases to 30°C at Δt=5min. This phenomenon is mainly caused by the variation of coolant height in the chamber, as the time exposed to high heat flux increases, the coolant filled has no contact with the plate anymore and the heat dissipation is weakened. In addition, through the comparison of figure 5 and 6, it can be found that at the same time intervals, the liquid water always exhibits a better cooling effect than hydrogels, because the effective mass of water is smaller for hydrogels at the same coolant volume.

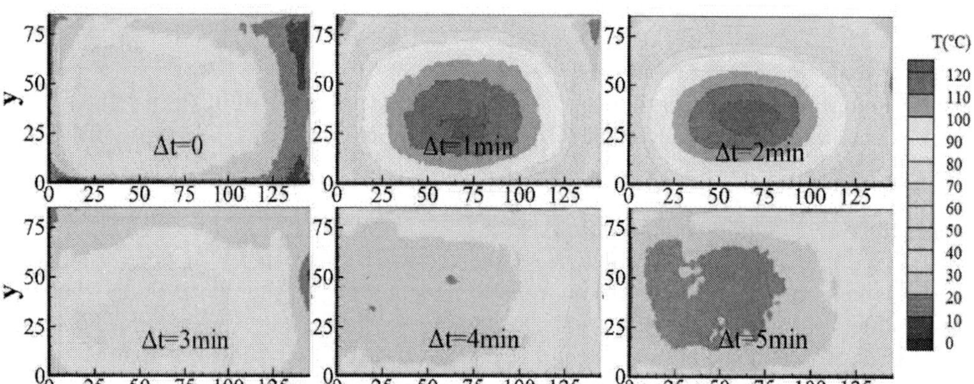

Figure 6. Temperature drop on the surface of porous plate with water and empty chamber

3.2. Local temperature on the surface and in the chamber

As the surface temperature and coolant temperature varies in different positions, the local rectangle area A is chosen (as shown in figure 2), and the corresponding average temperature is calculated.

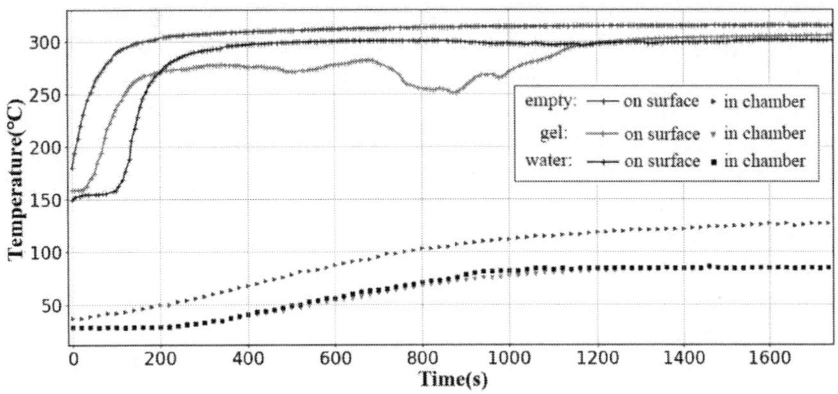

Figure 7. Variations of temperature averaged over area A and monitored by thermocouple 2 in the chamber

Figure 7 presents the transient variations of temperature within area A (curves) and the chamber temperature measured by thermocouple 2 under different chamber conditions. From figure 7, one can observe the following phenomena: 1) compared to the empty chamber, the temperatures on the surface and in the chamber are obviously lower when using hydrogels and water as coolant, especially at the initial stage. 2) In the case of hydrogels and water, there is a delay in temperature rising, and this delay period is longer when using hydrogels as coolant. Especially, during the intervals from 200s to 1000s, the hydrogels even has a better cooling effect than water. During this period, the hydrogels have a longer contact time with the porous plate due to the interstitial structure of stacked gels granule. 3) The chamber temperatures of gels and water change synchronously and increase gradually to 85°C. It indicates that under such experimental conditions, the monitor point is still be filled with coolant.

4. Conclusions

A series of exploratory experiments were conducted based on the newly proposed analogous transpiration cooling. From the analysis, the following conclusions can be drawn:

1) In comparison with the empty chamber, the filling of coolant, i.e. hydrogels and water, shows an excellent performance in slowing down the rising of surface temperature.
2) The delay period and the changing trend of surface temperature are different in analogous transpiration cooling with hydrogels and water. In general, water performs better within the first 200s, while the cooling effect of hydrogels lasts longer and is superior when the exposure time is more than 200s.
3) Hydrogels can provide considerable cooling performance with water and maintain a stable state without any coolant fluctuation and fluidity. Therefore the ATC using hydrogels as coolant holds great potential in the heat dissipation applications.

References.

[1] Van Foreest A, Gülhan A, Esser B, Sippel M, Ambrosius B and Sudmeijer K 2007 Transpiration Cooling Using Liquid Water *39th AIAA Thermophysics Conference* 39th AIAA Thermophysics Conference (Miami, Florida: American Institute of Aeronautics and Astronautics)
[2] Dong W and Wang J 2018 A New Model and its Application to Investigate Transpiration Cooling with Liquid Coolant Phase Change *Transp. Porous Media* **122** 575–93
[3] He F and Wang J 2014 Numerical investigation on critical heat flux and coolant volume required for transpiration cooling with phase change *Energy Convers. Manag.* **80** 591–7
[4] Huang G, Zhu Y, Liao Z, Ouyang X-L and Jiang P-X 2017 Experimental investigation of transpiration cooling with phase change for sintered porous plates *Int. J. Heat Mass Transf.* **114** 1201–13

[5] Huang Z, Xiong Y-B, Liu Y-Q, Jiang P-X and Zhu Y-H 2015 Experimental investigation of full-coverage effusion cooling through perforated flat plates *Appl. Therm. Eng.* **76** 76–85
[6] Wang J, Zhao L, Wang X, Ma J and Lin J 2014 An experimental investigation on transpiration cooling of wedge shaped nose cone with liquid coolant *Int. J. Heat Mass Transf.* **75** 442–9
[7] Bhutani U and Majumdar S 2017 Natural fibre envelope for cross-linked and non-cross-linked hydrogel-drug conjugates: Innovative design for oral drug delivery *Mater. Discov.* **8** 1–8
[8] Gennen S, Grignard B, Thomassin J-M, Gilbert B, Vertruyen B, Jerome C and Detrembleur C 2016 Polyhydroxyurethane hydrogels: Synthesis and characterizations *Eur. Polym. J.* **84** 849–62
[9] Yan B, Chen Z, Cai L, Chen Z, Fu J and Xu Q 2015 Fabrication of polyaniline hydrogel: Synthesis, characterization and adsorption of methylene blue *Appl. Surf. Sci.* **356** 39–47
[10] Cui S, Hu Y, Huang Z, Ma C, Yu L and Hu X 2014 Cooling performance of bio-mimic perspiration by temperature-sensitive hydrogel *Int. J. Therm. Sci.* **79** 276–82

Friction and Wear Properties of HfC-WC-Co Cemented Carbide

X Y Yan[1], S R Wang[1,*], S S Chen[2] and C Y Xue[1]

[1]School of Mechanical Engineering, University of Jinan, Jinan 250022, China
[2]School of Material Science and Engineering, Shandong University, Jinan 250022, China

*E-mail: me_wangsr@ujn.edu.cn

Abstract. HfC-WC-Co composites were prepared by cold press forming and hot press sintering, and their relative density, hardness and flexural strength were tested. The friction and wear experiments were carried out with the MMW-1 vertical universal wear tester. The friction coefficient and wear amount of the material were tested. The microstructure and phase of the wear surface were observed and analyzed by SEM, ultra depth of field microscope and XRD. The results show that the particles extraction, furrow and exfoliation caused by annular microcracks are the main forms of WC-Co composite wear. And appropriate amount of HfC can effectively improve the relative density, hardness and fracture toughness of materials. At the same time, it can effectively improve the sintering quality, inhibit the microcrack initiation and expansion, reduce the phenomenon of particle extraction and furrowing, and improve the friction and wear characteristics.

1. Introduction

WC-Co cemented carbide has excellent properties such as high hardness, high strength, high modulus of elasticity, low coefficient of thermal expansion, strong wear resistance and corrosion resistance. Therefore, it is widely used in the wear-resistant fields such as cutting tools, nozzles and drilling [1,2,3,4]. However, due to its inherent brittleness [5,6], its application and development in the engineering field is greatly limited. Therefore, the development of wear-resistant WC-Co hard alloy with high toughness and high hardness has practical application significance. The researchers found that adding appropriate amount of TaC [7], Cr_3C_2 [8], VC [9], TiC [10] to WC-Co cemented carbide can effectively improve the friction and wear properties and enhance the wear resistance. Previous studies have found that HfC may also be an ideal particle reinforcement for WC cemented carbide. HfC with NaCl-type face-centered cubic lattice structure has the characteristics of high melting point, good thermal conductivity, small thermal expansion coefficient (6.6x10-6/°C), high hardness (26.1Gpa) and good impact performance [11,12], what's more, It has strong phase stability, and has good compatibility with WC, which can change the state of Co binder, thereby improving the brittleness and wear resistance of WC-Co cemented carbide.

Therefore, in this paper, WC-Co-HfC cemented carbide is sintered by vertical vacuum hot-pressing sintering furnace, and then the relative density, hardness and fracture toughness of the material are tested. And friction and wear testing machine are used to test the friction coefficient and wear mass loss of the material. The super-depth three-dimensional microscope and scanning electron microscopy (FE-SEM) were used to observe the morphology after wear. The surface of the wear scar was detected and analyzed by XRD. The effect of HfC on the friction and wear characteristics of WC-Co cemented carbide was revealed.

2. Experiments

2.1. Material preparation

The base material of the experiment is WC, the binder is Co, and the grade is YG8. Adding HfC enhanced phase with mass fraction of 0%, 0.5%, 1.0%, 1.5%, 2%, 2.5% in the matrix. First, the matrix and additive materials were calculated according to the mass fraction ratio, weighed on an electronic analytical balance and placed in an agate tank of a planetary mill for ball milling. The ball mill tank is evacuated and then filled with argon. The argon concentration is 0.075 MPa, the ball milling time is 6 hours, the rotation speed is 360r/min, and the ball-to-material ratio is 5:1. Next, the above-mentioned ball-milled material is placed in a mold as shown in Figure.1 and cold-pressed at a cold pressure of 100 MPa, the pressure rate of 3 mm/min, and the pressure holding time of 120 s; Finally, the cold-formed sample is put into a graphite mold and sintered in a vacuum hot-pressing sintering furnace. The hot-pressing sintering process is shown in Figure 2. Samples cooled to room temperature were subjected to coarse grinding, fine grinding, fine grinding, polishing, ultrasonic cleaning, drying, and the like, and then subjected to relevant performance tests.

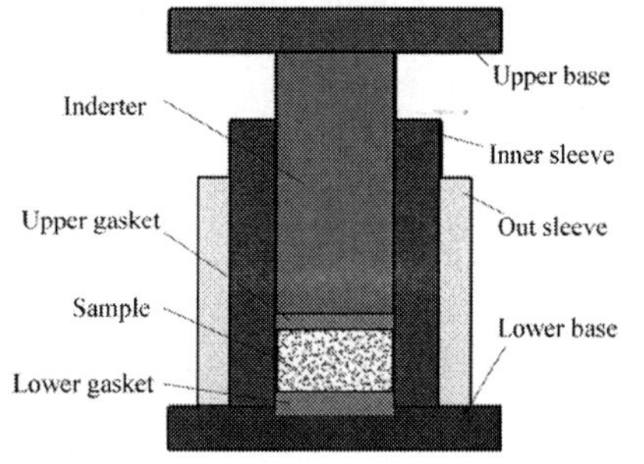

Figure 1. Schematic diagram of cold forming for sample in the mould.

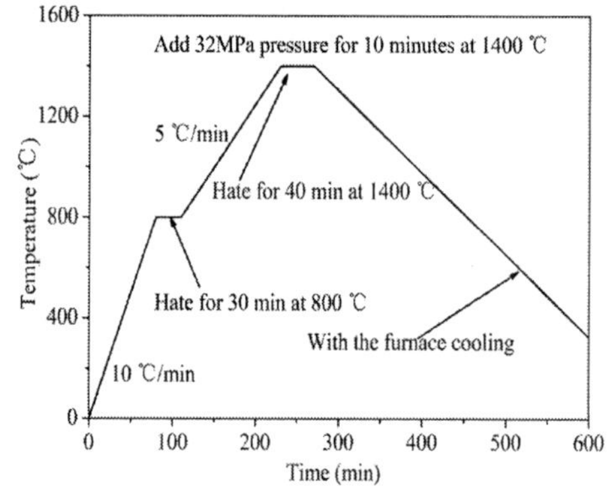

Figure 2. Sintering process.

2.2. Experimental methods

The hardness of the samples was measured using a 1000 N-dimensional microhardness tester, and each set of samples was averaged ten times. Before the measurement, first grind and polish on the pre-grinding machine with 200, 800, 1200 mesh sandpaper. The fracture toughness test was carried out on a CMT 5105 micro-control servo universal testing machine with a span of 20 mm using a three-point bending method. The experimental loading rate was 0.6 mm/min, and 10 samples were tested in each group, and the arithmetic mean value was obtained. The relative density of the sintered samples was tested by the Archimedes test method, and each group was measured 10 times to obtain an average value. The friction and wear test was carried out using the MMW-1 vertical universal wear tester. The test conditions were normal temperature and no lubrication. The load required for the experiment was applied by the loading rod P=200 N, and the experimental time was t=240 min. At the same time, the test sample was rotated by the motor to make the tested sample rub against the dual surface. The friction force F was measured by the experimental machine. The friction coefficient was obtained according to the formula $\mu=F/P$, and the friction coefficient was taken as the average value of the experimental results. An electronic scale with an accuracy of 0.1 mg was selected to measure the mass of the sample before and after the experiment. The difference in mass before and after was the amount of wear during the experiment. The wear amount was taken as the average value of 10 measurements. The product of the wear scar surface was detected and analyzed by means of a Rigaku D/Max-2550PC type diffractometer (XRD) the microstructure of the sample was observed and analyzed by S-4800 scanning electron microscopy (FE-SEM).

3. Results and analysis

3.1. Relative density and mechanical properties of the sample

The relative density of the materials is shown in Table 1. From the table we can see that with the addition of HfC, the relative density of the samples begins to increase. When the mass fraction of HfC is 1.5%, the relative density reaches the maximum, and HfC continues to be added, and the relative density begins to decrease. This indicates that the addition of HfC can effectively improve the sintering quality of the sample, increase the density, and reduce the generation rate of the pores.

Table 1. Relative density, Hardness and fracture toughness of composites

HfC(.wt%)	0	0.5	1.0	1.5	2.0	2.5
Relative density /%	81.23	94.60	94.70	95.67	95.05	93.28
Hardness (HV)	763.43	844.13	930	973.87	844.2	654.8
Fracture toughness(MPa·m$^{1/2}$)	13.34	14.67	15.56	16.45	15.12	12.45

The hardness and fracture toughness of the material are shown in Table 1. It can be seen from the figure that the hardness and fracture toughness of the composite are significantly changed when HfC is added. When the HfC mass fraction is 1.5%, the hardness and fracture toughness of the composites are increased by 11.84% and 59.05%, respectively, compared to the samples without HfC. With the continued addition of HfC, the flexural strength and fracture toughness of the composites have a decreasing trend.

3.2. Friction and wear properties

Figures 3 and 4 show the changes in friction factor and wear amount of samples with different HfC contents. It can be seen from Fig. 3 that under the same rotational speed and load, the friction coefficient of the sample material increases first and then decreases with the increase of HfC content, and the friction factor is between 0.33 to 0.51. It can be seen from Figures 4 that under the same rotational speed and load, the wear amount begins to decrease with the increase of HfC. When the HfC mass fraction is 1.5%, the wear amount is minimized, and the wear amount is 0.038%. The amount of wear was reduced by 24% compared to the sample without HfC added. However, the addition of HfC continued, and the amount of wear did not decrease.

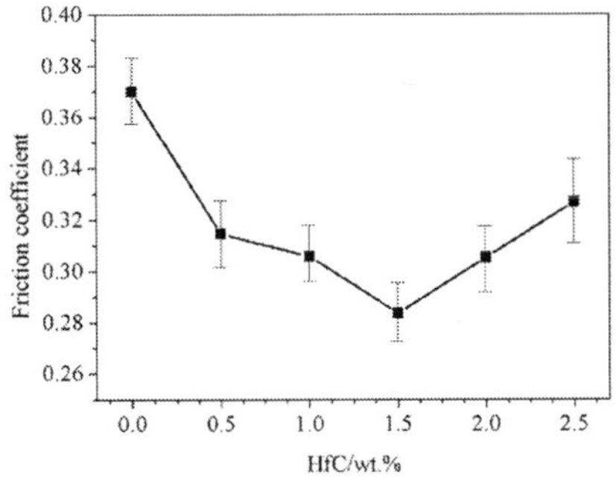

Figure 3. Sample friction coefficient diagram.

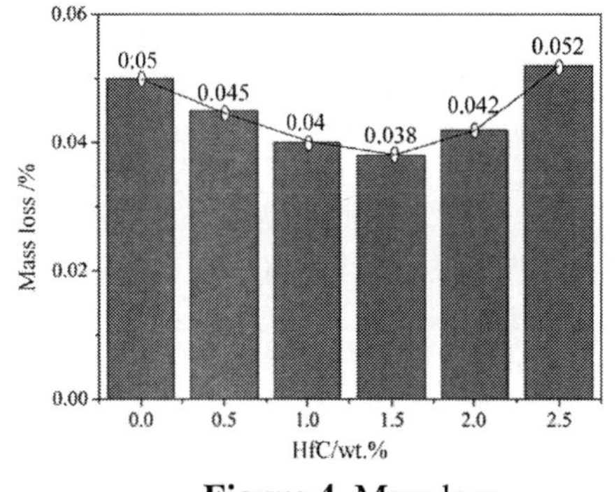

Figure 4. Mass loss.

It can be seen from Figure 5(b) that the surface of the sample also produces furrows and micro-pits, but the depth of the furrow is much smaller than that of Figure 5(a). This is because the addition of HfC increases the hardness of the sample, making it less prone to wear and improving the wear properties of the sample. At the same time, Figure 5(b) does not produce significant microcracks. This indicates that the addition of HfC can effectively improve the sintering quality of the sample and improve the crack resistance of the sample under alternating load. As can be seen from Figure 5(a) and (b), the wear performance of Fig. 5(b) is significantly better than that of Figure 5(a).

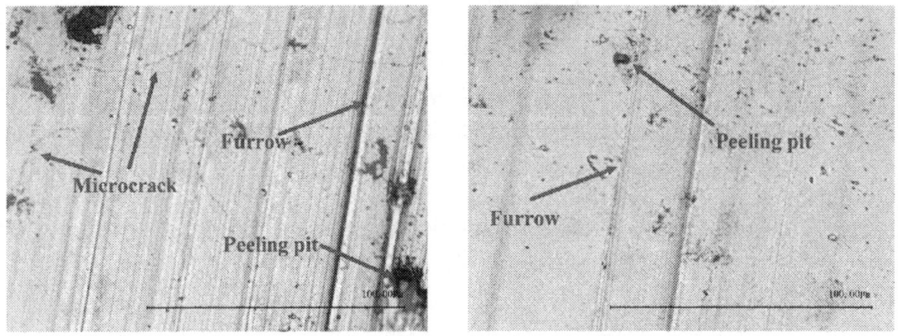

(a) WC-Co (b) WC-Co-1.5% HfC

Figure 5. Topography of the sample after wear.

Figure 6 is an XRD pattern of cemented carbide samples without HfC and 1.5% HfC added by Rigaku D/Max-2550PC diffractometer after abrasion. The surface of the sample to which HfC was not added (as Figure 6(a)) showed a strong WO_3 diffraction peak, while the peak of WC was weak, indicating that the hard phase WC in the composite had been decomposed by oxidation. The friction surface of the sample with 1.5% HfC added (as Figure 6(b)) also showed a strong WO_3 diffraction peak after abrasion,but the difference is that the peak of WC is stronger than the diffraction peak of WC in Figure 6(a), indicating that the hard phase WC in the sample to which 1.5% HfC is added is less decomposed. The diffraction peaks of CoO in Figure 6 (a) and (b) show that Co also undergoes an oxidation reaction during abrasion. During the friction process, the heat generated by the friction of the sample is much larger than the heat dissipated, so that the temperature of the sample rises and an oxidation reaction occurs, so that the surface is covered with an oxide film. During the friction process, the substrate is subjected to load, and the friction will be carried out on the oxide film. The elastic modulus and hardness of the WC oxidation products WO_3 and Co oxidation products CoO (or Co_3O_4) are lower than those of the matrix material. Therefore, the oxide film attached to the base material is easily broken. After the destruction, the new matrix exposed in the air acts on the surrounding environment to form a new oxide film. This repeated destruction and formation causes the surface wear of the sample to be intensified. The diffraction peak of HfO_2 does not appear in Figure 6(b), indicating that HfC does not undergo oxidation reaction, mainly because HfC has extremely strong phase stability from room temperature to melting point, and it is extremely difficult to react with oxygen. The elastic modulus of HfC and its hardness are also greater than that of the matrix oxide, and HfC replaces the matrix to wear, thereby improving the surface friction properties of the sample. And HfC can change the state of Co binder, making Co distribution more uniform in the matrix material, because the oxidation temperature of Co (350 °C) is much lower than the oxidation temperature of WC (500 °C). Therefore, the sample first generates an oxide of Co, and the oxide of Co in turn suppresses the formation of WO_3, thereby improving the friction properties of the samples.

Figure 7 shows the wear profile of the sample with 1.5% HfC added and no HfC added. As shown in Figure 7 (a) and (c), the sample to which 1.5% HfC was added had a good sintering quality. During the friction and wear process, the hard phase particles on the surface of the sample peel off from the friction surface under the action of the pressing force and the shearing force to form pits. At the same time, the peeled hard phase remains between the friction pairs to form abrasive grains, and the surface of the material forms a furrow during the relative sliding process. At the same time, the heat generated by the friction softens the binder phase of the sample, destroys the stability of the hard particles, and aggravates the abrasive wear. As shown in Figure 7(b) and 7(d), compared with Figure. 7(a) and 7(c), the sample to which no HfC was added was inferior in sintered quality, and a porous hole was formed on the surface. During the sliding contact with the stainless steel on the grinding head, the normal and tangential stresses on the friction surface cause the hard particles to collide with each other, resulting in microcracks inside the grain boundaries or grains. Under the action of alternating loads, the microcracks continue to expand until they intersect each other, forming a ring-shaped crack, and finally pulling out. At the same time, in the process of microcrack expansion to form a ring-shaped crack, two processes including the extraction of the particles to form pits and the wear of the abrasive grains to form the furrow are also accompanied.

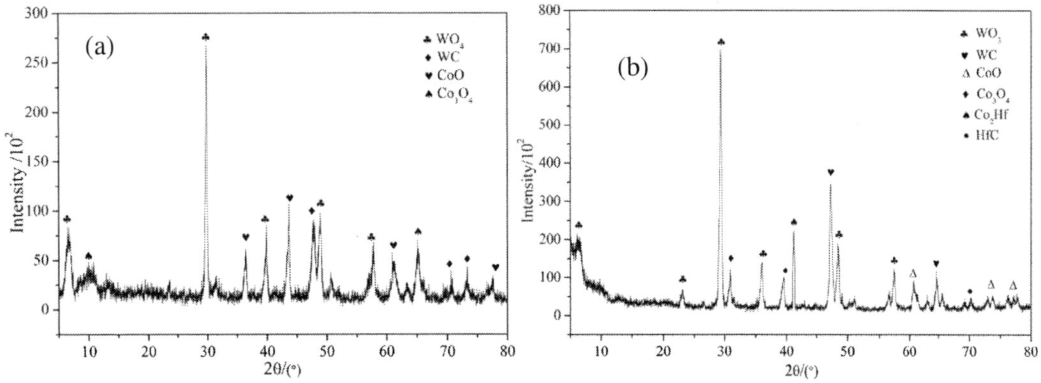

Figure 6. XRD pattern of two samples.

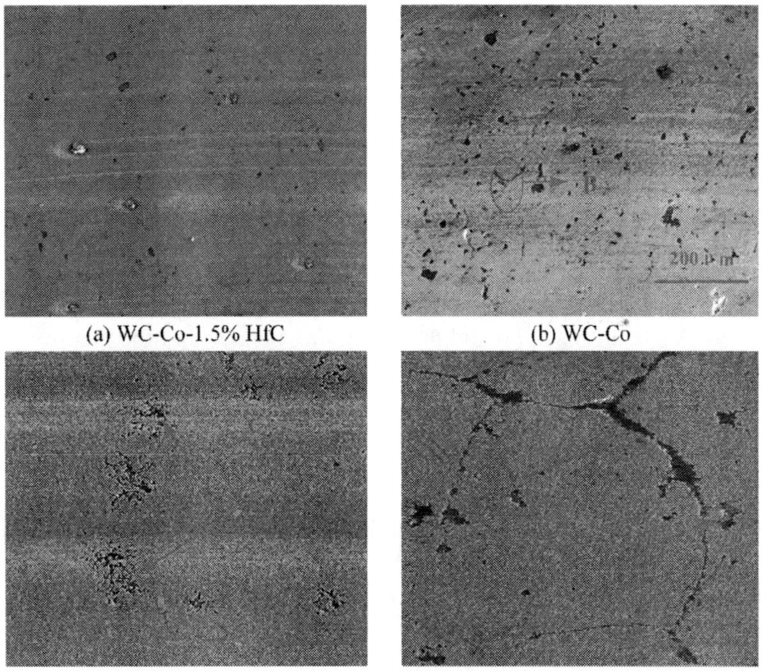

(a) WC-Co-1.5% HfC (b) WC-Co

(c) Amplified Region A (d) Amplified Region B

Figure 7. Topography of the sample after wear.

4. Conclusions

A study of the effect of HfC on Friction and Wear Properties of WC-Co Cemented Carbide have been conducted and the conclusions are listed as follows:

(1) Appropriate amount of HfC can effectively improve the friction coefficient and mass loss of WC-Co composites, both of which increase first and then decrease, and the friction factor is between 0.33 to 0.51. When the mass fraction of HfC is 1.5%, the friction coefficient and the mass loss of wear

are both at a minimum, and it has the advantages of high hardness and high fracture toughness, The furrow is shallow and the friction and wear performance is better.

(2) During the friction and wear process, the surface material of the sample is oxidized to form WO_3 and CoO oxide film, which aggravates the wear process, and the addition of HfC can inhibit the formation of WO_3, thereby improving the friction and wear properties.

(3) The wear of WC-Co composites is mainly achieved by particle extraction, furrows and exfoliation caused by annular microcracks. The addition of HfC can effectively improve the sintering quality, inhibit the microcrack initiation and expansion, reduce the phenomenon of particle extraction and furrowing, and improve the tribological properties.

Acknowledgments

This work was supported by the National Natural Science Foundation of P.R. China for the financial support (ID: 51872122), Distinguished Middle-Aged and Young Scientist Encourage and Reward Foundation of Shandong Province (ID: ZR2016EMB01) and Taishan Scholar Engineering Special Funding (2016-2020).

References

[1] Maneesh C, Fiona S, Eckart U, Rozalia A, Alon H, 2017, Diamond and related materials, wear performance of diamond coated WC-Co tools with a CrN interlayer, *Elsevier* **73**, pp 47-55

[2] Philipp H, Uta K, Nageswaran T A, Tomas B, Anders Wretland, 2017, Journal of Manufacturing Processes, Flank wear characteristics of WC-Co tools when turning Alloy 718 with high-pressure coolant supply, *ResearchGate* **30**, pp 116-123

[3] Zhang Y Y, Yakov E, Richard R. Chromik, 2018, Tribology International, Dry sliding wear behaviour of cold-sprayed Cu-MoS_2 and Cu-MoS2-WC composite coatings: The influence of WC, *Elsevier* **123**, pp 296-306

[4] Siavash M, Steven M, Michael A, Jan A, 2017, International Journal of Refractory Metals and Hard Materials, The wear of WC-Co drill bits during rotary-percussive drilling of reinforced concrete, **62(B)**, pp 202-209

[5] Yin G T, Wang Y Y, Cui H C, Fenggui Lu, Peiquan Xu, 2018, International Journal of Refractory Metals and Hard Materials, Effect of holding time and interlayer's thickness on the crack initiation and propagation and the dissolving behavior of the heat-treated facet WC grains, **71**, pp 45-60,

[6] Nahvi S.M., Jafari M., 2016, Surface and Coatings Technology, Microstructural and mechanical properties of advanced HVOF-sprayed WC-based cermet coatings, **286**, pp 95-102

[7] Vikas V, Manoj B.V., 2017, Wear, Sliding wear behavior of SPS processed TaC-containing Ti(CN)-WC-Ni/Co cermets against Silicon Carbide, **376-377(B)**, pp 1570-1579

[8] Siwakp P, Garbiec D, 2016, Transactions of Nonferrous Metals Society of China, Microstructure and mechanical properties of WC-Co, WC-Co-Cr_3C_2 and WC-Co-TaC cermets fabricated by spark plasma sintering, **26(10)**, pp 2641-2646

[9] Su Q D, Zhu S G, Ding H, Bai Y F, Di P, 2018, International Journal of Refractory Metals and Hard Materials, Effect of the additive VC on tribological properties of WC-Al_2O_3 composites, **75**, pp 111-117

[10] Guo Z X, Xiong J, Yang M, Dong G B, Wan W C, 2012, International Journal of Refractory Metals and Hard Materials, Tool wear mechanism of WC-5TiC-10Co ultrafine cemented carbide during AISI 1045 carbon steel cutting process, **35**, pp 262-269

[11] Ludovic C, Marianne B P, Sans J L, 2017 Corrosion Science, Effect of high temperature oxidation on the radiative properties of HfC-based ceramics , **126**, pp 55-264

[12] Yan X Y, Wang S R, Wang Y, Shandong Science,2018,Preparation and Mechanical Properties of Tungsten Carbide-based Cemented Carbides **31(03)**, pp 28-33

IOP Conf. Series: Materials Science and Engineering **491** (2019) 012017 doi:10.1088/1757-899X/491/1/012017

The Influence of Nanocomposite Carbon additive on Tribological Behavior of Cylinder Liner/Piston Ring

C Y Xue[1], S R Wang[1,*], Y Wang[2], G Q Wang[1] and X Y Yan[1]

[1]School of Mechanical Engineering, University of Jinan, Jinan 250022, China
[2]School of Physics and Technology, University of Jinan, Jinan 250022, China

[*]E-mail: me_wangsr@ujn.edu.cn

Abstract. Nanocomposite Carbon additive were dispersed at varying concentrations (1, 3, 5wt%) in SN/GF-5 lubricants for automotive engines. Raman spectroscopy is used to characterize the morphology of carbon.The effect of Nanocomposite Carbon additive on the friction and wear properties of the piston ring friction pair is investigated by using UMT friction tester. The tribological mechanism of Nanocomposite Carbon additive is further investigated by scanning electron microscopy,energy dispersive spectrometer and three-dimensional shape spectrometer. The results show that Nanocomposite Carbon additive exhibits excellent antifriction performance at high load and friction coefficient at low load. The effect of the nanocomposite carbon additive is the best when the mass fraction is 3%; however, the nanocomposite carbon additive concentrations is too high, which will lead to increased wear under high load. The nanocomposite carbon particles can smooth the micro-convex of the friction pair surface and reduce the surface roughness. The contact area of the friction pair is significantly increased, and the pressure is significantly reduced. The nano-composite carbon avoids the breakage of the bearing oil film when the metal surface peaks contact each other, ensuring the integrity of the oil film and uniformly carrying a large load.The nanocomposite carbon lubricant additive shown in this study can reduce energy consumption and extend engine life, resulting in significant cost savings.

1. Introduction

Reduction in friction and wear is one of the most important objectives of tribological research [1]. With the rise of nanotechnology, many studies have been conducted in recent years using carbon nanomaterials as additive in lubricating oil to improve the tribological properties of lubricants.There are few commercial additive containing carbon nanomaterials, mainly because the dispersion and stability of nanoparticles in lubricants are not well solved [2].The extremely fine grains of the nanoparticles,which have a large surface energy. Large self-aggregation tendency between particles, which form large aggregates after a period of time [3]. Nanocomposite carbon additive mainly contains Ultra-dispersed diamond graphite powder manufactured from explosives,which will not cause chemical pollution. Russia's Yu. Neverovskaya has successfully modified the surface of diamond with a coupling agent to improve the lipophilicity of the diamond and the suspension in the organic phase,which proposed a dispersion model for surface modification of diamond with silane coupling agent[4-5]. Red' kin et al. added a composite additive containing nanodiamond-graphite to the engine oil, and the antifriction and antiwear properties were significantly improved [6]. At the same time, the contents of elements such as sulfur, chlorine, phosphorus, and fluorine are lowered. The friction coefficient is lowered, the smoothness of the friction pair is increased, and the noise is lowered [7]. After mechanochemical modification of nanocomposite carbon it can form a stable colloidal system in various lubricating oils.and it can fill the micro-cracks and unevenness on the metal surface freely grind the sharp protrusion on the friction pair surface and remove the uneven peak of the friction pair. the addition of nanocomposite carbon additive to the lubricating oil can improve the working life of the engine and transmission and it can reduce surface wear[8].the authors used the umt tester to carry out the reciprocating friction and wear test. The effects of tribological properties of nanocomposite carbon additives on cylinder liner piston rings were investigated. The morphology and EDS analysis of the wear scar of the cylinder liner were used to explore the Nanocomposite Carbon. The tribological mechanism of Nanocomposite Carbon additive under the frictional pair of cylinder liner/piston ring is discussed.

2. Experimental section

2.1. Materials
Ultra-dispersed diamond graphite powder was purchased from Nanjing XFNANO Materials Tech Co., Ltd. SN/GF-5 lubricant were provided by Tianjin Co.,Ltd.T-161A High Molecular Weight Polyisobutylene Succinimide was supplied by Jinzhou Runda Chemical Co., Ltd.T-109 calcium alkyl salicylate was purchased from Xinxiang Richful Lube Additive Co., Ltd. Coupling agent of organic titanate were provided by Dalian Richon Chem Co., Ltd. Nanocomposite Carbon Lubricant fluid Density(15 °C) and Viscosity(40 °C) are 1.00g/cm^3 and 79.15 Cst. Ultra-dispersed diamond graphite nanoparticle size is 4nm. The upper and lower friction pair properties are 6×17×10mm , 45×34×8mm, R_a=0.43μm and hardness:55 HRC.

2.2. Characterizations and measurements

2.2.1. Configuring nano-composite carbon lubricant additive. Ultra-dispersed diamond graphite powder is mechanically chemically modified by a ball milling method to remove hard agglomeration of the nanoparticles. The nanomaterial is wrapped with a titanate coupling agent. The surface agglomerates are agglomerated by mechanical stress and activated to modify the surface crystal structure and physicochemical structure. The nanoparticle lattice undergoes displacement and internal energy increases. Under the action of external force, the titanate coupling agent acts as a surfactant to participate in the reaction and achieve the purpose of surface modification. The surfactant is coated with the surface of the particle by chemically adsorbing or chemically reacting the surface-active organic functional group with the surface layer of the nanoparticle. Nanocomposite Carbon additive with good lipophilic dispersion properties was prepared. The treated Nanocomposite Carbon additive was added to the SN/GF-5 lubricating oil, and the detergent T-109 and the dispersing agent T-161A were added. Different proportions of 1%, 3% and 5% Nanocomposite Carbon lubricant additive were prepared.

2.2.2. Tribotest rig description. The upper and lower specimens are obtained from the engine piston ring and the cylinder liner, and the dimensions are as shown in the table1.The main structural features of the test rig are shown in Fig. 1.The oil sample was dripped 1 ml before the test. The test was divided into a 30 s run-in and a 30-minute test phase. The run and phase load was 50 N. The test phase loads were 50, 100, 250 and 400 N, respectively, with a temperature of 100 ° C, a stroke of 1 mm and a frequency of 10 Hz. The cylinder liner was placed in petroleum ether before and after the test and ultrasonically cleaned. The weight loss percentage of the liner is measured on the balance. Then, the three-dimensional surface topography instrument was used to observe the wear profile of the liner. Scanning electron microscopy and energy spectrometer were used to observe the wear profile of the liner and analyze the surface element composition of the wear scar.

IOP Conf. Series: Materials Science and Engineering **491** (2019) 012017 doi:10.1088/1757-899X/491/1/012017

3. Results and Discussion

3.1. Raman spectroscopy

Raman spectroscopy is a means of studying carbon structure efficiently and can characterize the morphology of carbon. The Raman spectroscopy of the Nanocomposite Carbon additive are shown in Fig. 2. As can be seen from the Fig.2, the broad Raman peak near 1329 cm^{-1} is the characteristic peak of the sp^3 structure nano-diamond, and the Raman peak observed near 1580 cm^{-1} is the sp^2 structure nano-graphite. Since the Raman scattering cross section of diamond is 1/60 of graphite, this indicates that nano-composite carbon has both nano-diamond and sp^2 structure nano-graphite residue [9-10].

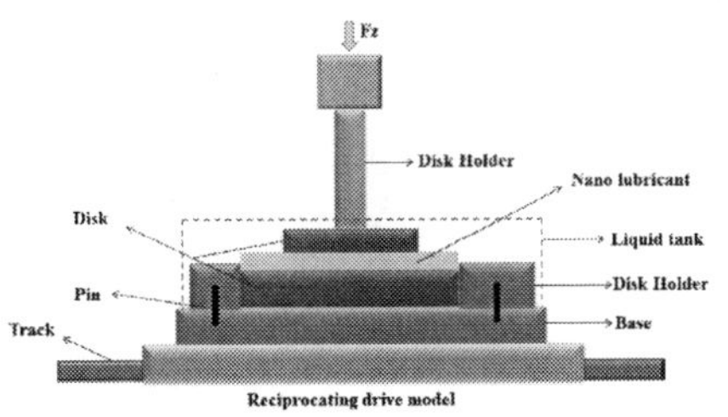

Figure. 1 Photograph of sample holders in UMT

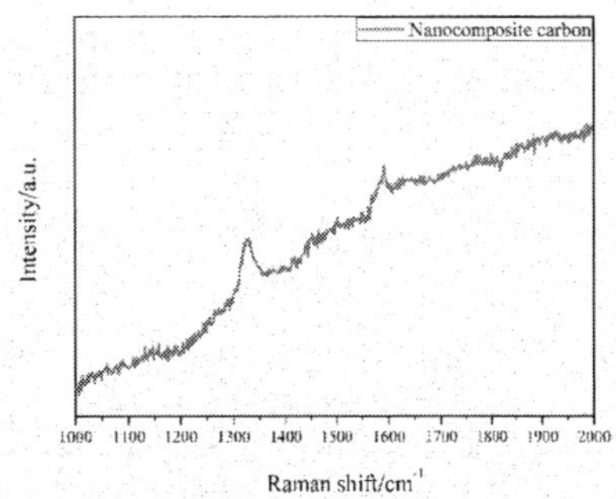

Figure. 2 Raman spectra of Nanocomposite Carbon

3.2. Effect of Nanocomposite Carbon additive Content on Friction and Wear Properties of Cylinder/Piston Ring Friction Pairs

Fig. 3 shows the friction coefficient curve of a SN/GF-5 engine oil using a nanocomposite carbon additive with different mass fractions for the cylinder liner piston ring friction pair. It is found through the friction curve that the addition of nanocomposite carbon additive reduces the friction coefficient. When the mass fraction of the nanocomposite carbon additive is 3%, the friction coefficient is the lowest. The friction coefficient of the friction pair decreases first and then increases with the increase of the nanocomposite carbon mass fraction. The excellent antifriction and anti-wear properties of nanocomposite carbon are demonstrated [11]. It can be seen from Fig. 4 that the wear percentage of the nanocomposite carbon mass fraction of 3% is the lowest. The nanodiamond in the nanocomposite carbon can micro-cut the micro-protrusion of the worn surface. Therefore, the metal surface is smoother and the coefficient of friction is low. Excessive mass fraction of nanocomposite carbon will increase wear and tear.

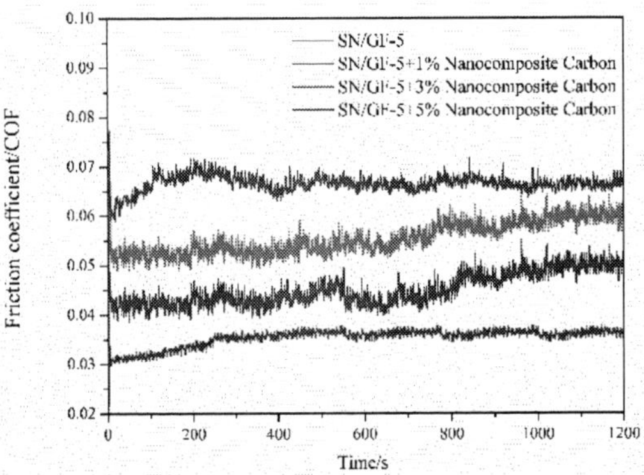

Figure. 3 Coefficient of friction curves

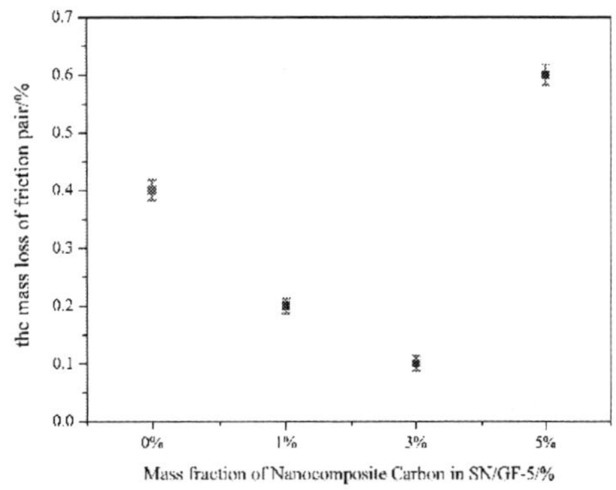

Figure. 4 The percentage of the friction pair mass loss

3.3. Effect of load on friction and wear properties of cylinder/piston ring friction pair

Fig. 5 shows the average friction coefficient of the cylinder/piston ring friction pair using different mass fractions (0%, 1%, 3%, and 5%) of nanocomposite carbon additive engine oils at different loads (50N, 100N, 250N, and 400N). At low load (50N, 100N), the contact area of the friction pair is small. The oil film is discontinuous, and the effect of the nanocomposite carbon additive is not prominent. As the load increases, the contact area of the friction pair increases, and the nanoparticles in the nanocomposite carbon additive adsorb on the contact surface to improve contact. And it can enhance the carrying capacity of the oil film. At high loads (250N, 400N), the average coefficient of friction is lowest when the mass fraction of nanocomposite carbon is 3%. Figure 6 shows the cylinder/piston ring friction wear rate of different mass fraction nanocomposite carbon engine oils under different loads. With the increasing of the load, the wear rate of the cylinder/piston ring friction pair is increased. The wear rate of the friction pair using nanocomposite carbon mass fraction 5% engine oil is the highest.

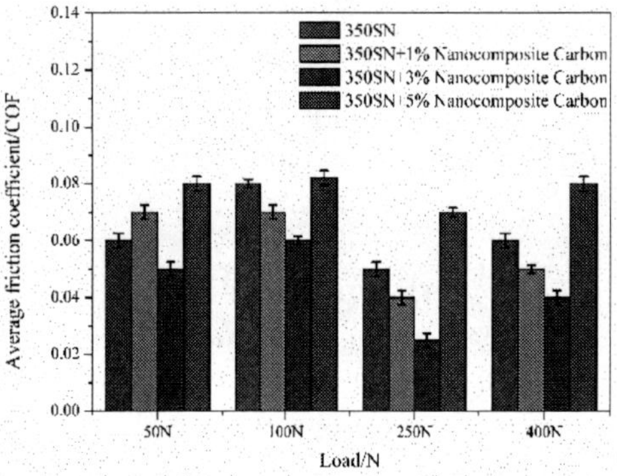

Figure. 5 Average friction coefficient

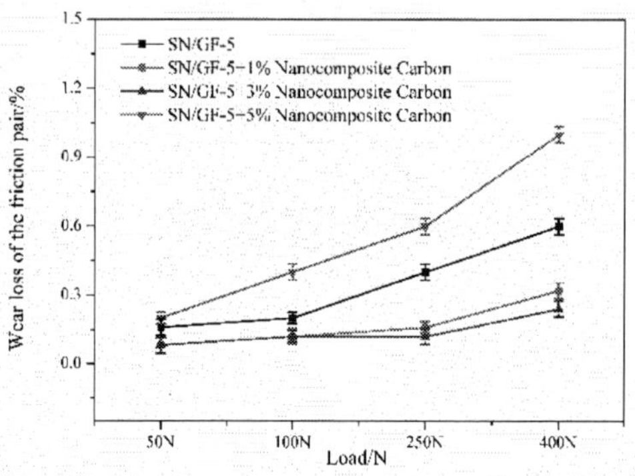

Figure. 6 Wear rate of the friction pair

3.4. Analysis of friction and wear mechanism

Fig. 7 shows the SEM and 3D morphology analysis of the wear scar at the 400 N Nanocomposite Carbon mass fraction of 0%, 1%, 3%, and 5% after the test. It can be seen from Fig. 7(a) that when the Nanocomposite Carbon additive with a mass fraction of 5% is added, the Nanocomposite Carbon in the contact zone of the piston ring of the cylinder liner increases the friction and wear, and a large number of pits and furrows appear, and the surface is uneven. The degree of wear is mainly due to abrasive wear. When there is no Nanocomposite Carbon additive in the contact zone of the cylinder/piston ring friction pair, the furrow still appears, which may be caused by the scratching of the cylinder liner in Fig. 7(b). It can be seen from Fig. 7(c) that when the mass fraction is 3%, Nanocomposite Carbon is added. After, furrows appear in the wear scar zone of the cylinder liner. The furrows are distributed throughout the wear scar zone, but the furrows are arranged neatly, the surface is smooth, and the cracks are significantly reduced. It can be seen from Fig. 7(d) that the cylinder liner wears is reduced. The surface is flat and the depth of the wear scar is shallow. In summary, the addition of an appropriate amount of Nanocomposite Carbon additive to the SN/GF-5 engine oil will greatly reduce the wear of the cylinder liner. Under different loads, the cylinder liner wear table is more distinct. Specifically, with the increase of load, the flatness of the wear surface decreases, and the depth of the furrow increases.

The 3D morphology of the wear surface of different concentrations of nanocomposite carbon lubricant after the friction and wear test is shown in Fig. 8. The three-dimensional topography mainly represents the distance between the peaks and valleys of the surface contour. In Fig. 8a, the three-dimensional topography of the mass fraction of nanocomposite carbon is 3%, and the distance between the peak and the valley is 3.49 μm. As the concentration of the additive increases, the depth of the wear scar increases significantly. Furrows and pits can be found on the worn surface. The maximum sample size of 5% is 12.07 μm (Fig. 8b).

The EDS analysis was performed on the region shown by the SEM in Table 1., and black cracks were found in the SEM image. The flake graphite confirmed as a matrix by spectrum 1.2. When 1%

and 3% Nanocomposite Carbon were added to SN/GF-5, the wear surface element composition was Fe, C, Mn, Si and other matrix elements, indicating that no other elements on the friction pair surface participated in the friction reduction. When the mass fraction of Nanocomposite Carbon in SN/GF-5 is 3%, the content of C, Mn, Si and other elements on the wear surface decreases(spectrum 3.4.). It is indicated that the addition of the Nanocomposite Carbon additive forms a reaction film[12].

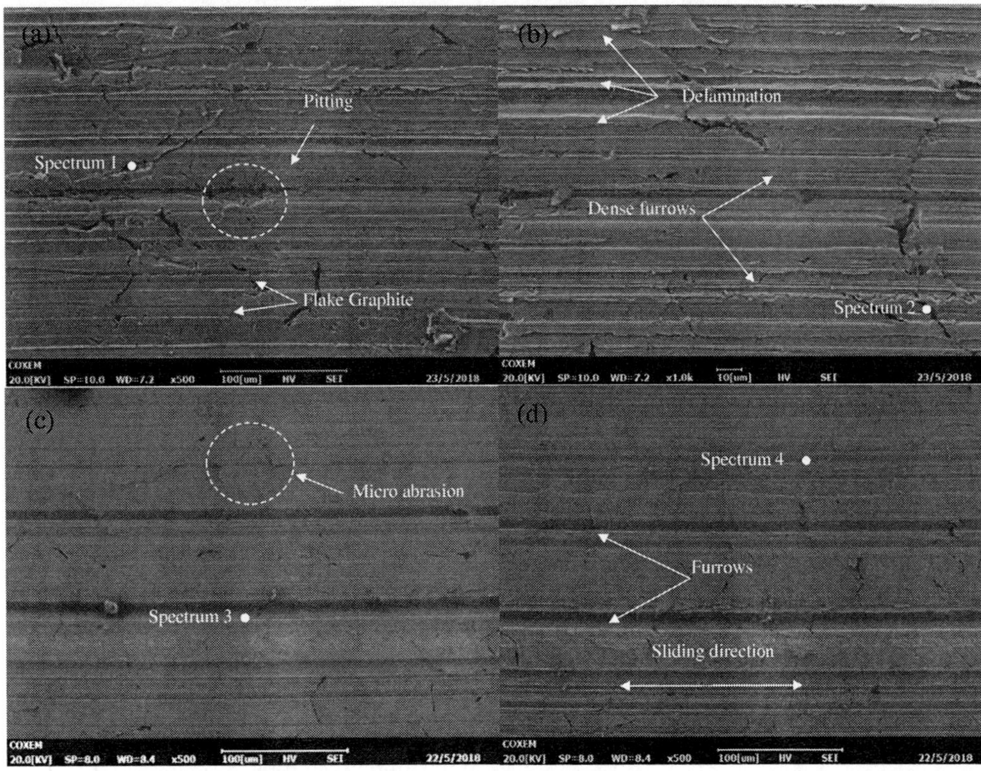

Figure. 7 SEM photograph of cylinder liner worn surfaces for (a) 5% (b) 0%, (c) 3% and (d)1%

Table 1. EDS spectra and elemental analysis of cylinder liner worn surfaces

Spectrum	Rlement mass fraction /%					
	Fe	C	O	Si	Mn	F
1	0.63	98.29	1.08	0	0	0
2	0	99.01	0.91	0	0	0
3	89.89	6.66	0	1.44	0	2.01
4	88.49	6.91	0	1.10	1.50	2.00

4. Conclusions

a At low load (50N, 100N), the anti-friction effect of the nanocomposite carbon additive on the cylinder/piston ring friction pair is not obvious. At high loads (250N, 400N), the nanocomposite carbon additive has a significant effect on the friction reduction of the cylinder/piston ring friction pair.

b Excessive Nanocomposite Carbon additive will aggravate the wear of the liner/piston ring friction pair, and the wear form is mainly abrasive wear. The amount of wear of the liner increases to the increase in the mass fraction of Nanocomposite Carbon additive in the SN/GF-5 engine oil.

c The addition of the nanocomposite carbon additive to the lubricating oil can enhance the bearing capacity of the lubricating oil film. The adsorption of nanoparticles in the nanocomposite carbon will adhere to the metal surface to reduce wear. The micro-cutting action of the nano-diamond in the nanocomposite carbon enables the surface of the friction pair to be smoother.

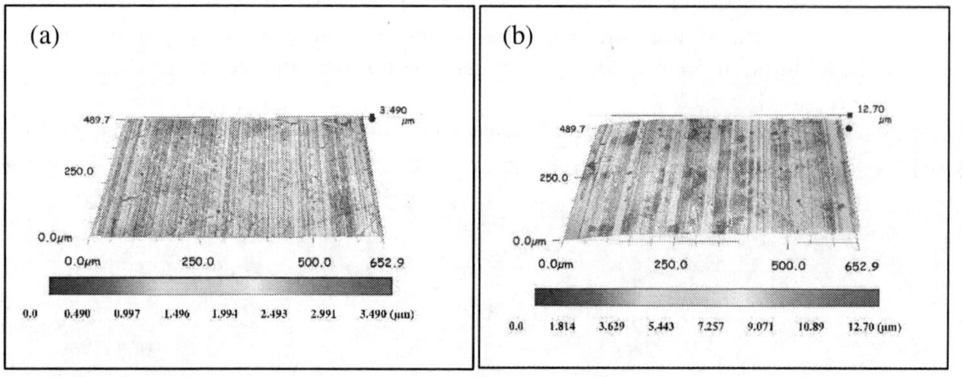

Figure. 8 3D morphologies of cylinder liner worn surfaces for (a) 3% and (b) 5%

Acknowledgment
This work was supported by the National Natural Science Foundation of P.R. China for the financial support (ID: 51872122), Distinguished Middle-Aged and Young Scientist Encourage and Reward Foundation of Shandong Province (ID: ZR2016EMB01) and Taishan Scholar Engineering Special Funding (2016-2020).

References

[1] Shen MW, Luo JB, Wen SZ, 2001. The tribological properties of oils added with diamond nano-particles. Tribol Trans;**44**:494-8.
[2] Shen, M; Luo, J; et al., 2001.The tribological properties of oils added with diamond nano-particles.Tribology Transactions, **44(3)**: 494-498.
[3] Ivanov, M.G.; Kharlamov, V.V.; et al., 2004.Tribological properties of the grease containing polytetrafluoroethylene and ultradispersed diamonds.Trenie i Iznos,,**25(1)**:99-103.
[4] A Yu Neverovskaya, A P Voznyakovski, V Yu Dolmatov, 2004.Structure of the Dispersive Medium and Sedimentation Resistance of Suspensions of Detonation Nanodiamonds JI.Physics of the Solid State, **46(4)**: 662-664.
[5] AP Voznyakovski,2004.Self-Organization in Nanocomposites Based on Detonation Nanodiamondsl JI .Physics of the Solid State, **46(4)**: 644-648.
[6] Red'kin,V.E.,2004.Lubricants containing ulradisperse diamond-graphite powder.Khimiya i Tekhnologiya Topliv i Masel,**(3)**:32-35.
[7] SW Hwang,SC Chang,TE Zhang, HK Kim, 2017.Tribology behavior of a lubricant with nano-diamond particles on steel. International Journal of Engineering & Technology,**9(1)**:169-174
[8] PH Tsai,HY Chu,Effects of the Nano-Diamond Additive on the Tribological Performance Improvement of Lubricating Grease. Key Engineering Materials, 2015,**642(6)**:298-302
[9] M Mermoux,A Crisci,T Petit,HA Girard,JC Arnault, 2014.Surface Modifications of Detonation Nanodiamonds Probed by Multiwavelength Raman Spectroscopy,Journal of Physical Chemistry C,**118(40)**:23415-23425
[10] M Mermoux S Chang HA Girard JC Arnault, 2018.Raman spectroscopy study of detonation nanodiamond,Diamond & Related Materials,**87**:248-260
[11] HT Hsieh ,GL Chen ,PH Tsai,HY Chu, 2015.Effects of the Ultra-Dispersed Nano-Diamond Additive on the Grease Boundary Lubrication Performance in the Reciprocating Journal Bearing Test, Key Engineering Materials, **642**:303-306
[12] M Marko,B Branson,E Terrell, 2015.Tribological Improvements of Dispersed Nanodiamond Additive in Lubricating Mineral Oil, Journal of Tribology,137 (1):01180

IOP Conf. Series: Materials Science and Engineering **491** (2019) 012018 doi:10.1088/1757-899X/491/1/012018

Proposals for Improving the Efficiency and Durability of the Turbines of Turbo-Pump Assemblies in Liquid-Propellant Rocket Engines by Using Double-Sided Crest-Type Radial Labyrinth Seals

V S Vasiliev[3, *], P S Levochkin[1], V K Chvanov[1] and S F Timushev[2]

[1] NPO Energomash named after academician V.P. Glushko, 1 Burdenko street, Khimki, Moscow Region, 141400, Russia.
[2] Moscow Aviation Institute (National Research University), 4 Volokolamskoe shosse, Moscow, 125993, Russia.
[3] Post-graduate student, NPO Energomash named after academician V.P. Glushko, 1 Burdenko street, Khimki, Moscow Region, 141400, Russia.

[*]Email: vasilev_vs@npoem.ru.

Abstract. Nowadays the improvement of the design of liquid-propellant rocket engines (LPREs) depends on a whole range of factors and activities, among which it is possible to distinguish, for example, the improvement of energy characteristics that are part of the LPREs, units and assemblies. The article compares the operation of an LPRE turbo-pump turbine using double-sided and single-sided labyrinth seals in Energomash's engines. Main geometrical parameters of each seal variant, images built for the calculations of 3D models of seals, the resulting calculation grid, as well as graphs of the pressure distribution and velocity fields are presented for a visual comparison. Sectoral leakage was calculated for each variant of the seal, a calculated assessment of the possibility of reducing the temperature of the working gas at the turbine inlet is carried out. In the future attention should be paid to a more detailed study of this type of seals using of modern computing power.

1. Introduction

Greater fuel efficiency is the key objective of activities to improve the design of liquid-propellant rocket engines (LPREs). It is achieved by enhancing the fuel efficiency of engine units and assemblies. Ancillary tasks include extending the service life, reliability and re-usability of units and assemblies, and reducing the time required for their design and manufacture. In order to succeed in all of these tasks, designers should address a range of issues at all stages in the creation of LPREs, from R&D to serial production [1, 2].

Optimal design of the turbo-pump units (TPUs) of LPREs is among the most important of these issues [3, 4]. Various approaches to modelling and analysis are used in TPU design. The present article focuses on the design of the radial labyrinth seals of TPU turbines and aims to give a conclusive assessment of how the geometry of the shroud turbine seal impacts the flow of working gas through the channels of the seal by numerical modelling of three-dimensional flow.

TPU turbines on LPREs designed by Energomash, with afterburning of the oxidizing generator gas, have fairly large gaps in the radial labyrinth seals, allowing substantial leakage of gas through the flow sections of the seals, which in turn has an impact on the operation of the turbine. The proposal is to reduce leakage by altering the design of the radial labyrinth seals. At present, single-sided crest-type radial labyrinth seals are widely used on the TPU turbines of Energomash LPREs. However, the use of double-sided seals could reduce leakage and increase the efficiency of the turbine. The use of double-sided seals has already been proposed in various engine designs, but the proposals have not been implemented in contemporary engines [5]. Figure 1 shows the visual form of the seals.

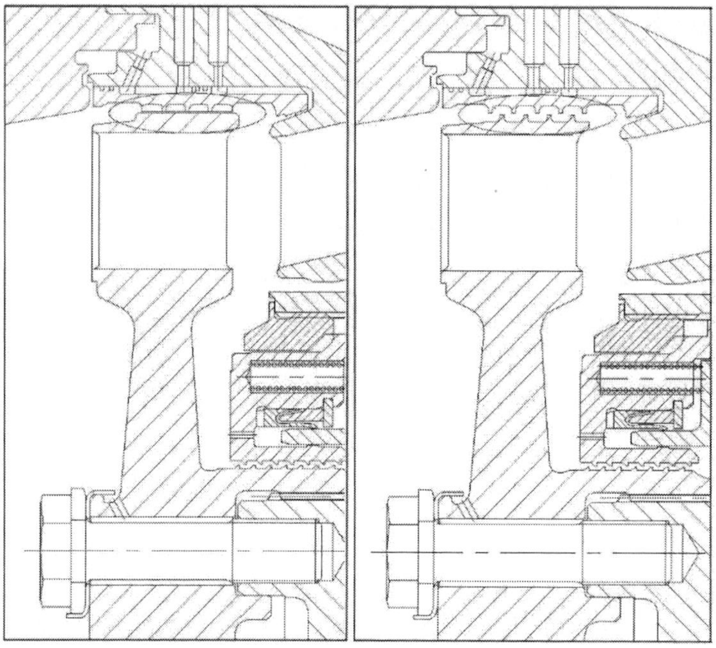

a) b)

Figure 1. A single-sided (a) and double-sided (b) crest-type radial shroud labyrinth seal.

2. Description of the approach

The present article compares the performance of a single-sided crest-type radial shroud labyrinth seal (currently used in the turbines of the TPUs of Energomash LPREs) with double-sided alternatives by numerical simulation of the three-dimensional flow in a stationary formulation.

Calculations are carried out for shroud seals with various geometries in order to identify the design, which ensures minimum leakage rate and, as a consequence, highest efficiency of the turbine. Therefore, the main evaluation criterion is the compaction leakage rate ("$\dot{m}$"). The reliability of modern LPREs, particularly those intended for multiple use, depends to a large extent on reliable operation of the TPU turbine, which is largely determined by the temperature of the working gas [6]. It is therefore important to find ways of reducing the gas temperature without power losses, and the present article also offers estimates of success in achieving this objective [7].

3D models of alternative designs for the flow sections of the seals were obtained using a CAD system. Since the problem is symmetrical to the axis of rotation, any sector can be used for calculation, such as for instance a 15-degree sector, which significantly reduces the calculation time in comparison with a full 3D model of the flow section. The geometry of the flow section is designed taking account of the main features of the working gas flow in the shroud seal channels, determining a specific leakage rate for a given effective pressure differential. The geometric parameters of the 3D models also take account of the so-called "umbrella effect", caused by low-temperature oxygen flows from the bearing support, which cool the surface of the turbine impeller, causing it to bend. As a result, the impeller blades move towards the blades of the nozzle apparatus. The Figure 2 below shows how the impeller bends during engine operation (the geometry scale is in millimeters).

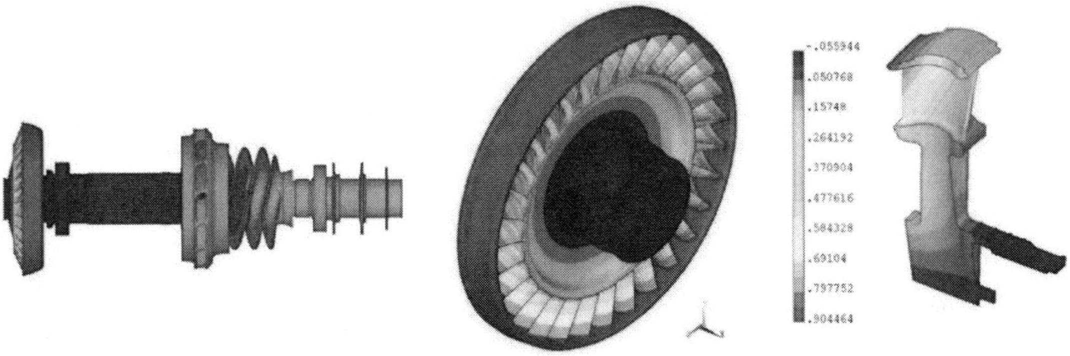

Figure 2. The umbrella effect in the 3D-model.

Calculations for the 3D models were carried out using a computational hydrodynamics program [8]. The grid is built automatically, with adaptation to the specified geometry, and the average grid size in each model is above 550,000 cells. An initial grid with parallelepiped cells is defined in the calculation area, and then geometrical sub-regions are selected for calculation using a denser grid. Each cell of the initial grid is divided in these sub-regions into 8 equal cells of a new adaptation level. If necessary, the cells are divided again, and the operation is repeated until the required level of accuracy is obtained. This sub-grid resolution of geometry approximates curvilinear boundaries on a rectangular grid. The cells through which the boundary passes lose their original parallelepiped form and are transformed into arbitrarily-shaped polyhedrons. This approach enables a sufficient degree of accuracy in calculations even with a relatively low number of grid nodes.

The calculation is carried out using an iterative method from zero initial conditions. At the beginning of the calculation, for the period of pressure growth, the time step is $1e^{-4}$ s, after which it is set automatically and amounts to approximately $1e^{-8}$ s.

The values of the boundary parameters are provided by the operation of a particular Energomash LPRE and are as follows:

- temperature of the working gas at turbine inlet = 581.4 °C;
- temperature of the working gas at turbine outlet = 467.9 °C;
- absolute pressure at turbine inlet = 539.37 kgf/cm^2;
- absolute pressure at the turbine outlet = 283.92 kgf/cm^2;
- rotation speed of the turbine impeller = 22,825 rpm.

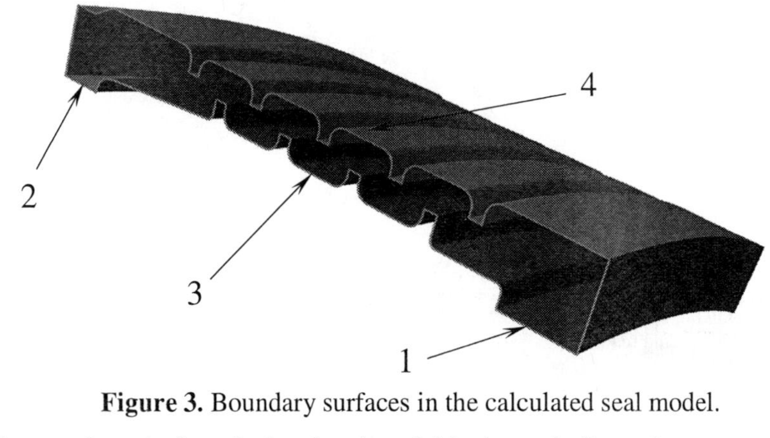

Figure 3. Boundary surfaces in the calculated seal model.

The boundary surfaces in the calculated seal model is shown in figure 3.

The calculation area consists of four sub-regions:
1 (blue) - working gas input surfaces (input parameters specified).
2 (green) - working gas output surfaces (output parameters specified).
3 (red) – turbine impeller surfaces (rotation condition specified).
4 (gray) - turbine stator surfaces (non-rotating wall).

The numerical modelling used mathematical models of computational gas dynamics that describe the motion of a working fluid at various speeds, taking account of the effects of compressibility, turbulence and heat transfer, and the k-ε model of turbulent flow for modelling of gas flow at large and small Reynolds numbers in a stationary arrangement for any Mach numbers.

The calculation results were used to generate graphs of pressure distribution (figures 7, 12, 17 and 22) and velocity fields (figures 8, 13, 18 and 23), and sectoral leakage was calculated for each variant of the seal. Also, for each seal option, main geometrical parameters (figures 4, 9, 14 and 19), 3D-model (figures 5, 10, 15 and 20) and calculation grid (figures 6, 11, 16 and 21) are given.

3. Calculation results.

Variant No.1. Initial geometry. The principal criteria for the selection of this design are the use of a single-sided crest-type seal to reduce the contact area of the seal's working surfaces, and the application of spiral design of the crest optimizing the downstream entrainment of microscopic particles.

The leakage value is $\dot{m}$ = 1.502062 kg/sec.

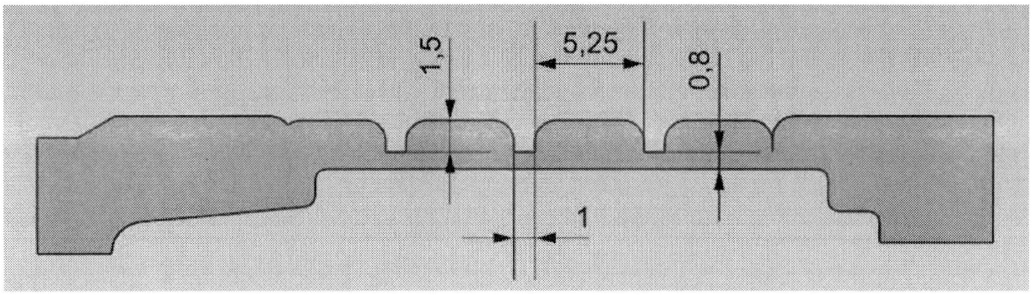

Figure 4. Main geometrical parameters of the seal.

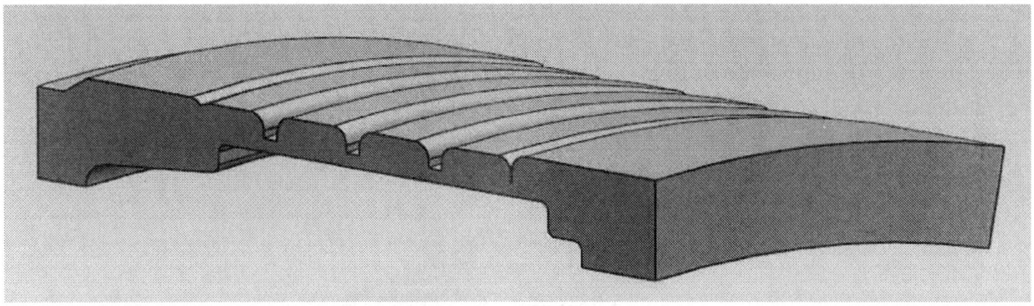

Figure 5. 3D-model of the seal.

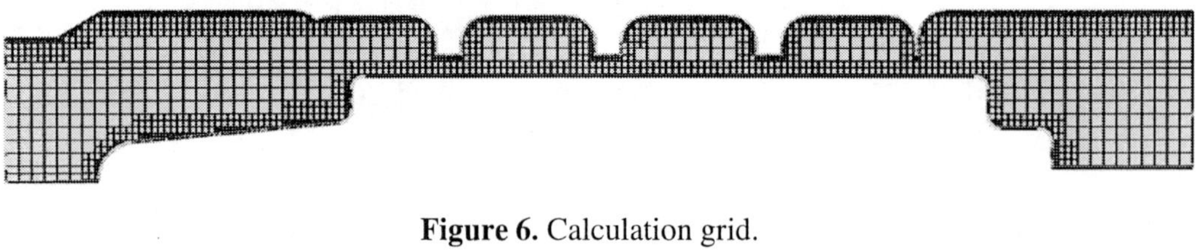

Figure 6. Calculation grid.

Figure 7. Pressure distribution in the seal.

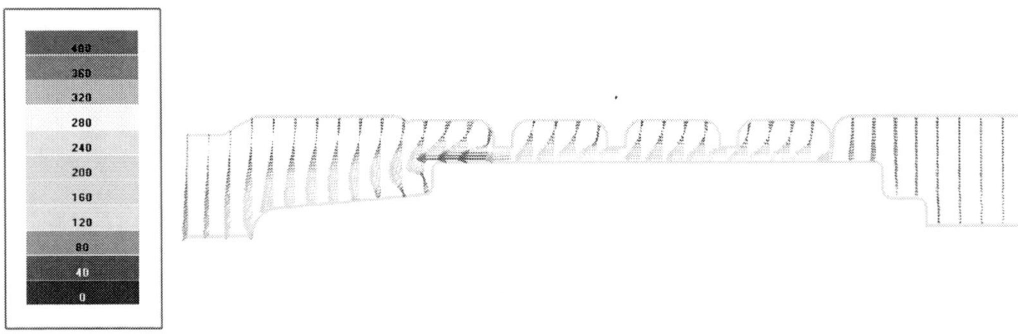

Figure 8. Velocity field in the seal.

Variant No.2. A double-sided crest-type radial labyrinth seal, retaining main geometrical parameters of the original seal.

The leakage value is $\dot{m}$ =1.622974 kg/sec.

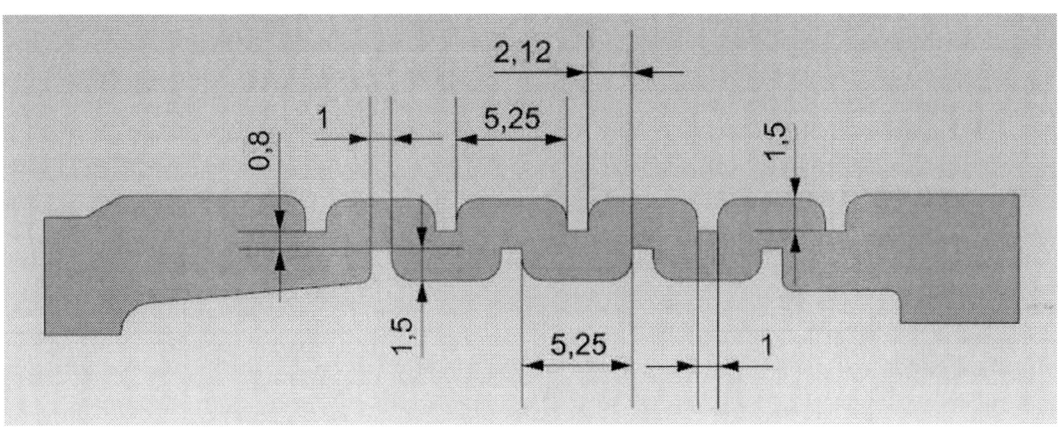

Figure 9. Main geometrical parameters of the seal.

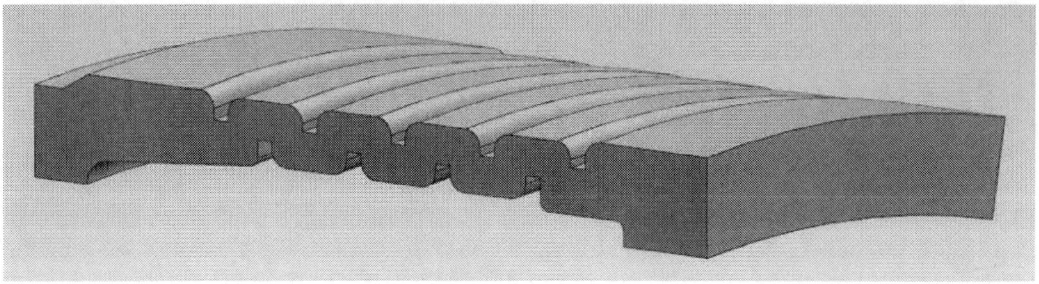

Figure 10. 3D-model of the seal.

Figure 11. Calculation grid.

Figure 12. Pressure distribution in the seal.

Figure 13. Velocity field in the seal.

Variant No.3. A double-sided crest-type labyrinth seal, retaining the basic geometric parameters of the initial seal, but reducing the gap between the crests of the turbine impeller and the crests of the stator wall by increasing the height of the crests. Development testing of engines at Energomash has shown that it is feasible to manufacture seals with gaps as small as 0.15 mm.

The leakage value is $\dot{m}$ =1.349547 kg/sec.

IOP Conf. Series: Materials Science and Engineering **491** (2019) 012018 doi:10.1088/1757-899X/491/1/012018

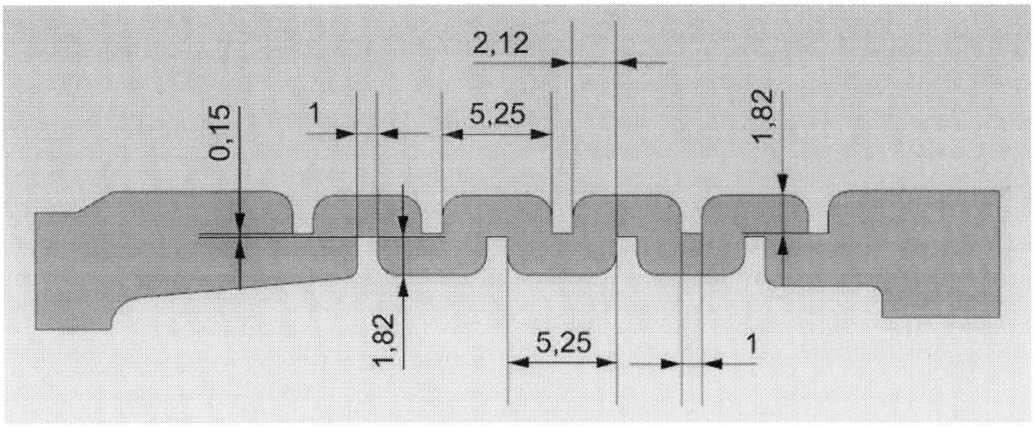

Figure 14. Main geometrical parameters of the seal.

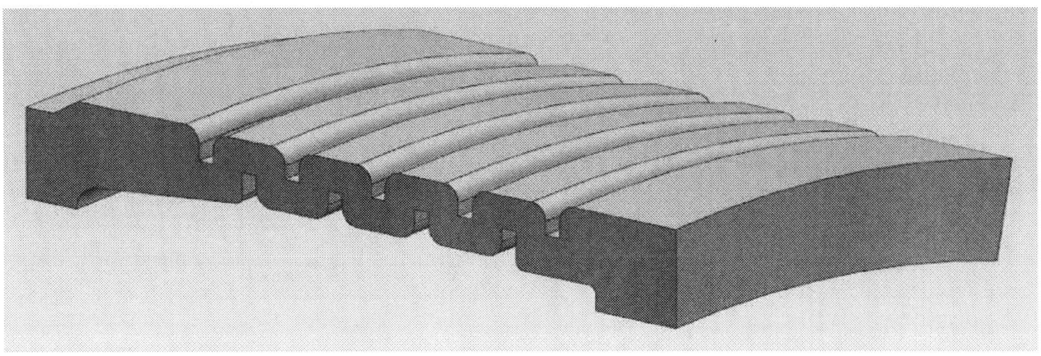

Figure 15. 3D-model of the seal.

Figure 16. Calculation grid.

Figure 17. Pressure distribution in the seal.

Figure 18. Velocity field in the seal.

Variant No.4. A double-sided crest-type labyrinth seal, retaining the basic geometric parameters of the initial seal, but reducing the gap between the crests of the turbine impeller and the crests of the stator wall by bringing the counterparts of the seal closer together.

The leakage value is ṁ=1.235992 kg/sec.

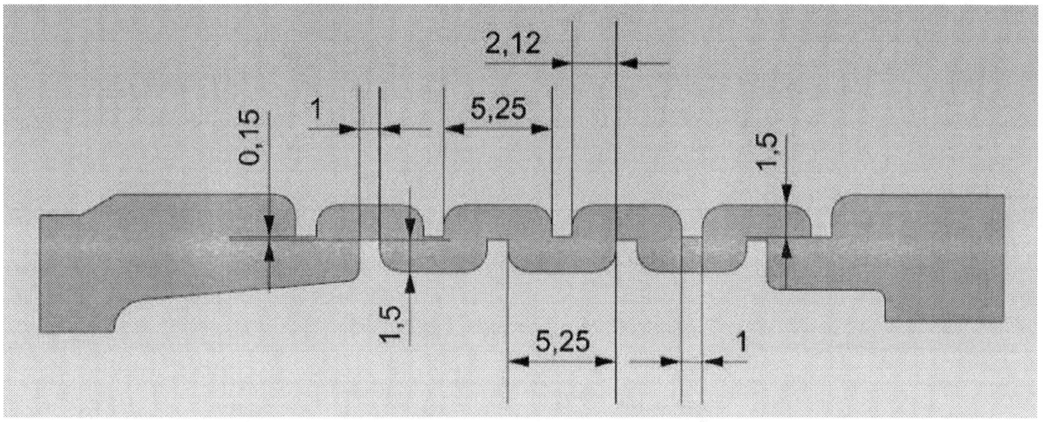

Figure 19. Main geometrical parameters of the seal.

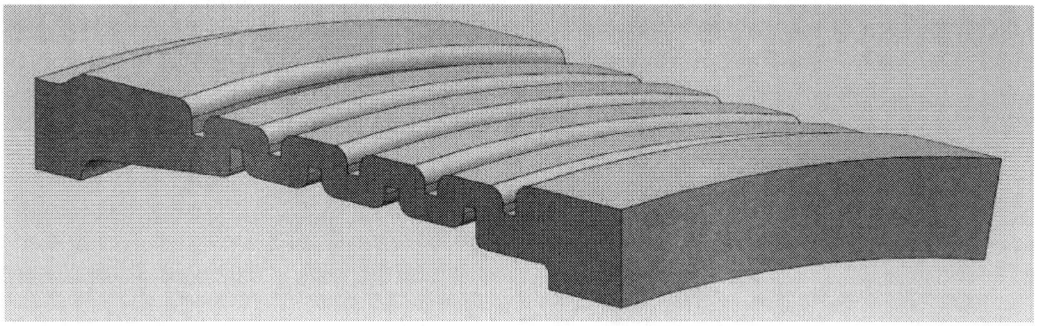

Figure 20. 3D-model of the seal.

Figure 21. Calculation grid.

Figure 22. Pressure distribution in the seal.

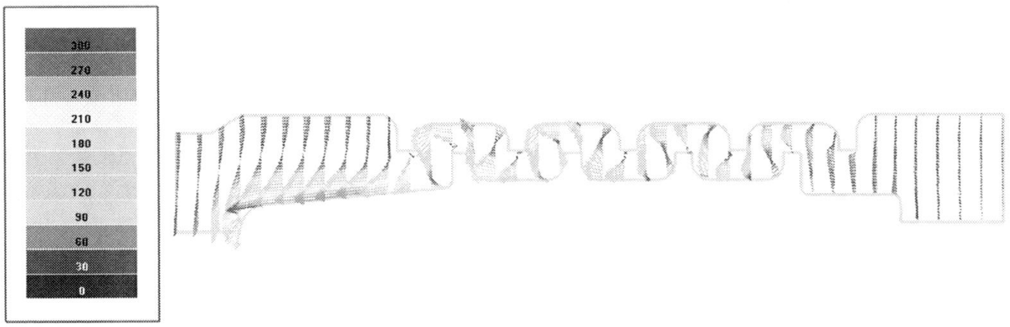

Figure 23. Velocity field in the seal.

4. Conclusions

The present article reports the results of numerical modelling of gas-dynamic processes in the flow section of a shroud seal in the TPU turbine of a liquid-propellant rocket engine.

The values of sectoral leakage obtained for the seal variants in question are given in Table 1.

Table 1. The values of sectoral leakage obtained for the seal variants in question.

Seal variant No.	Leakage, $\dot{m}$ (kg/sec).
1	1.502062
2	1.622974
3	1.349547
4	1.235992

As seen from the calculations, the best result in terms of reducing the flow rate of working gas in the shroud seal of the turbine, in comparison with the initial geometry, is offered by variant No. 4, which uses a double-sided crest-type radial seal with reduction of the radial clearance in the shroud seal, providing a reduction of the leakage rate.

Calculation of turbine efficiency variation due to the geometry of individual seal variants shows that a slight change in the temperature of the working gas is also obtained. By increasing the efficiency of the turbine, without changing the capacity, seal No.4 provides a decrease in the temperature of the working gas at turbine inlet by approximately 1.5-2% (8-12 °C). Lowering the temperature of generator gas entails greater reliability and efficiency of the TPU turbines of the LPRE.

Future research should focus on detailed studying of this type of seal, considering other possible changes in its geometry and conducting more detailed analysis of design parameters in order to determine the effect of geometric dimensions on the leakage value. Strength calculations should also be carried out in order to establish how leakage can be minimized. It will also be important to carry to numerical modelling of seal variants in a full-scale turbine model in order to study the effect of non-stationary and non-homogeneous properties of flow in the turbine on the leakage rate.

References

[1] Kirillov I.I. *The Theory of Turbomachines.* Leningrad: Mechanical Engineering, 1972.

[2] Lyovochkin P.S., Chvanov V.K., Romasenko E.N., Sidorenko A.S., Kanalin Yu.I., *Problems in the Development of Turbine Pump Assemblies and Booster Pumps for Modern Liquid-Propelled Rocket Engines.* Moscow: NPO Energomash, 2017.

[3] Belyaev E.N., Chvanov V.K., Chervakov V.V. *Mathematical Modelling of the Working Process of a Liquid-Propelled Rocket Engine.* Moscow: Moscow Aviation Institute Publications, 1999.

[4] Kanalin Yu.I., Skibin S.A., Chernysheva I.A. "Operational features of TPU turbines for the needs of an LPRE", *NTO NITs* No. 769, 131, 2014 (*Proceedings of NPO Energomash*), No. 32.

[5] Skibin S.A., Sternin L.E., Chernysheva I.A., Kanalin Yu.I., Poletaev N.P., Kazennov I.S., "Optimization of the construction of labyrinth turbine seals based on a spatial flow model". *NTO* No. 769, 92, 2013.

[6] Kasaev Kh.V., Trofimov R.S, "Reliability of Aircraft engines". *Mechanical Engineering*, 1982.

[7] Ovsyannikov B.V., Borovsky B.I., "Theory and calculation of fuel units for liquid rocket engines". *Mechanical Engineering*, 1986.

[8] Timushev S.F., Klimenko D.V., Firsov V.P., Antyukhov I.V., Numerical modelling of pressure pulsations and non-stationary loads in the radial turbine of a turbo-expander". *Proceedings of the Moscow Aviation Institute*, 82, 2015.

IOP Conf. Series: Materials Science and Engineering **491** (2019) 012019 doi:10.1088/1757-899X/491/1/012019

Experimental Study on Closed Loop Oscillating Heat Pipe for Different Filling Ratios

Shrekumar Soni[1] and Dr. Ajit Kumar Parwani[1,*]

[1]Department of Mechanical Engineering, Institute of Infrastructure Research and Management (IITRAM), Maninagar (East), Ahmedabad, Gujarat, India - 380026

[*]ajitkumar.parwani@iitram.ac.in

Abstract. An experimental study was performed on Oscillating Heat Pipe (OHP) with closed loop structure. The capillary tube made up of copper of this OHP has inner diameter of 2.15 mm and thickness of wall is 1 mm. Different set of experiments were performed with filling ratios ranging from 30% to 100%, heating power ranging from 115 to 430 W at evaporator section and free convection condition at condenser section. The water has been considered as a working fluid. The thermal resistance was calculated for measuring the thermal performance of OHP. The changes in thermal resistance with heat input for various filling ratios has been carried out. The thermal resistance achieved lowest for 70 % filling ratio.

1. Introduction

Nowadays, the world is experiencing many challenges in solving heat transfer issues in various engineering systems. Transferring heat efficiently is one of the most important tasks in various engineering applications. Conventional heat transfer equipment cannot transfer heat effectively and also have moving parts. As compared to the best-known conductors like copper the OHP can transmit up to one thousand times more thermal energy with small temperature drop [1]. OHP or Pulsating Heat Pipe (PHP) does not need a wicking structure and has no moving parts hence OHP has silent operation and also requires minimum maintenance. Hence OHP can directly impact on cost, reliability and performance of various applications.

Figure 1 (a) and Figure 1 (b) shows schematic and experimental set up of closed loop OHP. A closed loop OHP is typically a meandering tube with small diameter and comprised of a serpentine channel [2]. A closed loop structure is formed when the two ends of tube are connected to each other. The tube is fully evacuated initially and then filled with the suitable working fluid partially. The working fluid taken in this study is water. The effects of surface tension force cause the formation of liquid slugs and vapour plugs. The liquid slugs of working fluid starts to evaporate as the heat is transferred from the evaporator section which causes the rise of vapour pressure inside the tube. This increase of vapour pressure in the evaporator section causes the vapour bubbles to expand that pushes the liquid towards the condenser. On the other side in the condenser section there is condensation of vapour bubbles which reduces the vapour pressure. This process is continuous between the condenser section and the evaporator section and an oscillating motion is set up within the OHP. The amplitude and the frequency of the oscillating motion within the tube will be dependent on the mass fraction of the liquid [3].

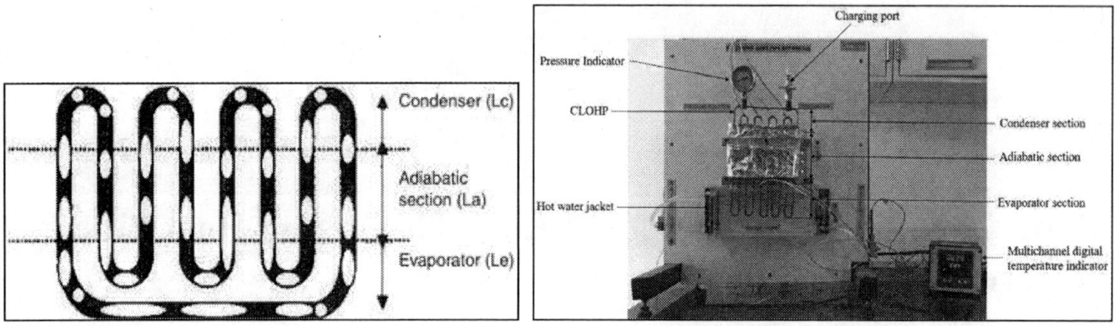

(a) (b)

Figure 1. (a) Closed loop Oscillating Heat Pipe schematic [2] **(b)** Experimental setup

Various articles on experiment studies of OHP have been reviewed. Tong et al. [4] used charge coupled device (CCD) cameras for flow visualization in experiments of closed loop OHP having pyrex glass tubes with methanol as a working fluid. They found that the operation of closed loop OHP required a minimum critical power input which initiates fluid flow in tube. With the filling Ratio (FR) of 0.6 or 60 % the circulation of the working fluid is possible which can transport heat from the evaporator to condenser section. During the start-up period large amplitude oscillations from the evaporator to the condenser occur and thereafter working fluid circulation is continuous. They also found the increase of circulation velocity with power input. Xue et al. [5] studied the full visualization and startup performance of an ammonia OHP. They found that for the operation of closed loop OHP with bottom heating mode and ammonia as a working fluid a very small startup power is required compared to water due to particular characteristics of ammonia. The OHP can start even when the evaporator temperature is only four centigrade higher than the temperature of condenser. Muraleedharan [6] performed experiments for flat heat pipes of different metals. He observed maximum capillary limit equal to 177 W for Aluminum flat heat pipe with axial grooves and composite wick with acetone as a working fluid. For stainless steel flat heat pipe, maximum capillary limit for acetone is 125 W and with water is 650 W. Experimental studies were performed on closed loop OHP by Zhang et al. [7] for heating powers ranging from 5 to 60 W and FR from 60% to 90% with pure natural convection condition and working fluid as water. The best overall thermal performance was determined to be 70% FR. Tanshen et al. [8] conducted experimentation on OHP charged with aqueous Al_2O_3 nanoparticles, MWCNTs and their hybrids. They studied phenomenon of heat transfer in terms of frequency of pressure fluctuation in OHP. They concluded that the best thermal performance for the OHP is obtained when the FR ranges from 50% to 60% and power input in evaporative section influences frequency of pressure fluctuation. The lowest thermal resistance is accomplished by the hybrid of aqueous Al_2O_3 and purified MWCNT. Mozumder et al. [9] conducted experiments on heat pipe using working fluids of three different types i.e. water, methanol and acetone and varied FR as 35%, 55%, 85% and 100%. They concluded that the heat transfer coefficient increases with FR of working fluid greater than 85% of volume of evaporator. They had shown with FR greater than 85% decreases the thermal resistance and it also reduces temperature difference across the evaporator and condenser. Baitule and Pachghare [10] carried out experimental analysis of closed loop OHP with variable filling ratio ranges from 0% to 100% in steps of 20%. The OHP was tested on four working fluids namely methanol, ethanol, acetone and water. They have found that the thermal resistance of closed loop OHP decreases with the increase of heat input. For the input power condition up to 48 W the thermal resistances are lowest for acetone and highest for water and above the 48 W the thermal resistance of acetone increases slightly. They have also found that 60% filling of OHP give the optimum result i.e. lower thermal resistance for each working fluid.

Some numerical studies have also been reported in the literature. Gupta and Parwani [11] performed CFD modeling of OHP in bottom heated mode. They have presented to achieve the lowest thermal resistance of 0.295 K/W during the working of OHP. Shafii et al. [12] presented analytical models for looped OHP with multiple liquid slugs and vapor plugs. They concluded that by increasing the diameter of the OHP, the total average heat transfer increases. The total average heat transfer significantly decreases with decreasing the temperature of the heating wall sections. Singh [13] performed CFD modelling and numerical analysis of performance of closed-loop OHP with alcohol as working fluid. He analysed that frequent oscillation with alcohol comes earlier than the water because alcohol having low specific heat will give cooling effect much earlier than water. He also observed that the oscillation will starts earlier with high value of heat flux input at evaporator section.

In this study, the water was used as a working fluid for OHP. The values for filling ratio taken were 30 %, 50 %, 70 %, 90 %, 100 %. The fill ratio is defined as the fraction by volume of the heat pipe, which is initially filled with the liquid [14]. The values of heat input have been taken by different hot water temperatures (i.e. 50 °C, 60 °C, 70 °C, 80 °C). The whole experimental setup is fabricated at Thermal lab of Institute of Infrastructure Technology Research and Management (IITRAM), Ahmedabad.

2. Experimental setup details

The experimental apparatus of OHP fabricated for the current study is shown in Figure 1(b), with an overall width and height of 180 mm and 280 mm respectively. The OHP has a long capillary copper tube with an outside diameter of 4.15 mm and inside diameter of 2.15 mm. The bending radius of serpentine channel is 9 mm and total loop length is 2913 mm. The OHP set up is divided into three sections as shown in Figure 2 namely evaporator, adiabatic and condensation sections. The length of adiabatic section, evaporator section and condenser section can be changed. The arrangements for water in and out are given. The setup was levelled by level screws. The thermal load in the heating section was applied by acrylic flexiglass made hot water jacket of 300 mm × 65 mm × 120 mm. The thickness of the glass is 8 mm. The hot water bath is covered by a lid also. So, there is no such heat losses from the bath. The water is heated by heater rated 500 kW with different temperatures.

The OHP is cooled down by natural convection. The condenser section is open to atmosphere. Below the condenser section, the tube is well heat-insulated by the thick glass wool covered by aluminium foil, forming the adiabatic sections. The OHP has contained a closed-loop pipe, with a charging port and a pressure gauge.

The K type thermocouple wires are used with multichannel digital temperature indicator which has inaccuracy of ±1 % of Full Scale Deflection (FSD). The thermocouples i.e. T1 and T2 are connected to the water in and water out respectively. Four thermocouples i.e. T3, T4, T5, T6 are attached to the condenser section of OHP as shown in figure 2. The mass flow rate can be regulated by valve arrangement.

Some amount of heat would have been lost from adiabatic section since there is not possible to achieve perfect insulation. Heat would be also lost from water bath, inlet and outlet pipes and other small leakages. Cumulative heat losses were estimated to be approximately 2-3 % of power input. Percent uncertainty in heat input was estimated to be approximately 1.0 %. The filling ratio uncertainty of an OHP was estimated to be ± 0.10 %. The dimensional accuracy was assessed to be ±1.0 mm. The accuracy of various measuring instruments is given in Table 1.

Table 1. Accuracy of various measuring instruments

Sr. No.	Measuring Instrument	Accuracy
1	Resistance Heater	± 0.5 °C
2	Thermocouple	± 1.0 %
3	The Steel Ruler	± 1.0 mm

3. Experimental procedure

The experiments are performed with working fluid used as water. The evaporator section and adiabatic section length are same 100 mm while condenser section length is 80 mm. The volumetric flow rate is kept constant during the experiments. The water is heated by the heater. The hot water is then pumped in the hot water bath. The heat transfer takes place between the OHP and the hot water. The hot water in and hot water out temperatures are measured. Also, temperatures at condenser and evaporator section are measured. The transient tests on the heat pipe were conducted, in which heater is switched on and at regular intervals the increase of temperature readings are observed till the steady state is obtained. After reaching steady state condition, the temperatures reading at the six locations are measured. This procedure is repeated for other heat inputs and FR values. The level of filling ratios taken into consideration are 30%, 50%, 70%, 90 % and 100 %.

For the above mentioned FR the temperature readings at the six different locations on the surface of heat pipe has been noted after the steady state condition. Following equations are used for this study:

$$Q_i = \dot{m} C_p \, \Delta T \qquad \text{(W)} \tag{1}$$

$$R = (\overline{T}_e - \overline{T}_c) / Q_i \quad (°C/W) \tag{2}$$

Where, Q_i (W) is heat input by the hot water, C_P is water specific heat = 4.184 kJ/kgK, $\dot{m}$ (kg/s) is water mass flow rate = $\rho \times \dot{V}$, ρ and $\dot{V}$ are density (kg/m^3) and volumetric flow rate (m^3/s), ΔT = Temperature drop between water in and water out, °C, R (°C /W) is OHP thermal resistance, , $\overline{T}_c$ is condenser average temperature, °C, $\overline{T}_e$ = evaporator average temperature, °C, $\overline{T}_e$ and $\overline{T}_c$ can be calculated by,

$$\overline{T}_e = (T1+T2)/2, °C \tag{3}$$

$$\overline{T}_c = (T3+T4+T5+T6) / 4, °C \tag{4}$$

4. Results

The results are shown by plotting graphs of thermal resistance (R) variations with heat input (Q_i) for different FR.

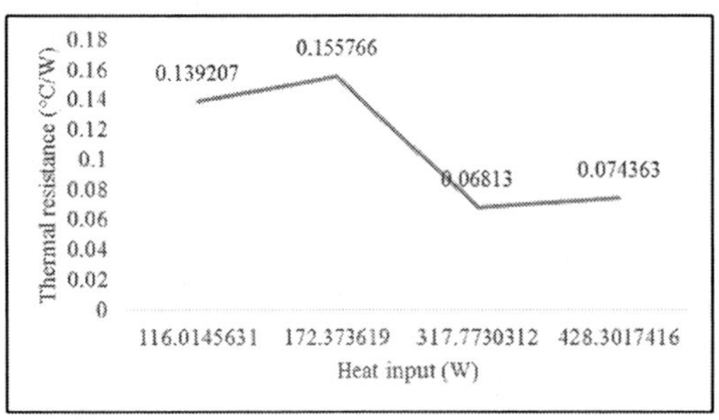

(a)

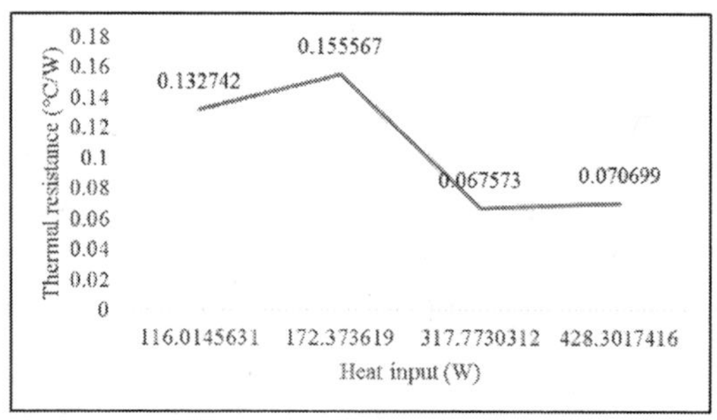

(b)

Figure 2. Thermal resistance variation with different heat inputs for (a) 30 % filling ratio (b) 50 % filling ratio

Figure 2 (a) shows that the values of thermal resistance ranging from 0.06813 °C/W to 0.15578°C/W with 30 % filling ratio, the minimum thermal resisrance achieved is 0.06813 °C/W at 317.773 W of heat input. The thermal resistance is maximum for 172.374 W for 30 % filling. Figure 2 (b) shows that the values of thermal resistance ranging from 0.06757 °C/W to 0.1556 °C/W with 50 % filling ratio. The minimum thermal resisrance achieved is 0.06757 °C/W at 317.4363 W heat input. The thermal resistance is decreased for each heat input with 50 % filling ratio than 30 % filling ratio.

IOP Conf. Series: Materials Science and Engineering **491** (2019) 012019 doi:10.1088/1757-899X/491/1/012019

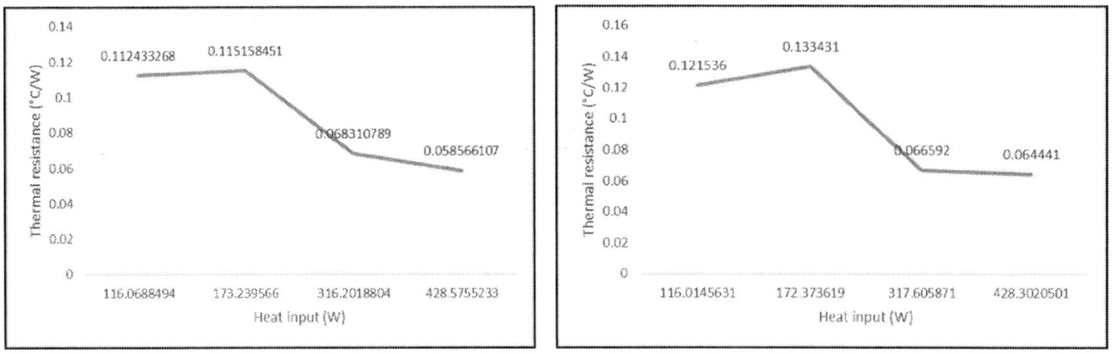

(a) (b)

Figure 3. Thermal resistance variation with different heat inputs for (a) 70 % filling ratio (b) 90 % filling ratio

Figure 3(a) shows that the values of thermal resistance ranging from 0.05857 °C/W to 0.1152 °C/W with 70 % filling ratio, the minimum thermal resisrance achieved is 0.005857 °C/W at 428.0264 W heat input. So, the thermal resistance is further decreased with increasing FR. Figure 3 (b) shows that the thermal resistance increased with FR varied from 70 % to 90%. The minimum thermal resistance achieved is 0.0644 for 90 % filling with 428.3021 heat input. For the case of 90% FR, the working fluid inside the OHP do not get sufficient space to generate bubbles, hence the performance of the OHP deteriorates. The thermal resistance is increasing from 116 W to 172.374 W for 30 % to 90 % filling ratio. Due to smaller increment of the heat input, it may not be able to increase the internal pressure of an OHP to circulate the working fluid flow efficiently. Because of this, there is the dominance of flow hindrance forces like surface tension forces, gravity forces etc. So, there is lesser heat transfer to the condenser hence lower values of condenser section temperature. Therefore, there is increase of thermal resistance for heat input from 116 W to 172.373 W.

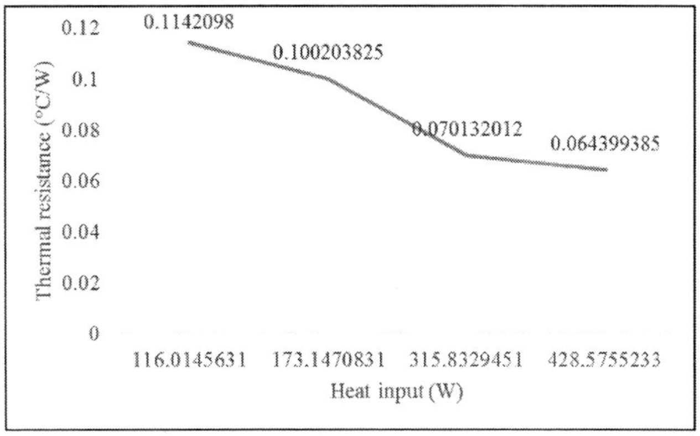

Figure 4. Thermal resistance variation with different heat inputs for 100 % filling ratio

At FR 100% the OHP operate as a single phase thermosyphon. Though oscillations do not occur under this condition, but substantial amount of heat transfer will still occurs due to circulation of liquid into the pipe by thermally induced buoyancy [12]. With increase in a heat input, the thermal resistance is decreased. The thermal resistance is minimum for 428.5756 W as shown in figure 4.

5. Conclusions

Experimental study was successfully conducted on a closed loop OHP. The water was used as a working fluid with FR 30 %, 50 %, 70 %, 90 %, 100 %. The thermal performance was measured in terms of thermal resistance. Following are the conclusions drawn from this study:

- The values of thermal resistance decrease with increase in filling ratio from 30 % to 70 % for each heat input. The thermal performance is maximum for 70 % filling. After 70 %, the performance decreases with increase in filling.
- The experimentally determined optimal filling ratio is 70% for the water as a working fluid. In short, 70 % filling ratio gives better result in terms of low thermal resistance and decrease in temperature difference across the condenser and evaporator.
- The thermal resistance is increasing from 116 W to 173 W and then it starts decreasing for higher values of heat input (i.e. for 317 W and 428 W).
- The performance of OHP with 100 % filling (or thermosyphon) has shown better result than 90 % filling in vertical position. The thermal resistance is continuously decreasing with increase in heat input.

References

[1] "Econotherm." [Online]. Available: http://affinitas.com.sg/images/The Heat Pipe Advantage - 0 7.

[2] M. Kutz, "Mechanical Engineers' Handbook: Energy and Power", Third Edition, vol. 4, John Wiley & Sons, Inc., 2005.

[3] D. A. Reay, P. A. Kew, and R. J. McGlen, "Heat Pipes: Theory, Design and Applications", Fifth Edition, 2006.

[4] B. Tong, T. Wong, and K. Ooi, "Closed-loop Pulsating Heat Pipe," *Appl. Therm. Eng.,* vol. 21, no. 18, pp. 1845–1862, 2001.

[5] Z. Xue, W. Qu, and M. Xie, "Full Visualization and Startup Performance of an Ammonia Pulsating Heat Pipe," *Propuls. Power Res.,* vol. 2, no. 4, pp. 263–268, 2013.

[6] C. Muraleedharan, "Heat Transfer and Fluid Flow Studies on Flat Heat Pipe," University of Calicut, 2001.

[7] X. M. Zhang, J. L. Xu, and Z. Q. Zhou, "Experimental Study of a Pulsating Heat Pipe Using fc-72, Ethanol, and Water as Working Fluids," *Exp. Heat Transf.,* vol. 17, no. 1, pp. 47–67, 2004.

[8] M. R. Tanshen et al., "Pressure Distribution Inside Oscillating Heat Pipe Charged with Aqueous Al_2O_3 Nanoparticles, MWCNTs and Their Hybrid," *J. Cent. South Univ.,* vol. 21, no. 6, pp. 2341–2348, 2014.

[9] A. K. Mozumder, A. F. Akon, M. S. H. Chowdhury, and S. C. Banik, "Performance of Heat Pipe for Different Working Fluids and Fill Ratios," *J. Mech. Eng.,* vol. M, no. 2, pp. 96–102, 2010.

[10] A. Baitule D and P R Pachghare, "Experimental Analysis of Closed Loop Pulsating Heat Pipe with Variable Filling," *Int. J. Mech. Eng. Robot. Res.,* Vol. 2, no. 3, 2013.

[11] A. Gupta and A. Parwani, "CFD Modeling for Thermal Performance of Closed Loop Pulsating Heat Pipe in Bottom Heated Mode," *Advances in Civil, Structural and Mechanical Engineering,* 2017, pp. 46–50.

[12] M. B. Shafii, A. Faghri, and Y. Zhang, "Thermal Modeling of Unlooped and Looped Pulsating Heat Pipes," *J. Heat Transfer,* vol. 123, no. 6, p. 1159, 2001.

[13] A. K. Singh, "Numerical Analysis of Performance of Closed- Loop Pulsating Heat Pipe", M.Tech. thesis, Department of Mechanical Engineering, National Institute of Technology Rourkela, June 2013.

[14] C. Fasula, "Oscillating Heat Pipes (OHP)," A Special Problems Paper, University of Rhode Island, May, 2009.

IOP Conf. Series: Materials Science and Engineering **491** (2019) 012020 doi:10.1088/1757-899X/491/1/012020

Determination Methodology for Stability Domain of Hybrid Electric Power System with Multiple Excitation Sources

Zhongyan Li[1,2] and Donghai Hu[3,*]

[1]School Of Mechanical Engineering, Shanghai Jiao Tong University, Shanghai, 200240, China
[2]Jinlong United Automotive Industry (Suzhou) Co., Ltd. Suzhou, 215123, China
[3]School of Automotive and Traffic Engineering, Jiangsu University, Zhenjiang 212013, China

[*]Corresponding E-mail: 1000004735@ujs.edu.cn

Abstract. Irregular internal excitation (engine excitation and motor excitation) and external load excitation can cause torsional vibration in hybrid electric propulsion system which even leads to the break of shaft. The objective of this paper is to determinate the stability domain of hybrid electric power system under different drive mode with multiple excitation sources. To achieve the goal, the simplified three-mass torsional model of hybrid electric power system is established. Then we apply the multi-scale method to solve the nonlinear torsional model. Finally we set different parameters of engine speed to simulate torsional vibration characteristics of hybrid electric power system. The simulation results exhibit that the torsional vibration characteristics of hybrid electric power system under two drive mode are different because the directional relationship of the moment of engine and permanent magnet synchronous motor (PMSM) is distinctive. It's hard to eliminate the torsional vibration of hybrid electric propulsion system due to phase difference between the engine excitation and the load excitation.

1. Introduction

In recent years, deteriorating energy and environmental problems promote the development of electric power systems and hybrid electric power systems [1,2]. Nevertheless, the application of electric power systems have encountered shortcomings, such as long charging time, limited operating range and lifespan. On the other hand, hybrid electric power systems have been applied increasingly for their flexible working mode, wide operating range and low emission [3,4,5]. Usually, the hybrid electric power system is composed of two power sources: one is the combination of engine and generator, and the other is the combination of motor and battery [6,7]. However, due to the complexity of the configuration of hybrid electric power system and strong nonlinearity of engine and permanent magnet synchronous motor(PMSM), irregular engine excitation and electromagnetic excitation can cause torsional vibration which even leads to the break of shaft [8,9]. So it's essential to analyze the torsional characteristics of hybrid electric power system to determine the stability domain with the engine and PMSM participated in.

The hybrid electric power system is a multi-degree of freedom(DOF) nonlinear system with multiple excitation sources. It is considered that the mass distribution of the hybrid electric power system is lumped [10]. Tang et al. [11] established a 16-DOF torsional vibration model of hybrid electric vehicle power system. The model is applied to analyze the torsional vibration characteristics and give the theoretical prediction of the natural frequency and corresponding vibration modes. The dynamic model was also used to present the torsional modes of hybrid electric power system in hybrid driving mode. The results indicate that the low frequency vibration is concentrated on the vehicle and wheels. Engine noise is the main noise in the hybrid driving mode. Tang et al. [12] developed a simplified three-mass torsional model of the hybrid electric power system to determine the dominant frequency. The results show that the simplified model can be used to describe the low frequency vibration characteristics of the hybrid electric power system accurately. However, the simplified three-mass model concentrates on the unbalanced torque of engine and does not consider the electromechanical coupling relationship between the engine and motor. Chen et al. [13] established the two-mass torsional vibration model of hybrid electric power system considering electromechanical coupling. Yue et al. [14] specifically analyzed the amplitude-frequency characteristic of torsional vibration in two different modes of the hybrid electric power system (boosting mode and generating mode). The corresponding results of the engine were compared to reveal that the coexistence of electromagnetic torque of motor and the engine torque changes the torsional vibration characteristics of the original engine shafting and intensifies the amplitude of the torsional vibration. Zuo et al. [15] built a multi-body dynamic model based on a simplified structure of a parallel hybrid system. The incentive model of engine was established in MATLAB, the dynamic simulation results show that the change of engine cylinder pressure is the main reason that affects the torsional vibration of the shaft system. However, many researches focus on the influence of engine excitation, as well as the analysis of natural frequencies and corresponding modalities of hybrid electric power system.

The purpose of this paper is to determine the stability domain of hybrid electric power system under two typical drive mode. A simplified three-mass nonlinear torsional model of hybrid electric power system under various drive mode was developed using the lumped parameter method in Section 2. The first-order approximate solution of the model was solved by multi-scale method in Section 3. The nonlinear dynamics theory was applied to reveal the torsional vibration characteristics of hybrid electric power system. Section 4 introduces the results of the influences of variable engine speed on the nonlinear torsional vibration characteristics of hybrid electric power system. Finally, conclusions are drawn in Section 5.

2. Nonlinear torsional model of hybrid electric power system

2.1. Electromagnetic torque model of PMSM

Considering the torsional vibration angle of PMSM, the rotor synthesis fundamental wave magnetomotive force is given below [16,17]:

$$F(\alpha,t)=F_{1m}\cos(\omega t-\alpha)+F_{2m}(\omega t-\alpha+(p\phi+\psi+\pi/2)) \tag{1}$$

Electromagnetic torque of PMSM is given below when the torsional vibration angel is ϕ.

$$T_M=F_m(\sin\psi\sin(p\phi)-\cos\psi\cos(p\phi)) \tag{2}$$

Where, $F_m=p\pi Rl\Lambda_0F_{1m}F_{2m}$; R is inner radius of stator; l represents effective length of rotor; Λ_0 is air gap permeability of PMSM.

The simplified equation of electromagnetic torque of PMSM is shown using Taylor approximation.

$$T_M=k_0+k_1\phi-k_2\phi^2-k_3\phi^3 \tag{3}$$

Where, $k_0=-F_m\cos\psi, k_1=pF_m\sin\psi, k_2=p^2F_m/2\cos\psi, k_3=p^3F_m/6\sin\psi$.

2.2. Mechanical torque model of engine

The unbalanced force sources of a four-cylinder engine are mainly composed of reciprocating moment of inertia T_j and combustion gas moment T_g. The combustion gas moment T_g can be expressed by the following infinite series using the Fourier series theory [18,19].

$$T_g = T_0 + \sum_{r=1}^{\infty} c_k \sin(k_\alpha + \theta_k) \tag{4}$$

Since the fourth-order and higher-order unbalanced moments are only 2.5% of that of second-order, it is mainly necessary to consider the second-order imbalance torque.

$$\begin{cases} T_j = 2\lambda m_j r^2 \omega^2 \sin 2\alpha \\ T_g = -r\pi D^2 (a_2 \sin 2\alpha + b_2 \cos 2\alpha)/4 \end{cases} \tag{5}$$

Where, m_j represents reciprocating equivalent mass of piston connecting rod group; r is crank radius; D is piston diameter; λ represents connecting rod ratio.

In actual analysis of engine torsional vibration, only the interference torque T_g generated by changeable gas pressure is generally considered. Furthermore, the cosine component in combustion gas moment T_g is small compared with the sine component. Considering torsional vibration angel, simplified mechanical torque model of engine is given below.

$$T_E = T_0 + \frac{\pi r D^2}{4} a_2 \sin 2\alpha \tag{6}$$

2.3. Nonlinear torsional model under boosting mode

A schematic representation of hybrid electric power system is shown in figure 1. The hybrid electric power system gets into the compound driving mode when the clutch is locked, and the PMSM drives, together with the engine [20,21].

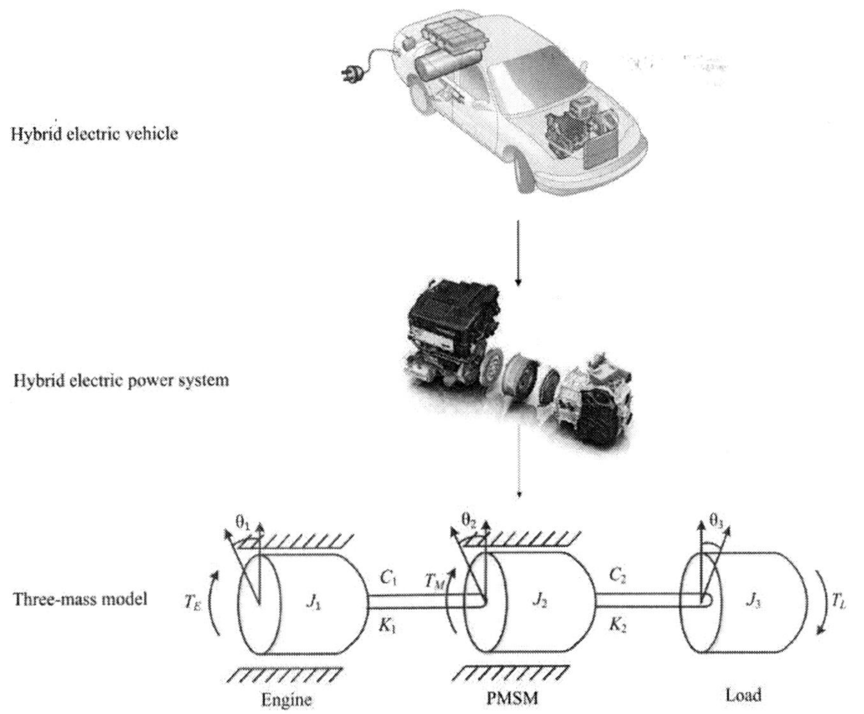

Figure 1. Schema of hybrid electric power system

The mechanical rotation equations is given below:

$$\begin{cases} J_1\ddot{\theta}_1 + C_1(\dot{\theta}_1 - \dot{\theta}_2) + K_1(\theta_1 - \theta_2) = T_E \\ J_2\ddot{\theta}_2 + C_1(\dot{\theta}_2 - \dot{\theta}_1) + C_2(\dot{\theta}_2 - \dot{\theta}_3) + K_1(\theta_2 - \theta_1) + K_2(\theta_2 - \theta_3) = T_M \\ J_3\ddot{\theta}_3 + C_2(\dot{\theta}_3 - \dot{\theta}_2) + K_2(\theta_3 - \theta_2) = -T_L \end{cases} \tag{7}$$

Where J_1, J_2 and J_3 represent the moment of inertia of engine, PMSM and load respectively; θ_1, θ_2, θ_3 represent rotation angles at the end of the two shafts; C_1 and C_2 are damping coefficient; K_1 and K_2 are stiffness of the two shaft respectively.

Suppose that ϕ_{10}, ϕ_{20} and ϕ_{30} are the torsion angles at the ends of the shaft of the hybrid electric propulsion system under the action of constant torques T_{E0}, T_{M0} and T_{L0} respectively. ϕ_1, ϕ_2 and ϕ_3 are the torsional vibration angles at the end of the shaft under different excitation, respectively. We can get equations shown below:

$$\begin{cases} \theta_1 = \phi_{10} + \phi_1, \theta_2 = \phi_{20} + \phi_2, \theta_3 = \phi_{30} + \phi_3 \\ \dot{\phi}_{10} = \dot{\phi}_{20} = \dot{\phi}_{30} = 0, \ddot{\phi}_{10} = \ddot{\phi}_{20} = \ddot{\phi}_{30} = 0 \end{cases} \tag{8}$$

The equivalent torsional vibration equations of the hybrid electric propulsion system under the engine torque disturbance ΔT_E, electromagnetic torque disturbance term ΔT_M and the load torque disturbance term ΔT_L is given as below:

$$\begin{cases} J_1\ddot{\phi}_1 + C_1(\dot{\phi}_1 - \dot{\phi}_2) + K_1(\phi_1 - \phi_2) = \Delta T_E \\ J_2\ddot{\phi}_2 + C_1(\dot{\phi}_2 - \dot{\phi}_1) + C_2(\dot{\phi}_2 - \dot{\phi}_3) + K_1(\phi_2 - \phi_1) + K_2(\phi_2 - \phi_3) = \Delta T_M \\ J_3\ddot{\phi}_3 + C_2(\dot{\phi}_3 - \dot{\phi}_2) + K_2(\phi_3 - \phi_2) = -\Delta T_L \end{cases} \tag{9}$$

We substitute the following equations into Eqs.9 $\Delta T_E = r\pi D^2 a_2 \sin 2\alpha / 4$, $\Delta T_M = k_1\phi_2 - k_2\phi_2^2 - k_3\phi_2^3$, $\Delta T_L = F\cos(\omega t)$. New variables are set $x_1 = \phi_1 - \phi_2, x_2 = \phi_2 - \phi_3$. The nonlinear torsional vibration model of hybrid electric propulsion system is given below:

$$\begin{cases} \ddot{x}_1 + A_1x_1 + \mu_1\dot{x}_1 + \mu_3\dot{x}_2 + (A_2x_1^2 + B_2x_2^2 + A_3x_1^3 + B_3x_2^3 + D_1x_1x_2 + D_2x_1^2x_2 + D_3x_1x_2^2) = f_1\sin(\Omega_1 t) \\ \ddot{x}_2 + H_1x_2 + \mu_4\dot{x}_1 + \mu_2\dot{x}_2 - (A_2x_1^2 + B_2x_2^2 + A_3x_1^3 + B_3x_2^3 + D_1x_1x_2 + D_2x_1^2x_2 + D_3x_1x_2^2) = f_2\sin(\Omega_2 t) \end{cases} \tag{10}$$

Where, $A_1 = \frac{k_1m_2}{J_2} + (\frac{K_1}{J_1} + \frac{K_1}{J_2}), A_2 = \frac{k_2m_2^2}{J_2}, A_3 = -\frac{k_3m_2^3}{J_2}, B_2 = \frac{k_2m_1^2}{J_2}, B_3 = \frac{k_3m_1^3}{J_2}$, $D_1 = -\frac{2m_1m_2}{J_2}$, $D_2 = \frac{3n_3m_1m_2^2}{J_2}, D_3 = -\frac{3n_3m_1^2m_2}{J_2}, \mu_1 = \frac{C_1}{J_1} + \frac{C_1}{J_2}, H_1 = \frac{K_2}{J_2} + \frac{K_2}{J_3} - \frac{k_1m_1}{J_2}, \mu_2 = \frac{C_2}{J_2} + \frac{C_2}{J_3}$.

3. Theoretical analysis of stability domain

First the small parameter ε and the following scale transformation [22] are introduced.

$$\begin{cases} A_i \to \varepsilon A_i, i = 2,3; B_i \to \varepsilon B_i, i = 2,3; D_i \to \varepsilon D_i, i = 1,2,3 \\ E_1 \to \varepsilon E_1, G_1 \to \varepsilon G_1, \mu_i \to \varepsilon\mu_i, f_i = \varepsilon f_i, i = 1,2 \end{cases} \tag{11}$$

The two-degree-of-freedom nonlinear model with the combination of parameter excitation and external excitation is shown as follows.

$$\begin{cases} \ddot{x}_1 + \omega_1^2x_1 + \varepsilon(A_2x_1^2 + B_2x_2^2 + A_3x_1^3 + B_3x_2^3 + D_1x_1x_2 + D_2x_1^2x_2 + D_3x_1x_2^2 + \mu_1\dot{x}_1 + \mu_3\dot{x}_2) = \varepsilon f_1\sin(\Omega_1 t) \\ \ddot{x}_2 + \omega_2^2x_2 - \varepsilon(A_2x_1^2 + B_2x_2^2 + A_3x_1^3 + B_3x_2^3 + D_1x_1x_2 + D_2x_1^2x_2 + D_3x_1x_2^2 - \mu_4\dot{x}_1 - \mu_2\dot{x}_2) = \varepsilon f_2\sin(\Omega_2 t) \end{cases} \tag{12}$$

We use the method of multiple scales to find the first-order uniform asymptotic solution of Eqs. 12.

$$\begin{cases} \omega_1(t,\varepsilon) = x_{10}(T_0,T_1) + \varepsilon x_{11}(T_0,T_1) \\ \omega_2(t,\varepsilon) = x_{20}(T_0,T_1) + \varepsilon x_{21}(T_0,T_1) \end{cases} \tag{13}$$

Where, $T_0 = t, T_1 = \varepsilon t$.

We introduce two tuning parameters σ_1 and σ_2.

$$\omega_1^2 = \Omega^2 + \varepsilon^2\sigma_1, \omega_2^2 = \Omega^2 + \varepsilon^2\sigma_2 \tag{14}$$

Eqs. 13 and 14 are substituted into Eqs. 12 to make the coefficients of the same power of ε on both sides of the equation equal.

ε^0:

$$\begin{cases} D_0^2 x_{10} + \omega_1^2 x_{10} = 0 \\ D_0^2 x_{20} + \omega_2^2 x_{20} = 0 \end{cases} \tag{15}$$

ε^1:

$$\begin{cases} D_0^2 x_{11} + \Omega^2 x_{11} = -2D_0D_1x_{10} - (A_2x_{10}^2 + B_2x_{20}^2 + A_3x_{10}^3 + B_3x_{20}^3 + D_1x_{10}x_{20} + D_2x_{10}^2x_{20} + D_3x_{10}x_{20}^2) \\ -\mu_1\dot{x}_{10} - \mu_3\dot{x}_{20} + f_1\sin(\Omega_1 t) \\ D_0^2 x_{21} + \Omega^2 x_{21} = -2D_0D_1x_{20} + (A_2x_{10}^2 + B_2x_{20}^2 + A_3x_{10}^3 + B_3x_{20}^3 + D_1x_{10}x_{20} + D_2x_{10}^2x_{20} + D_3x_{10}x_{20}^2) \\ -\mu_4\dot{x}_{10} - \mu_2\dot{x}_{20} + f_2\sin(\Omega_2 t) \end{cases} \tag{16}$$

The solution of the complex form of Equation 15 can be expressed as follows:

$$\begin{cases} x_{10} = A(T_1,T_2)e^{i\omega_1 T_0} + CC \\ x_{20} = B(T_1,T_2)e^{i\omega_2 T_0} + CC \end{cases} \tag{17}$$

Where CC represents the complex conjugate.

The elimination conditions for secular term is shown below.

$$\begin{cases} -2D_1A\Omega_1 - 3A_3A^2\bar{A} - 3B_3B^2\bar{B} - D_2\bar{A}^2B - D_3\bar{A}B^2 - \mu_1A\Omega - \mu_3B\Omega = 0 \\ -2D_1B\Omega_1 - 3A_3A^2\bar{A} - 3B_3B^2\bar{B} - D_2\bar{A}^2B - D_3\bar{A}B^2 - \mu_2B\Omega - \mu_4A\Omega = 0 \end{cases} \tag{18}$$

We let $A=(x_1+iy_1)/2, B=(x_2+iy_2)/2$ and bring to Eqs. 18 and separate the real imaginary part to get the following average equations.

$$\begin{cases} x_1' = 0.5\mu_1x_1 + 0.5\mu_3x_2 + Na_1 \\ y_1' = -0.5\mu_1y_1 - 0.5\mu_3y_2 + Na_2 \\ x_2' = 0.5\mu_4x_1 + 0.5\mu_2x_2 + Na_3 \\ y_2' = 0.5\mu_4y_1 - 0.5\mu_2y_2 + Na_4 \end{cases} \tag{19}$$

Where, $Na_i, i = 1,2,3,4$ is nonlinear functions and the symbol ' represents the derivative of the slow time scale T_1.

The characteristic equation at the zero solution has the following form:

$$\lambda^4 + R_1\lambda^3 + R_2\lambda^2 + R_3\lambda + R_4 = 0 \tag{20}$$

According to the Routh-Hurwitz principle [23], the stability conditions is given as follows.

$$\begin{cases} R_1 > 0 \\ R_1 R_2 - R_3 > 0 \\ R_3(R_1 R_2 - R_3) - R_1^2 R_4 > 0 \\ R_4 > 0 \end{cases} \tag{21}$$

4. Numerical simulation results of stability domain

4.1. Under boosting mode

Time history diagram and phase portraits of x_1 obtained by simulation when ω_1=150, ω_1=200, ω_1=250 are shown in figure 2. The initial values are set to be $x_{10}=x_{20}=\dot{x}_{10}=\dot{x}_{20}=0$, and the load excitation frequency is set to be $\omega_2=20$.

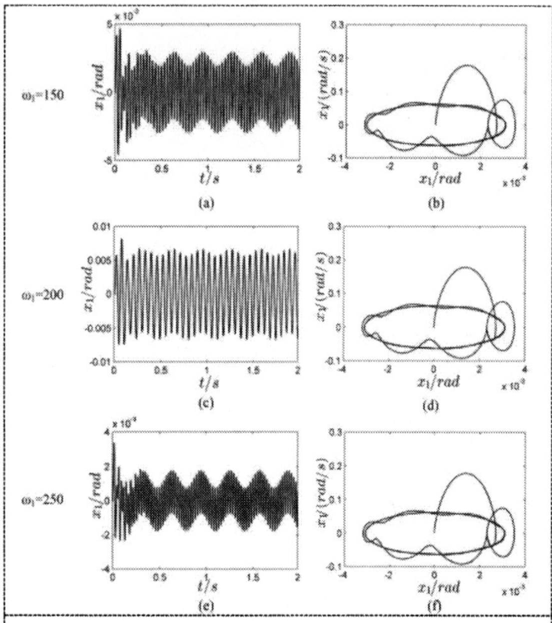

Figure 2. Time history and phase portraits of x_1 with different engine speed

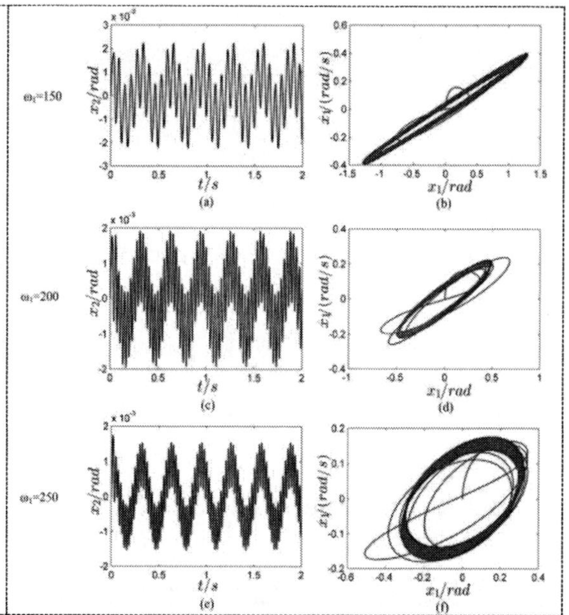

Figure 3. Time history and phase portraits of x_2 with different engine speed.

The Poincaré bifurcation diagram of x_1 with variation of parameter the rotational speed of engine ω_1 is shown in figure 4. When $150 < \omega_1 < 200$, the torsional vibration of shaft between engine and PMSM gets gradually stable, while the torsional vibration of the shaft gradually loses stability with increasing engine speed when $200 < \omega_1 < 250$.

Time history diagram and phase portraits of x_2 obtained by simulation when ω_1=150, ω_1=200 are shown in figure 3. The initial values are set to be $x_{10}=x_{20}=\dot{x}_{10}=\dot{x}_{20}=0$, and the load excitation frequency is set to be $\omega_2=20$. The Poincaré bifurcation diagram of x_2 with variation of parameter the rotational speed of engine ω_1 is shown in figure 5.

IOP Conf. Series: Materials Science and Engineering **491** (2019) 012020 doi:10.1088/1757-899X/491/1/012020

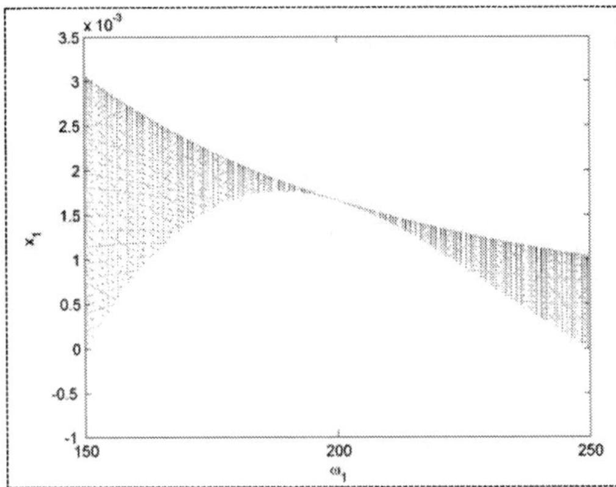

Figure 4. The Poincaré bifurcation diagram of x_1 with variation of parameter the rotational speed of engine ω_1

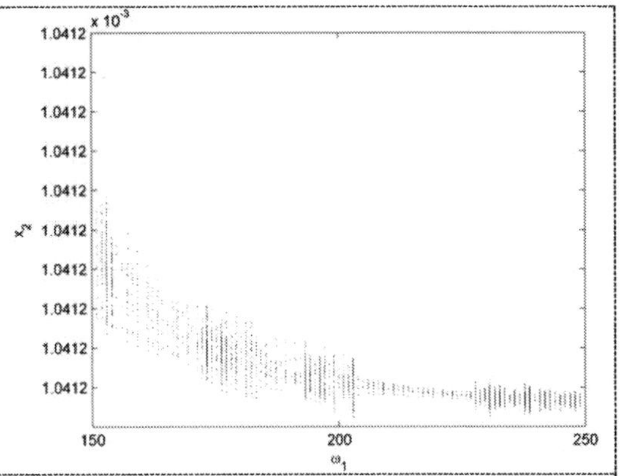

Figure 5. The Poincaré bifurcation diagram of x_2 with variation of parameter the rotational speed of engine ω_1

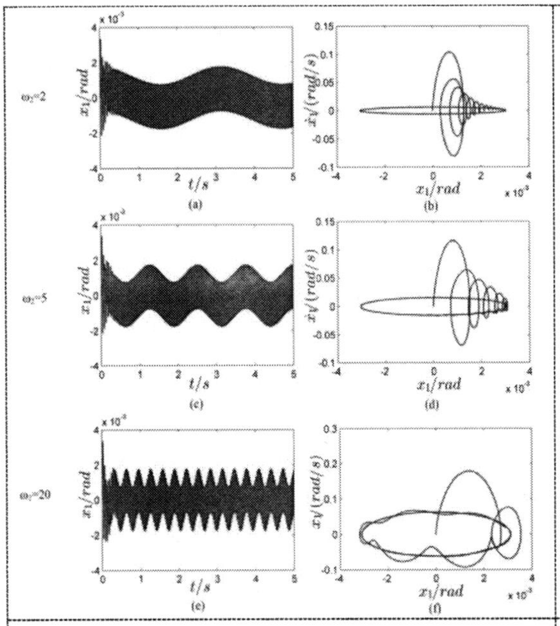

Figure 6. The Poincaré bifurcation diagram of x_1 with variation of load excitation frequency.

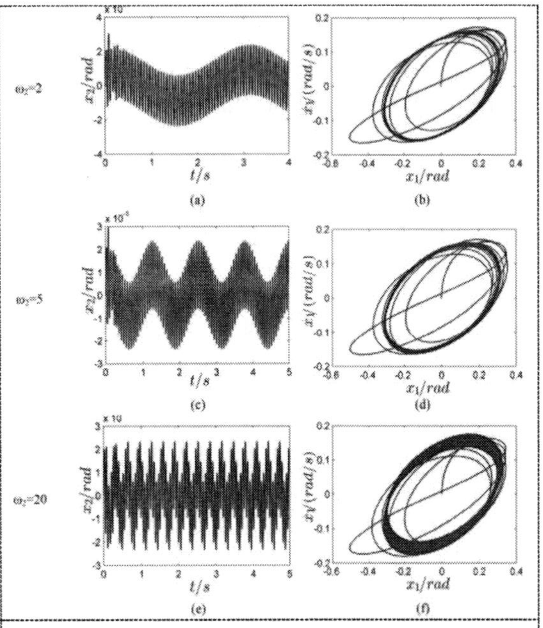

Figure 7. Time history and phase portraits of x_2 with different load excitation frequency.

4.2. Under generating mode

Time history diagram and phase portraits of x_1 obtained by simulation are shown in figure 6. The initial values are set to be $x_{10}=x_{20}=\dot{x}_{10}=\dot{x}_{20}=0$, and the engine speed is set to be $\omega_1=200$.The Poincaré bifurcation diagram of x_1 with variation of load excitation frequency ω_2 is shown in figure 8.

Time history diagram and phase portraits of x_2 obtained by simulation when $\omega_2=2, \omega_2=5, \omega_2=20$, are shown in figure 7. The initial values are set to be $x_{10}=x_{20}=\dot{x}_{10}=\dot{x}_{20}=0$, and the load excitation frequency is set to be $\omega_1=200$. The Poincaré bifurcation diagram of x_2 with variation of load excitation frequency ω_2 is shown in figure 9.

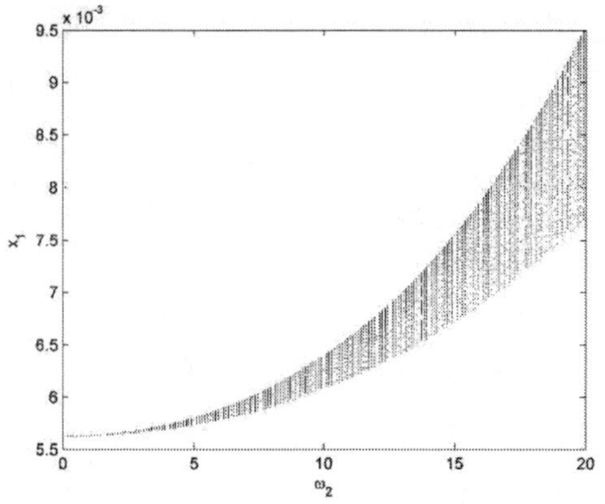

Figure 8. The Poincaré bifurcation diagram of x_1 with variation of load excitation frequency ω_2.

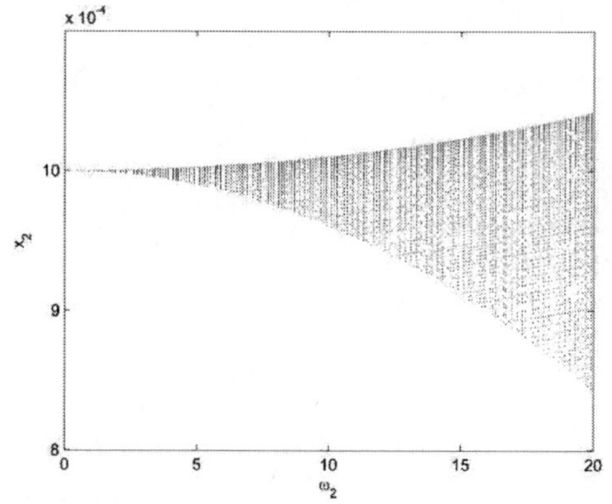

Figure 9. The Poincaré bifurcation diagram of x_2 with variation of load excitation frequency of engine ω_2.

5. Conclusion

In this paper, a three-mass torsional vibration model of engine-PMSM-load of hybrid electric propulsion system is established by using lumped parameter method. Using nonlinear dynamics theory, the influence of different engine speed and load excitation frequency on nonlinear torsional vibration stability of hybrid electric propulsion system is studied. We get the following conclusions:

(1) There is a phase difference between the engine excitation and the load excitation in the hybrid electric propulsion system, which makes the hybrid electric propulsion system hard to eliminate the torsional vibration;

(2) With the engine speed as the control parameter, the torsional vibration characteristics of the two shaft are different because the directional relationship of the moment of engine and PMSM is distinctive from that of PMSM and load. When $150 < \omega_1 < 200$, the torsional vibration of shaft between engine and PMSM gets gradually stable, while the torsional vibration of the shaft gradually loses stability with increasing engine speed when $200 < \omega_1 < 250$.

(3) With the load excitation frequency as the control parameter, as the load excitation frequency increases, the torsional vibration of the hybrid electric propulsion system gradually loses stability.

Funding

This work was supported by the National Natural Science Foundation of China (grant no. 51275291), National Natural Science Foundation of China (grant no. 51705208) and China Postdoctoral Science Foundation (grant no. 2018M632240).

References

[1] Liang L, Wang X, and Song J 2016 *MECH SYST SIGNAL PR* **87** 17-29.
[2] Chan C.C 2007 *PROC IEEE* **95** 704-718.
[3] Chao Y, Song J, Li L and Cao D 2016 *MECH SYST SIGNAL PR* **77** 649-664.
[4] Bradley T H, and Frank A A 2009 *Renew. Sust. Energ. Rev* **13** 115-128.
[5] Inman S, El-Gindy M, Haworth D C 2003 *INT J HEAVY VEH SYST* **10** 167-187.
[6] Yang Y, Hu X, Pei H and Peng Z 2016 *Appl. Energy* **168** 683-690.
[7] JonasFredriksson, HenrikWeiefors and BoEgardt 2002 *Veh. Syst. Dyn* **37** 359-376.
[8] Guo X and He H 2016 *MATEC Web of Conferences* **81** p08002.
[9] Abouobaia E, Bhat R and Sedaghati R 2015 *J INTEL MAT SYST STR* **1** 1-13.
[10] Ebrahimi M, Farshidianfar A and Bartlett H 2003 *Proc. Inst. Mech. Eng. Part D-J* **215** 217-229.
[11] Tang X, Jin Y, Zhang J, Zou L and Yu H 2014 *Proc. Inst. Mech. Eng. Part D-J* **228** 94-103.

[12] Tang X, Yang W, Hu X and Zhang D 2017 *MECH SYST SIGNAL PR* **85** 329-338.
[13] Chen K, Hu J and Peng Z 2017 *Energy Procedia* **105** 3164-3172.
[14] Yue D, Miao D and Zhang J 2008 *Auto. Eng* **3** 211-214.
[15] Zuo J, Liu S, Huang Z and Hu Q 2012 *Appl. Mech. Mat* **190** 825-831
[16] Yu Y and Mi Z 2013 *J Appl. Math* 1-11.
[17] Chen X, Yuan S and Peng Z 2015 *Nonlinear Dyn* **80** 541-552.
[18] Burston, Amanda, Ward B and Davies R 2001 *MON NOT R ASTRON SOC* **326** 403-415.
[19] Ying G, Meng G and Jing J 2009 *Arch. Appl. Mech* **79** 287-299.
[20] Chen L, Xi G and Sun J 2012 *IEEE T VEH TECHNOL* **61** 2936-2949.
[21] Du C and Ding X 2015 *IAEAC* 905-910.
[22] Cacan, Martin R, Stephen L and Leamy M 2014 *Nonlinear Dyn* **78** 1205-1220.
[23] Cermak, J and Nechvatal L 2017 *Nonlinear Dyn* **87** 939-954.

IOP Conf. Series: Materials Science and Engineering **491** (2019) 012021 doi:10.1088/1757-899X/491/1/012021

Investigation of Parameters Affecting Heat Transfer and Fluid Flow of a TEFC Electric Motor by Using Taguchi Method

Huan-Sen Peng[1,*] and Feng-Hsiang Lai[1]

[1]TECO Group Research Institute, TECO Electric & Machinery Co., Ltd., Taipei, Taiwan

*Corresponding author e-mail: devin.peng@teco.com.tw

Abstract. The thermal issue is an important criterion for the design of high efficiency motor. In this study, the numerical simulation on the thermal performance of TEFC electric motor is proposed. The parameters such as fin pitch ratio, fin dimensions and fan cover location relative to the fin are discussed to investigate the effects of them and decrease the winding temperature. In order to examine the various parameters quickly without suffering from a great number of parametric computations, Taguchi method is introduced to reduce the computational effort. The case studies results show the effects of fin and fan cover configuration on average temperature rise of winding. And it is found that the fin pitch ratio has a significant influence on average temperature rise of winding. These case studies not only provide physical insight into the thermal-fluid phenomenon but also find out the optimal combination of parameters from given levels by using SN ratio. An additional simulation is performed and shows a good agreement with predicted value. The new design proposed in this study shows 4.8℃ decrease in average temperature rise of winding.

1. Introduction

The thermal management of electric machine becomes more important because of the requirements for miniaturization, energy efficiency and cost reduction. Good thermal management reduce the danger of irreparable failures and increases electric motor life. Heat losses generated from the stator and rotor are transferred to the frame and end shields and dissipates through the outer fins to the atmosphere. Hence, the improvement of thermal and hydraulic performances are very important.

In recent years, some researchers devoted their efforts in investigating the thermal management of electric motor. Valenzuela and Tapia [1] performed the evaluation of the heat transfer capacity of a finned frame under the different velocities of the cooling fan. The empirical evaluation of the heat transfer coefficient of two finned frames at different speeds of the cooling fan was developed. Illagolla [2] analyzed a fin for examining the effects of the convective heat transfer for temperature distribution of the electric water pump motor by using MATHCAD. Importance of the convective heat transfer was also discussed. The results show that the temperature gradient of the fin increase along the length of the fin from base to the end when increasing the convective heat transfer coefficient. Guo [3] employed CFD modelling to obtain the distribution of air flow rate, air velocity and temperature of a 55 kW surface permanent magnet synchronous motor (SPMSM). The helpful empirical formula of heat transfer coefficients of fins were proposed for electric machines design with the same structure. A

IOP Conf. Series: Materials Science and Engineering **491** (2019) 012021 doi:10.1088/1757-899X/491/1/012021

numerical investigations have been carried out on thermal-flow characteristics of a totally enclosed fan cooled induction motor by Moon et al [4]. Four design parameters such as fin attached to frame, channel embedded in frame, duct through rotor-core and fan-cover surrounding external self-rotating fan were examined. Air flow rate of fan and temperature rise of insulated winding were considered to determine the fin array-height, thickness and pitch-attached to frame surface. And the number of channels embedded in frame was examined to look into how effective does inner-circulated-air cool down. Ulbrich et al. [5] proposed a novel approach on conjugate heat transfer problem in 2-D. In their study, a constant heat transfer coefficient between fin surface and fluid was assumed in order to describe the heat flow through rectangular fins analytically. And the optimal heat dissipation is achieved by having more but shorter fins under the condition of constant total surface area.

Taguchi method is a frequently used method to find out the optimal combination of parameters with minimum effort and cost. Many scholars use this method to optimize the electromagnetic performance, thermal performance or reduce the noise in the electric motor design [6-8]. Taguchi method shows salient usefulness in optimization design of electric motor applications.

The purposes of this study are to examine the effect of fin and fan cover geometry on the thermal and fluid characteristic of TEFC electric motor (as shown in figure 1) and minimizing the average temperature rise of winding. The obtained results from original design were validated by comparing them with experimental data. Parametric studies through CFD and Taguchi method were performed for improving thermal performance. Furthermore, the optimal case was selected using SN ratio and confirmed by an additional simulation.

Figure 1. TEFC electric motor used in the analysis.

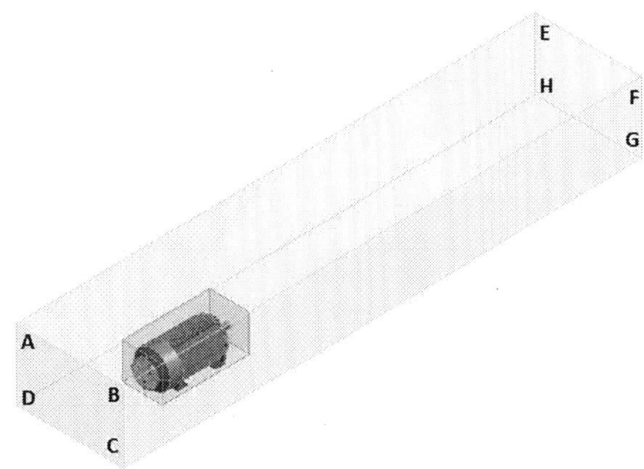

Figure 2. Computational domain and boundary conditions.

2. Mathematical model and numerical method

2.1. Simulation

The governing equations to be considered are the time-averaged continuity, momentum, and energy equations. An eddy viscosity model is used to account for the effect of turbulence phenomena. In solid region, solid energy equation is employed. The boundary conditions for this problem are state as figure 2. For fluid field, no-slip boundary condition was employed for the ground (Plane CDHG). And entrainment boundary conditions were employed in other five surrounding planes. For thermal field, adiabatic boundary condition was assumed for the ground. And the other five surrounding planes were assumed as ambient temperature.

2.2. Loss sources

Power losses in electric motor are composed of resistive losses in stator and rotor conductors, iron losses in the magnetic circuit, additional losses and mechanical losses. In this thermal model, heat source is the average losses obtained by measurement at rated operation.

2.3. Numerical method

The thermal and fluid fields of computational domain are solved by the commercial software ANSYS FLUENT™. This conjugate heat transfer problem is assumed to be steady-state and run with an implicit and segregated solver. The multiple reference frame model (MRF) is employed to deal with the rotation parts (fan, shaft, and rotor). Unstructured mesh mainly composed of tetrahedral cells is applied for complicated geometries like frame, end shields and winding while structural mesh is applied to sweepable parts. Fine grids are deployed where high velocity or temperature gradient takes place, including air gap between the stator and rotor, blades, fins and so on.

2.4. Taguchi method

The Taguchi method was developed by Genichi Taguchi during the 1950s. It's a methodology for applying statistics to improve the quality of manufactured good, determine optimal design parameters of engineering, and so on. Taguchi method employs orthogonal array to investigate several design parameters with a low number of case studies. The Taguchi method is established based on the signal-to-noise ratio (S/N ratio). For analyzing the S/N ratio, three types of quality characteristics are defined, i.e. the nominal-the-better, the smaller-the-better, and the larger-the-better. In this study, the smaller-the-better quality characteristic was chosen for minimizing the average temperature rise of winding. For the smaller-the-better cases, the definition of SN Ratio can be expressed as

$$\mathrm{SN} = -10\log[\bar{y}^2 + S^2] \tag{1}$$

where $\bar{y}$ is mean value of observations, S is standard deviation. In simulation case studies, $\bar{y} = \mathrm{y}$, and $\mathrm{S} = 0$.

And the selection of L9(3^4) orthogonal array using Taguchi method shown in table 1. We define four factors to affect the average temperature rise of winding. The factors include fin pitch ratio (fin pitch/fin thickness), fin thickness, fin height and fan cover location relative to the fin (-a/b). Figure 3 shows the details of configuration. The factors with their values and their levels considered in this study were given in table 2.

Table 1. L9(3^4) orthogonal array.

Experiment	A	B	C	D
1	1	1	1	1
2	1	2	2	2
3	1	3	3	3
4	2	1	2	3
5	2	2	3	1
6	2	3	1	2
7	3	1	3	2
8	3	2	1	3
9	3	3	2	1

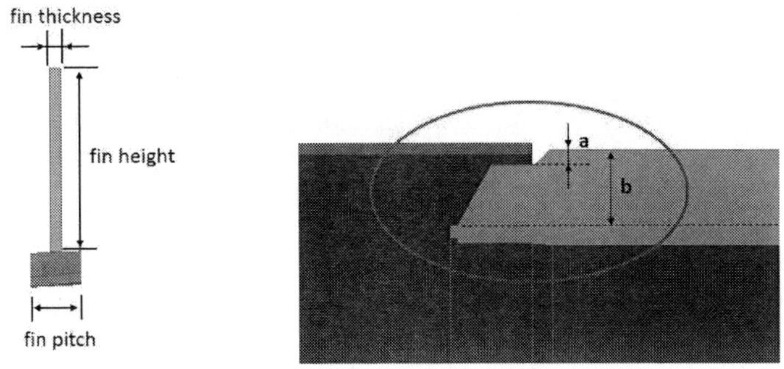

(a) (b)

Figure 3. Details of fin and fan cover configuration (a) fin dimensions (b) fan cover location relative to the fin.

Table 2. Control factors and their levels.

Factors	Levels		
	1	2	3
A: fin pitch ratio (-)	9.17	3.4	1.45
B: fin thickness (mm)	3	5	7
C: fin height (mm)	44	55	77
D: fan cover location relative to the fin (-)	0%	-20%	-40%

3. Results and discussion

3.1. Validation

In thermal management of the induction motor, winding temperature is a more critical component than others. Hence, the average temperature rise of winding is compared with measurement data to verify the present numerical model, as shown in table 3. The numerical prediction for the average temperature rise of winding shows a good agreement with measurement data. The validation indicated that the thermal model and assumptions are reliable.

Table 3. Comparison of the results between measurement and simulation.

		Average temperature rise
F#449T	CFD	68.8
400HP	Measurement	71.6
T_{amb} = 25.6°C	Error	-2.8

3.2. Case studies

The thermal performance of the electric motor can be calculated for each simulation of the orthogonal array by using the observed values of the average temperature rise of winding from table 4. The results for average temperature rise of winding are presented as SN ratio for each factor at each level in figure 4. From the figure, it is clear that the average temperature rise of winding decrease with all the three factors, i.e. fin pitch ratio, fin thickness, and fin height. The rate of decrease for the average temperature rise of winding is higher for fin pitch ratio when compared with the fin thickness and fin height. It can be observed that SN ratio will increase sharply as the fin pitch ratio decreases from 9.17 to 1.45. It reveals that the increment of amount of fins improves the thermal performance. Although the flow resistance will increase as the amount of fins increase, the heat transfer area still compensate for the decrease of air flow rate under the factor setting level ranges.

Figure 4 also shows that the fin height does not have a significant influence on the SN ratio. The heat transfer area will increase as the fin height increases. However, the fin efficiency decreases with increase in fin height. And the convective heat transfer coefficient isn't a constant from the bottom to the tip of fins. Hence, the increment of heat dissipation doesn't proportional to the heat transfer area. The heat dissipation rate gradually slow down. The thermal phenomenon mentioned in previous discussed also show in figure 5.

The influence of fan cover location relative to the fin shows different tendency from previous factors. As the factor decrease from 0% to -20%, the diameter of fan cover outlet decrease and increase the flow resistance. The air flow rate decreases. However, the change of fan cover configuration reduce the circulation between fan cover and end shield. And the air flow will concentrate near the bottom of fins. The convective heat transfer coefficient increases and improve the heat transfer rate in high temperature region. When the factor decrease from -20% to -40% shows the opposite result. The diameter of fan cover outlet decrease and more closer to the frame. The flow resistance becomes too high and reduce the more air flow rate and flow velocity in the near frame region. It caused the negative effects on thermal performance. And the result reveals that the optimal fan cover location relative to the fin is between -20% and -40%. The flow fields also show in figure 6.

Table 4. Orthogonal array L9(3^4) with calculated average temperature rise of winding and SN ratio.

Case	A	B	C	D	y_t	S/N ratio
1	9.17	3	44	0%	80.88	-38.157
2	9.17	5	55	-20%	77.36	-37.770
3	9.17	7	77	-40%	73.57	-37.334
4	3.4	3	55	-40%	74.72	-37.468
5	3.4	5	77	0%	73.44	-37.318
6	3.4	7	44	-20%	70.52	-36.966
7	1.45	3	77	-20%	68.93	-36.768
8	1.45	5	44	-40%	70.75	-36.995
9	1.45	7	55	0%	65.88	-36.376
				Ave.	72.894	-37.2391

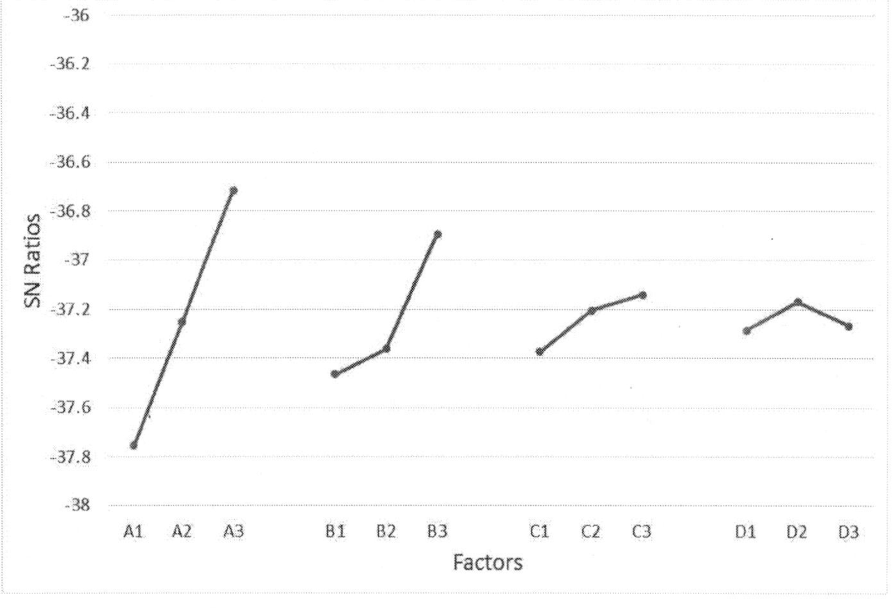

Figure 4. SN ratios of factor level for average temperature rise of winding.

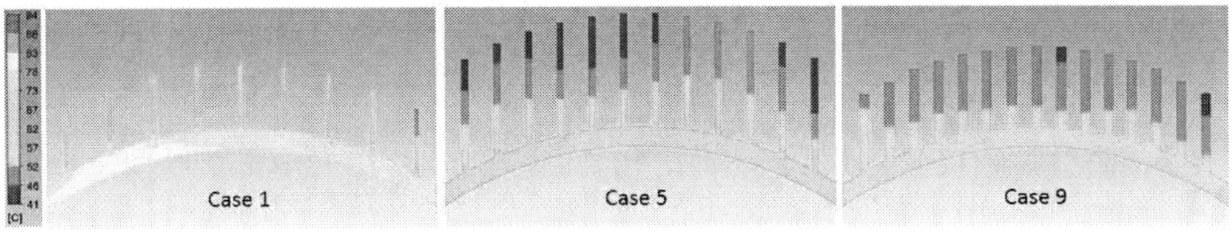

Figure 5. Temperature distributions of fins.

Figure 6. Flow field of near fin regions.

3.3. Optimal combination of control factors

The case studies results give the whole picture of the effects of fin and fan cover configurations on the average temperature rise of winding and an optimal combination of these factors can be selected. Based on the influence of each factor shown in figure 4, it is seen that the factor-level combination (A3, B3, C3, D2) contributes to minimization of average temperature rise of winding. The optimal case can be predicted by using the empirical model:

$$SN_{optimal} = \overline{SN} + (\overline{SN}_{A3} - \overline{SN}) + (\overline{SN}_{B3} - \overline{SN}) + (\overline{SN}_{C3} - \overline{SN}) + (\overline{SN}_{D2} - \overline{SN}) = -36.1957 \quad (2)$$

Therefore, the predicted average temperature rise of winding in optimal case is 64.53℃. Finally, an additional simulation of optimal case was performed for confirmation. The average temperature rise of winding obtained from CFD is 63.94℃. The result shows a good agreement with prediction.

4. Conclusions

In this paper, the numerical simulation on the thermal performance of TEFC electric motor is proposed. Various parameters such as fin pitch ratio, fin dimensions and fan cover location relative to the fin were discussed in order to investigate the effects of them and decrease the winding temperature. Instead of a number of possible combinations for four parameters and three levels for each of them (3^4=81), Taguchi method was used to reduce the computational effort. With this technique, there are only 9 patterns defined by L9(3^4) orthogonal array need to be calculated. The case studies results show the effects of fin geometry and fan cover location on average temperature rise of winding and provide physical insight into the heat transfer and fluid flow phenomenon. And it is found that the fin pitch ratio has a significant influence on average temperature rise of winding. Moreover, the optimal case from given levels was selected using SN ratio and confirmed by an additional simulation. The new design proposed in this study shows 4.8℃ decrease in average temperature rise of winding.

Acknowledgments

The support of TECO Electric & Machinery Co., Ltd. is gratefully acknowledged. The authors also want to thank Green electric machine division for technical advices and sharing experimental data for this research. Truly thanks to Mr. William Hu, deputy general director of TECO Group Research Institute, for providing me with comments on the research.

References

[1] Valenzuela M A and Tapia J A 2006, *Proc. Conf. Rec. IEEE IECON* pp 4835–4840.
[2] Illagolla N J B 2014, *Scientific Research Journal* Vol. II, Issue X pp 34-39.

[3] Guo B C 2014, *17th International Conference on Electrical Machines and Systems (ICEMS) (Hangzhou, China)* pp 3264-3267.
[4] Moon S H, Jung Y H and Kim K W 2016, *XXII International Conference on Electrical Machines (ICEM) (Lausanne, Switzerland)* pp 1928-1933.
[5] Ulbrich S, Kopte J and Proske J 2018, *IEEE Transactions on Industrial Electronics* 65 (2).
[6] Shin W G and Lee S H 2018, *J. Mech Sci Technol* 25 (2), pp 317-322.
[7] Azizi H and Vahedi A 2015, *Turk J Elec Eng & Comp Sci* pp 17-27.
[8] Shirazi A N, Yousefi B, Gholamian S A and Rashidaee S 2013, *Int. J. Comp. Sci. & Appl.* 3 (2) pp. 31-38.

IOP Conf. Series: Materials Science and Engineering **491** (2019) 012022 doi:10.1088/1757-899X/491/1/012022

Numerical Simulation of Transpiration Cooling within Variable Property of Water and Steam

Meng Wang[1], Fei He[1,*], Nan Wu[1], Jianhua Wang[1] and Guangqi Dong[2]

[1] Department of Thermal Science and Energy Engineering, University of Science and Technology of China, Jinzhai Road 96, Hefei 230027, PR China
[2] Beijing Power Machinery Research Institute, Beijing 100074, PR China

hefeihe@ustc.edu.cn

Abstract. Transpiration cooling with coolant phase change provides a wide range of applications in effective thermal protection. Water is usually employed as the coolant and experiences huge temperature and pressure changes during the transpiration cooling process. Mathematical model and numerical approach of transpiration cooling with phase change in consideration of the real variation of water thermodynamic properties are established in this paper, and compared with the constant property model. The IWAPS Industrial Formulation 1997 is used to calculate the thermodynamic properties of water in wide ranges of temperature and pressure. The numerical results obtained indicate some thermodynamic properties vary significantly with temperature and pressure, leading to higher phase change temperature and coolant driving force. With the increase of mass flow rate, the two property models exhibit the similar changing trend in temperature and pressure and there is a lag in the start and end positions of phase change.

1. Introduction

During hypersonic flight in the near space, aerodynamic heat is one of the severest problems, which leads to extremely high temperature of the vehicle surface, especially for the nose cone and leading edge [1]. For instance, when a hypersonic vehicle cruises with Mach 20 at 47km, the surface temperature may exceed 4000K [2]. The extremely high temperature causes degradation in performance of material, and the structural strength will be affected. Therefore, it is very important to develop effective thermal protection systems to ensure the normal flight of hypersonic vehicles.

Transpiration cooling has been recognized as one of the most effective thermal protection methods for hypersonic vehicles exposed to extremely high heat flux. The coolant absorbs a lot of heat due to the large specific surface area of the porous matrix, and at the same time, the coolant flowing out of the porous matrix forms a continuous, even and stable film on the surface to separate hot flow and structure [3]. And it has been validated that the transpiration cooling with liquid coolant is more effective than gaseous coolant because of the huge latent heat of vaporization [4].

There has been a lot of researches on the transpiration cooling with liquid coolant, including experimental investigation and numerical simulation. In the experimental aspect, a series of investigations on the transpiration cooling with liquid coolant have been carried out by Wang et al. and Zhao et al [5,6]. under subsonic and supersonic conditions. These experiments show the high cooling efficiency of transpiration cooling with liquid coolant. In the numerical simulation aspect, two-phase mixture model (TPMM) and separated phase model(SPM) under local thermal equilibrium(LTE) assumption and non-equilibrium(LTNE) condition have been used to study transpiration cooling with liquid coolant [7–10]. In these researches, purified water is used as the coolant because of the huge latent heat of vaporization, high density and low price. But almost in all the existing numerical simulations, thermodynamic properties of water and steam are simplified to constant. Actually in the cooling process, when coolant flows from the tank at ordinary temperature to the surface of the vehicle at high temperature, coolant temperature increases and pressure decreases, which will lead to the change of thermodynamic properties. Therefore, it is meaningful to do further researches on the transpiration cooling with liquid water considering the real properties of water.

To calculate the thermodynamic properties of water and steam, IAPWS-IF97 that published by the International Association for the Properties of Water and Steam (IAPWS) is applied. It provides internationally accepted formulations for the properties of light and heavy steam, water and selected aqueous solutions for scientific and industrial applications in 1997 [11].

In this paper, numerical simulations on the two-phase flow of transpiration cooling with liquid water are carried out in real property model and constant property model. The two-phase flow characteristics and the effect of coolant mass flow rate are also discussed and compared.

2. Model

2.1. Physical Model

The physical model considered in this study is an infinite porous plate with a thickness of L, which is sketched in Figure 1. The upper side of the plate is exposed in hot main flow, where the heat flux (Q in the figure) boundary condition is exerted. Liquid coolant is injected into the pores from the bottom of the plate at temperature T_c with mass flow rate $\dot{m}_c$. The coolant flows through the plate and absorbs heat from porous matrix. With enough heat exchange, the plate can be classified into three regions in the vertical direction, i.e. liquid region, two-phase region and vapor region. If the heat flux is low, the vapor region and even two-phase region will disappear.

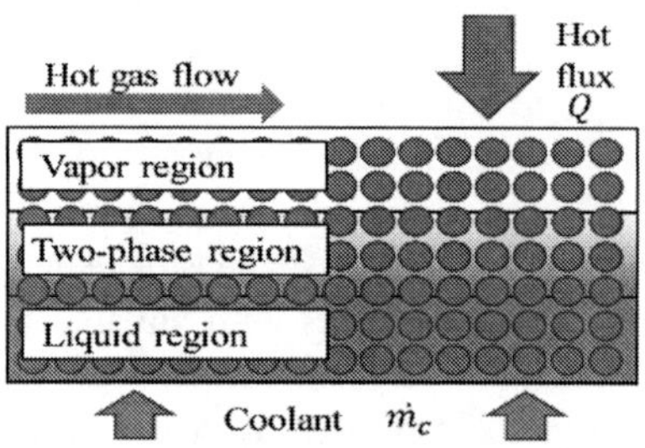

Figure 1. Physical model of transpiration cooling with liquid coolant

2.2. Mathematical Model

Based on TPMM-LTNE model, the coolant flow and heat exchange are described by the same conservation equations and constitutive relationships in all regions, which are listed in Table 1 [7,8].

In mass and momentum equations, the two-phase flow is described by a mass-average mixture velocity. Fluid and solid energy equations are solved separately, and heat exchange between fluid and solid within the porous matrix is considered. And in constant property model, Fluid energy equation is simplified.

Table 1. Model equations

	Variable property (IAPWS-IF97) model	Constant property model
Mass	$\nabla \cdot (\rho \vec{u}) = 0$	
Momentum	$\vec{u} = -\frac{K}{\mu}(\nabla p - \rho_k \vec{g})$	

Fluid energy	$\nabla\cdot(\gamma\vec{u}H_2)+2\rho\vec{u}\nabla h_{v,sat}=\nabla(k_{f,eff}\nabla T)+\nabla\cdot(h_{fg}\vec{j})+Q_{sf}$ $\vec{j}=-D(s)\nabla s+\frac{f(s)K\Delta\rho}{\nu_v}\vec{g}$	$\nabla\cdot(\gamma\vec{u}H_2)=\nabla\cdot(\Gamma_h\nabla H_2)+\nabla\cdot\left[f(s)\frac{K\Delta\rho h_{fg}}{\nu_v}\vec{g}\right]+Q_{sf}$
Solid energy	$\nabla\cdot(k_{s,eff}\nabla T_s)=Q_{sf}$	
Constitutive relationships	$\gamma=\frac{[(\rho_v/\rho_l)(1-s)+s][h_{v,sat}(1+\lambda_l)-h_{l,sat}\lambda_l]}{(2h_{v,sat}-h_{l,sat})s+(\rho_v h_{v,sat}/\rho_l)(1-s)}$ $\nu(s)=\frac{1}{\frac{k_{rl}(s)}{\nu_l}+\frac{k_{rv}(s)}{\nu_v}}$ $\lambda_l(s)=\frac{\nu(s)k_{rl}}{\nu_l},\lambda_v(s)=\frac{\nu(s)k_{rv}}{\nu_v}$ $f(s)=k_{rv}(s)\lambda_l(s)$	
Capillary pressure	$p_c=\left(\frac{\epsilon}{K}\right)^{1/2}\sigma J(s)$ $J(s)=1.417(1-s)-2.120(1-s)^2+1.263(1-s)^3$	

2.3. Thermodynamic Properties Formulation

In the numerical simulations of transpiration cooling with phase change, thermophysical properties of coolant are obtained by two different models, i.e. the variable property model by IAPWS-IF97 formulation, and the constant property model to set properties as those of saturated water and saturated steam at 101325Pa. Different from the constant property model, properties in the variable property model are the functions of temperature and pressure and calculated by IAPWS-IF97. The IAPWS-IF97 consists of a set of equations which cover the following valid ranges of temperature and pressure:

$$273.15\ \text{K}\leqslant T\leqslant 1073.15\ \text{K}\quad p\leqslant 100\ \text{MPa}$$

$$1073.15\ \text{K}\leqslant T\leqslant 2273.15\ \text{K}\quad p\leqslant 10\ \text{MPa}$$

Specifically, the variations of some main parameters with temperature and pressure are displayed in Figure 2.

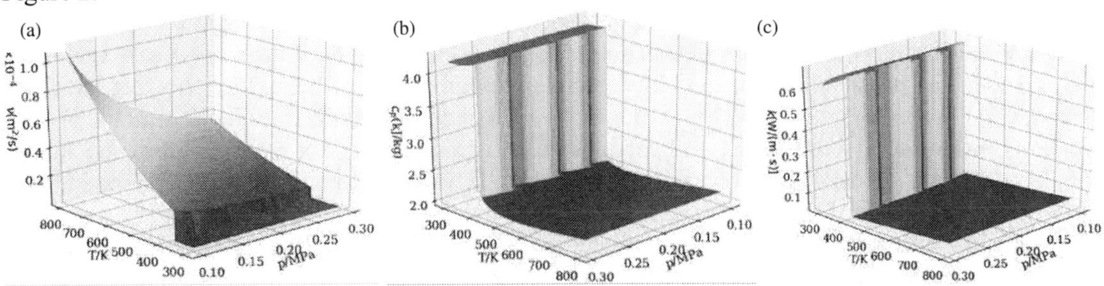

Figure 2. Properties of water and steam at pressure between 0.1 to 0.3MPa, temperature between 293.15 to 800K

3. Numerical Solution

The above mathematical models are solved by limited volume method. The convergence of the solving process is achieved when the residual error drops below 1×10^{-8}. For one-dimensional problem, the computational domain is divided into uniform control volumes. The grid independency is verified by using three meshes with grid numbers of 800, 2500 and 10000. The difference between the results is not significant. Finally, the mesh with a grid number of 2500 is chosen. The detailed solution procedure and algorithmic process are given as follows:

1) Calculate initial corrected enthalpy distribution from temperature and pressure.

2) Solve mass and momentum equation.
3) Update properties dependent pressure.
4) Solve solid and fluid energy equation.
5) Update properties.
6) Repeat from step 2 until convergence.

4. Results and Discussions

In all the simulations, the pressure and heat flux applied on the hot side of the plate is 101325Pa and $2 \times 10^6\ \mathrm{W/m^2}$, respectively. The porous plate has a thickness of 10cm and a thermal conductivity of $20\ \mathrm{W/(m \cdot K)}$. The permeability of the porous matrix is $1 \times 10^{-12}\mathrm{m}^2$ when the porosity is 0.3 and average sphere diameter is 5×10^{-4} m.

4.1. Distributions of Temperature and Pressure

When the coolant mass flow rate is set as $0.5\ \mathrm{kg/(m^2 \cdot s)}$, the transpiration cooling performances are investigated by the IAPWS-IF97 model and constant property model respectively. Figure 3 shows the fluid temperature and pressure distributions in the porous plate.

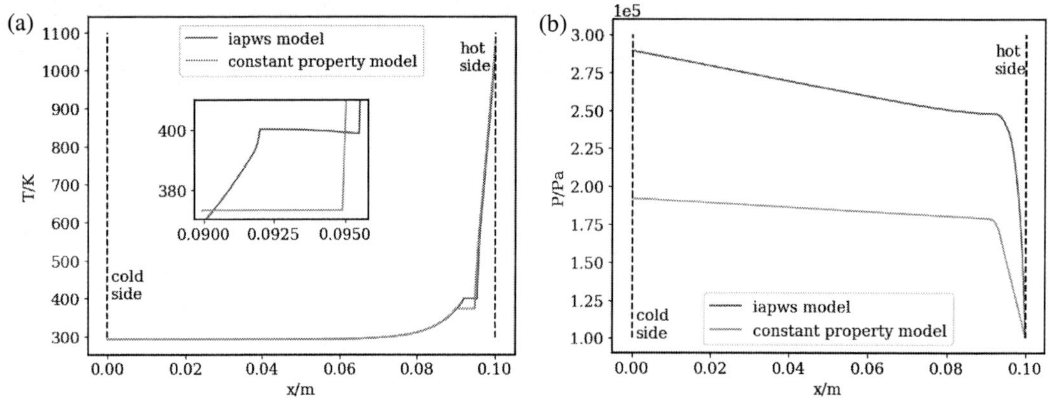

Figure 3. Temperature (a) and pressure (b) distributions in two property models

Two models show the same change trend of temperature in liquid and vapor regions, i.e. temperature increases slightly in liquid region and increases rapidly in vapor region. But in two-phase region, the fluid temperature distributions obtained by the models are different, including the following interesting phenomena: 1) compared to the constant property model, the fluid temperature in the variable property model is much higher and the phase change location is closer to the hot side; 2) a decrease of fluid temperature along the flow direction can be observed in the two–phase region, as shown in the subfigure (a) of Figure 3. These differences can be explained by the pressure distribution in 3(b). Due to the kinematic viscosity varying with temperature and pressure, the pressure drops obtained by the variable property model is higher than that obtained by constant property model, which indicates that the driving force of coolant required is much underestimated using the traditional constant property model. A higher pressure corresponds to a higher phase change temperature. In IAPWS-IF97 model, the variations of coolant thermal properties with pressure are considered, so the phase change temperature predicted decreases along the coolant flow direction. Furthermore, higher phase change temperature means more latent heat is absorbed when phase change happens, therefore a thicker liquid region and thinner two-phase region are obtained using the IAPWS-IF97 model.

4.2. Influence of Coolant Mass Flow Rate on Temperature and Pressure Distributions

In this section, the temperature and pressure distributions are calculated using the two models when coolant flow rate varies from $0.5\ \mathrm{kg/(m^2 \cdot s)}$ to $1.0\ \mathrm{kg/(m^2 \cdot s)}$. And the results are shown in Figure 4.

For the same heat flux, the specific enthalpy becomes smaller as the coolant injected increases, and thereby the coolant temperature at the hot side drops. When the coolant injected further increases, the specific enthalpy is lower than saturated enthalpy, and the temperature keeps at phase change temperature. The difference between the temperatures obtained by two models is small because the change of specific heat is tiny in the range of temperature and pressure considered here. In both models, the driving pressure decreases firstly then increases with the increase of coolant mass flow rate due to the reduction of vapor region and two-phase region. The driving pressure obtained by IAPWS-IF97 model is always higher than that by constant property model. The maximum difference occurs at the mass flow rate of $0.5\ \mathrm{kg/(m^2 \cdot s)}$, where driving pressure by IAPWS-IF97 model is 57% higher.

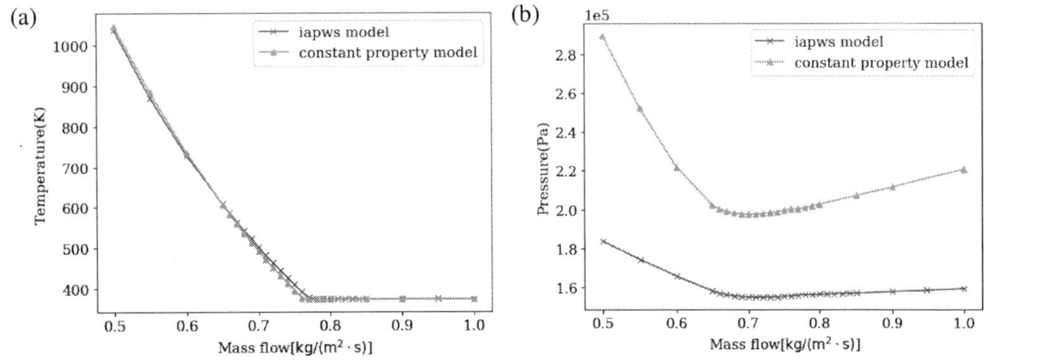

Figure 4. The variation of temperature at hot side (a) and driving pressure (b) with coolant mass flow rate

4.3. Influence of Coolant Mass Flow Rate on Phase Change Location

Pressure loss of coolant occurs mainly in two-phase and vapor regions, which greatly affects the coolant distribution. The phase change locations at different mass flow rates are investigated here. Figure 5 shows the starting and ending positions of two-phase region. As coolant mass flow rate increases, the starting location of two-phase region moves obviously towards the hot side of the plate. It is noted that the starting position of two-phase region obtained by IAPWS-IF97 model is always behind that by constant property model. That's because the phase change temperature is higher in IAPWS-IF97 model due to a higher pressure. When mass flow rate increases up to $0.77\ \mathrm{kg/(m^2 \cdot s)}$, the vapor region disappears, leading to a step change of starting location of two-phase region there. The ending locations predicted by the two models both move towards the hot side with the coolant mass flow rate and maintain at the hot side when vapor region disappears. Thickness of two-phase region means the ending location of phase change minuses starting location. It can be concluded that the two-phase region predicted by IAPWS-IF97 model is smaller than that by constant property model.

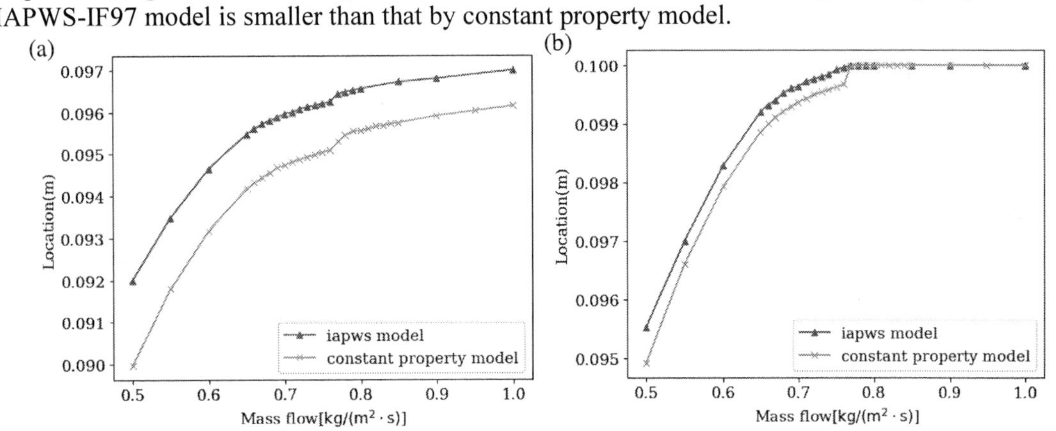

Figure 5. Starting (a) and ending (b) locations of two-phase region with coolant mass flow rate

5. Conclusion

In this paper, the real thermal properties of water and steam are considered in the simulation of transpiration cooling with liquid water as coolant. Then the numerical investigations on transpiration cooling using the variable property model based on IAPWS-IF97 formulation and the constant property model were carried out. Through the analysis and comparison, the following conclusions can be drawn:

(1). Comparing to the constant property model, the phase-change temperature predicted by variable model is higher and decreases in the two-phase region due to the pressure change there, and the driving force of coolant estimated by IAPWS-IF97 model is much higher and shows more significant variation trend along the flow direction.

(2). When the coolant mass flow rate increases, coolant temperature at the hot side drops and the driving force firstly decreases and then increases. Comparing the results from the two models, the temperatures at hot side are always very close, but the driving forces show remarkable difference.

(3). With the increase of coolant mass flow rate, the phase change location move toward the hot side and two-phase region decreases, but the starting and ending positions predicted by the variable property model have a lag in comparison with constant property model.

Reference

[1] T. Reimer, M. Kuhn, A. Gülhan, B. Esser, M. Sippel and A. Foreest 2011 Transpiration Cooling Tests of Porous CMC in Hypersonic Flow 17th AIAA International Space Planes and Hypersonic Systems and Technologies Conference (San Francisco, California: American Institute of Aeronautics and Astronautics)

[2] M. Sippel 2007 Introducing the SpaceLiner Vision 7th International Symposium on Launcher Technologies (Barcelona, Spain) p 10

[3] A. Laganelli 1970 A comparison between film cooling and transpiration cooling systems in high speed flow 8th Aerospace Sciences Meeting (West Germany: American Institute of Aeronautics and Astronautics)

[4] J. Bellettre, F. Bataille, A. Lallemand and H. Y. Andoh 2005 Studies of the transpiration cooling through a sintered stainless steel plate *Exp. Heat Transf.* **18** 33–44

[5] J. H. Wang, L. J. Zhao, X. C. Wang, J. Ma and J. Lin 2014 An experimental investigation on transpiration cooling of wedge shaped nose cone with liquid coolant *Int. J. Heat Mass Transf.* **75** 442–9

[6] L. J. Zhao, J. H. Wang, J. Ma, J. Lin, J. L. Peng, D. J. Qu and L. Z. Chen 2014 An experimental investigation on transpiration cooling under supersonic condition using a nose cone model *Int. J. Therm. Sci.* **84** 207–13

[7] C. Y. Wang and C. Beckerma 1993 A two-phase mixture model of liquid-gas flow and heat transfer in capillary porous media-l. Formulation *Heat Mass Transf.* **36** 2747–58

[8] J. X. Shi and J. H. Wang 2011 A Numerical Investigation of Transpiration Cooling with Liquid Coolant Phase Change *Transp. Porous Media* **87** 703–16

[9] F. He, J. H. Wang, L. C. Xu and X. C. Wang 2013 Modeling and simulation of transpiration cooling with phase change *Appl. Therm. Eng.* **58** 173–80

[10] W. J. Dong and J. H. Wang 2018 A New Model and its Application to Investigate Transpiration Cooling with Liquid Coolant Phase Change *Transp. Porous Media* **122** 575–93

[11] W. Wagner, J. R. Cooper, A. Dittmann, J. Kijima, H.-J. Kretzschmar, A. Kruse, R. Mareš, K. Oguchi, H. Sato, I. Stöcker, O. Šifner, Y. Takaishi, I. Tanishita, J. Trübenbach and Th. Willkommen 2000 The IAPWS Industrial Formulation 1997 for the Thermodynamic Properties of Water and Steam *J. Eng. Gas Turbines Power* **122** 150

IOP Conf. Series: Materials Science and Engineering **491** (2019) 012023 doi:10.1088/1757-899X/491/1/012023

Grid Convergence Study for Detached-Eddy Simulation of Flow over Rod-Airfoil Configuration Using OpenFOAM

Siti Ruhliah Lizarose Samion[1, a], Nur Haziqah Shaharuddin[1, b] and Mohamed Sukri Mat Ali[1, c]

[1]Malaysia-Japan International Institute of Technology, University of Technology Malaysia, 54100 Kuala Lumpur, Malaysia

E-mail: [a]lializarose@gmail.com; [b]haziqahshaharuddin11@gmail.com; [c]researchsukri@gmail.com

Abstract. A rod-airfoil is a benchmark configuration for simulating Airfoil-Turbulence Interaction Noise (ATIN). The numerical simulation is modelled using unsteady Detached Eddy Simulation (DES). The grid refinement involves two stages of assessments. The first compares the sensitivity of two reasonable estimated cells to the flow behaviours of the case. Based on the finding from the first stage, further grid refinement is proceed in stage two. In stage two, a minimum of three different grid resolutions are considered; the fine, medium and coarse grids in order to investigate the grid independency. Richardson extrapolation and Grid Convergence Index (GCI) are introduced to quantitatively evaluate the grid independency. Based on the results between those three different grids, a monotonic convergence criterion has been achieved. The reduction in GCI value indicates that the grid convergence error has been significantly reduced, in which the fine grid has a GCI value is around less than 0.5%.

1. Introduction

An airfoil is a common engineering geometry. Its application especially in the wind turbines, turbofan engines and the helicopter rotors, however induces generation of airfoil- turbulence interaction noise (ATIN) [1,2,3]. The airfoil experiences undulation of lift as a result of unsteady pressure fluctuations produced by the interaction of turbulence in the upstream flow and the airfoil leading edge, and consequently it is responsible for the ATIN [4]. The ATIN is mostly noticeable at lower frequency because larger turbulent structures are the strongest in its noise generation mechanism [5].

The turbulence physics upstream of airfoil is an important feature in the ATIN study as it governs the noise generating mechanism. Thus, ATIN reduction should be correlated with the upstream turbulent characters. However, the correlation is not yet been discussed comprehensively I the open literature.

In most ATIN investigations, the upstream turbulence is generated and conveyed downstream to the airfoil by vortex generator so that the same condition of ATIN can be mimicked as in real condition[6, 7, 8]. Extensions of unsteady CFD techniques to the prediction of ATIN generated by high Reynolds number flows in complex geometries have first to be benchmarked on relevant test cases. Such a test case must be based on geometry that contains some of the aerodynamic mechanisms encountered in ATIN applications, but remain simple enough from the computational point of view in order to achieve parametric study purposes. The rod-airfoil configuration is a relevant benchmark case for the ATIN investigation. Jacob et al [9] was among the pioneer to introduce this rod-airfoil configuration. This is because at high Reynolds numbers, the rod sheds the well-known von Karman vortex street which acts as an oncoming disturbance onto the airfoil. Jacob et al. [9] highlighted three strong dimensional effects responsible for spectral broadening around the rod vortex shedding frequency in the subcritical regime, and identified that the airfoil leading edge was the main contributor to the noise emission in a rod-airfoil configuration due to vortex-structure interaction. Moreover, further understanding on the details of the rod-airfoil interactions have been gained through numerical simulations too. Previous studies [10,11,12,13,14,15] found good agreement between numerical calculation and experiment of ATIN. Hence, the current study aims to provide a systematic approach for grid convergence study of flow around a rod-airfoil using Grid Convergence Index (GCI) that is based on Richardson extrapolation.

2. Numerical model description

2.1 Test case and mesh description

The investigated case is a rod-airfoil configuration in a three-dimensional uniform incompressible flow at constant free stream velocity. The test case is as illustrated in Figure 1. A rod with diameter D with a downstream airfoil of chord length $9.5D$ immersed in a fluid of constant free stream velocity U_∞. The geometrical parameters and flow dynamic quantities are non-dimensionalized by D and U_∞ respectively. Gap distance between the rod and airfoil is set $3.5D$ as according to Yong Li et al. [16] the airfoil will experience fully developed vortex in the rod wake. The upstream, downstream, up and bottom boundary distance are also mentioned in the schematic diagram of computational domain. The span distances are $10D$ for the early stage grid refinement and $3.5D$ in the later.

The grid convergence performance was assessed by two stages. Table 1 presents the number of grid for different cases investigated in current study.

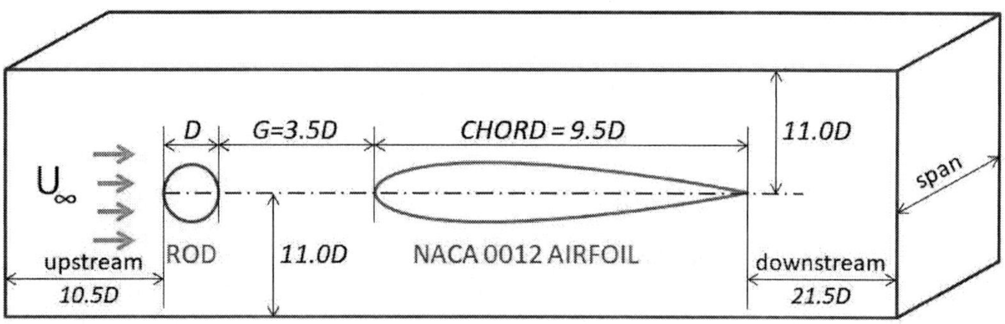

Figure 1. Model configuration in the simulation domain with the parameters. (Figure not in scale.)

Table 1. Number of grids of different cases.

Grid Refinement Stage	CASE	$N_x \times N_y \times N_z$
Stage 1 (span$10D$)	A	17,040
	B	180,680
Stage 2 (span $3.5D$)	C	1,042,720
	D	1,783,740
	E	3,153,200

2.2 Solution Methodology

The rod-airfoil study involves turbulence structure investigation because it was primarily aimed to investigate the noise emission. The most effective simulation model to incorporate fairly well with

IOP Conf. Series: Materials Science and Engineering **491** (2019) 012023 doi:10.1088/1757-899X/491/1/012023

turbulence is the Large Eddy Simulation (LES) if compared to the Reynolds Average Navier-Stokes (RANS). However, current study implemented Detached Eddy Simulation (DES) computations on the flow over the rod-airfoil due to the high demand of computational cost of the LES. All flow conditions and settings are summarized in Table 2.

Table 2. Flow conditions and setting.

Reynolds Number, Re $= \frac{U_\infty D}{\nu}$	20,000
Mach Number, Ma	0.044
Turbulent Model	Spallart-Allmaras DDES
Time step, Δt (s)	1.5 ~ 2.0$\times$ 10^{-4}

Computations in this work were performed using OpenFOAM. In particular, the merged piso-simple algorithm known as pimpleFoam solver is used. The convergence criterion for pressure and velocity solutions is set so the residuals fall below the tolerance of 10^{-9} and 10^{-8} respectively at each time step. 2nd order backward scheme is used for the temporal discretisation. The convection term is discretized using Gauss linearUpwind grad (U) and the viscous term is discretized using Gauss Gamma scheme. The time step is set accordingly as to keep the CFL value less than unity.

3. Grid Refinement Stage 1

Firstly at this stage, two cases are compared to obtain the grid refinement reliability. Two cases were prepared – one with the coarsest grids with smallest cell height as $\mathbf{0.2D}$ and the other is with smallest cell height approximately $\mathbf{0.05D}$. The visualisation of flow and the $\boldsymbol{y^+}$ values are assessed at this stage from each case.

3.1 Flow visualisation

Figure 2 depicts the grid distribution of the case with the $\boldsymbol{y^+}$ values of the geometry. The wall $\boldsymbol{y^+}$ is a non-dimensional wall distance often used in CFD defining the ratio between turbulent and laminar influences in a cell. Near-wall regions have bigger gradients in variables and in the flow physical momentum. The simulation adopts wall function in the calculation thus the $\boldsymbol{y^+}$ values are bound to meet in the range of $\mathbf{30 \le y^+ \le 60}$ [17]. Values of $\boldsymbol{y^+} \approx \mathbf{30}$ are most desirable for wall functions [18]. Hence, based on the $\boldsymbol{y^+}$ values comparison, the grids of case B is more preferable due to the rod's $\boldsymbol{y^+}$ value is nearer to 30.

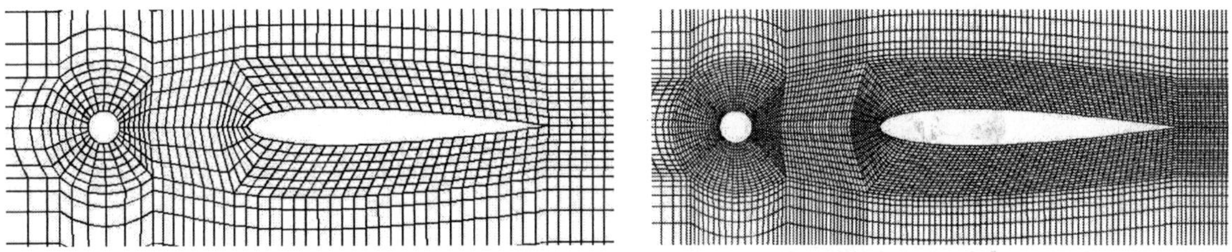

(a) Case A with $y^+_{rod} \approx 60$ and $y^+_{airfoil} \approx 38$ (b) Case B with $y^+_{rod} \approx 41$ and $y^+_{airfoil} \approx 32$.

Figure 2. Cell distribution around the rod-airfoil.

The flow visualization is then assessed in Figure 3. Flow separations are observed in both cases. However, the shear layers formation is not visible in case A. This instantaneous streamwise velocity results of case A shows not-properly formed vortex shedding if compared to that of Case B's. The alternation of the shear layers in the downstream of the rod is clear in Case B. This proves the fine grids reliability in Case B is able to reproduce the expected vortex shedding.

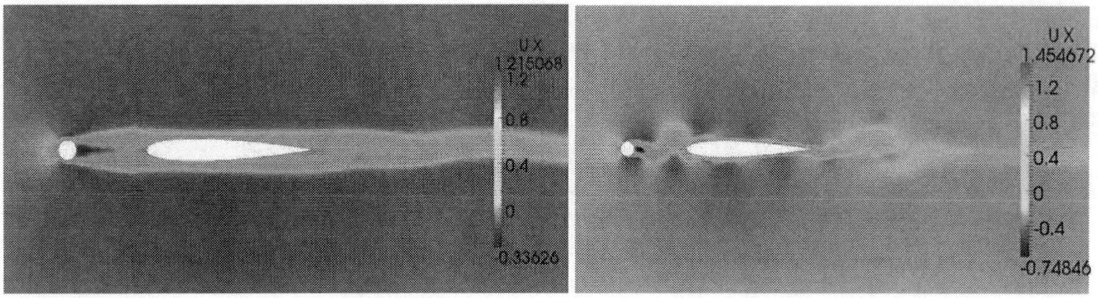

Figure 3. Instantaneous streamwise velocity contours at $z = 0$. LHS: Case A. RHS: Case B.

The flow visualization of the spanwise plane right upstream ($0.5D$ in front) from the airfoil is also compared as shown in Figure 4. The flow behaviour in spanwise direction is useful for inspection of spanwise correlation length.

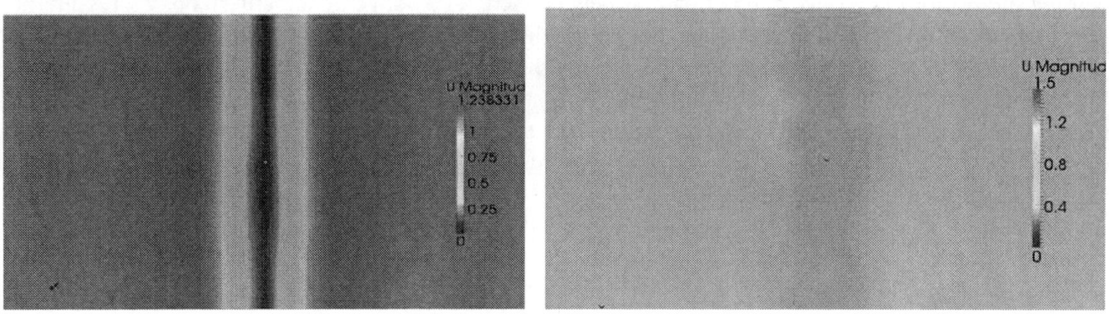

Figure 4. Instantaneous streamwise velocity contours crossectional at $= -0.5D$.
LHS: Case A. RHS: Case B.

4. Grid Refinement Stage 2

Three different grid resolutions were used in this stage. The coarsest is Case C, case D has medium grid and the finest is Case E. The upstream , top and bottom is $10D$ away from the boundary, while the outlet is $20D$ away from the boundary and the span length is $3.5D$.

4.1 Grid convergence study by Richardson extrapolation

Richardson extrapolation, introduced by Richardson [19] is also known as "the differed approach to the limit $(h \to 0)$". It defines a higher-order estimate of flow fields from a series of lower-order discrete values $(f_1, f_2, \dots, f_n)$. A convergence study needs three grid resolutions at minimum [20]. Roache [21] has generalized Richardson Extrapolation by introducing the p^{th}order methods:

$$f_{exact} \approx f_1 + |(f_1 - f_2)/|(r^p - 1)| \tag{1}$$

The grid refinement ratio r in this study is fixed, and defined as $r = \Delta_C/\Delta_D = \Delta_D/\Delta_E = 1.7$

From equation (1), the extrapolated value is varied by p^{th}order decision. As stated in Stern [19], the order of accuracy can be estimated by using the following equation:

$$p = \frac{ln(\varepsilon_{32}/\varepsilon_{21})}{ln(r)} \tag{2}$$

$$\varepsilon_{i+1,i} = f_{i+1} - f_i \tag{3}$$

To evaluate the extrapolated value from these solutions, the convergence of the system must be first determined. The possible convergence conditions are: (1) Monotonic convergence : $0 < R < 1$ (2) Oscillatory convergence : $R < 0$ and (3) Divergence : $R > 1$, where R is the convergence ratio and it is determined by the $R = \varepsilon_{32} / \varepsilon_{21}$.

Table 3 summarizes the order of accuracy for root mean square lift coefficient CL_{rms} and mean drag coefficient CD_{mean} from the simulation results of the three different grids. The convergence is monotonic for both variables assessed. Strouhal number St results are not included in this analysis due to the differences between all the cases are not significantly big.

Table 3. Order of accuracy Grid Convergence Index for the flow variables.

	ε_{32}	ε_{21}	R	p	GCI_{32} (%)	GCI_{21}(%)
CL_{rms}	0.1091	0.0076	0.0696	4.6763	4.71	0.34
CD_{mean}	0.1515	0.0015	0.0099	8.1008	0.25	0.0024

The Grid Convergence Index (GCI) defines a uniform measure of convergence for grid refinement studies as stated in Roache [21]. The GCI is derived from estimated fractional error obtained from the generalization of Richardson extrapolation. The GCI value represents the resolution level and how much the solution approaches the asymptotic value. The GCI for thefine grid resolutions can be calculated by the following:

$$GCI_{i+1,i} = F_s \frac{|\varepsilon_{i+1,i}|}{f_i(r^p-1)} \tag{4}$$

The safety factor F_s selected for the study is 1.25 as followed from Wilcox [22]. As observed from Table 3 previously, there is reduction in GCI values for the three successive grids ($GCI_{21} < GCI_{32}$). The GCI for finer grid GCI_{21} is relatively low compared to the GCI of the coarser one GCI_{32}. This implies that the dependency of the numerical simulation on the cell size has been reduced. In addition, the grid independent solution is acceptably achieved due to the GCI reduction from coarser grid to the finer grid. In other words, further refinement will not result in much change.

The variables obtained are compared with the extrapolated value using Equation (1). The comparisons are then plotted in Figure 4. The extrapolated value of CL_{rms} is just slightly lower than that of finer grid results (h/D=0.001D). Therefore, it is proved that the solution is converged within the refinement from coarser to the finer grid. Also, in this paper the discrepancy between the simulation value and this extrapolated value is defined as the error

$$E_i = \frac{f_i - f_{RE}}{f_{RE}} \tag{5}$$

Figure 5 shows that the successive grid refinement has nearly achieved the asymptotic value at the finest grid resolution where the relative error compared with the RE is only 0.002%, hence it is grid independent.

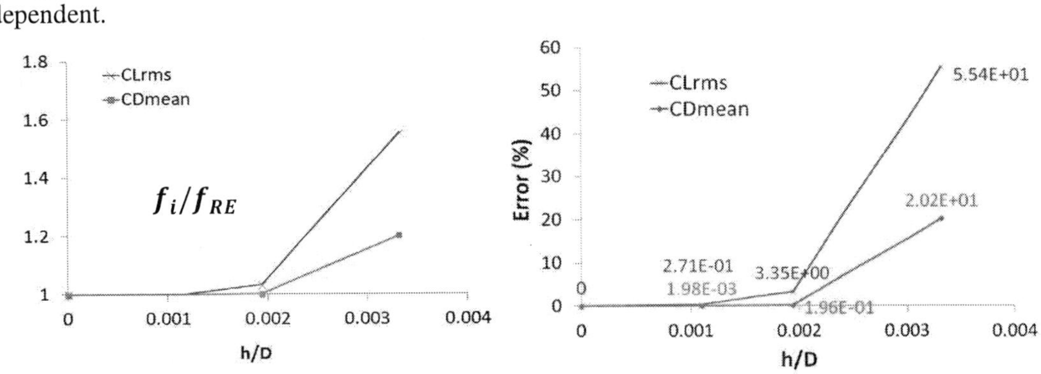

Figure 5. LHS: Comparison of global parameters normalized by the extrapolated value, among the three grid solutions. Extrapolated values are based on Richardson Extrapolation estimation. RHS: Error percentage for the global parameters of rod.

4.2 *Results comparison with previous studies*

Table 4 compares the results of current DES with previous studies of the same case. The Strouhal number is in excellent agreement with the literature. The root mean square lift coefficient and mean drag coefficient are slightly lower, but they are still in good agreement.

Table 4. Comparison of global flow dynamic quantities with previous literature.

Cases	Year	Reynolds number	Method	CL_{rms}	CD_{mean}	St
Cantwell & Coles	1983	140,000	Exp. (rod only)	0.3-0.4	1.0-1.35	0.19
Casalino et al.	2003	49,000	URANS	-	0.8	0.24
Greschner et al.	2008	48,000	k-ε DES	0.42	0.81	0.185
Giret et al.	2015	48,000	LES	0.41	0.92	0.2
Current study (case E)	**2018**	**20,000**	**DES**	**0.209**	**0.756**	**0.19**

5. Conclusion

Two stages of grid assessments were carried out in this paper to obtain the best grid suitable for a rod-airfoil simulation in a Detached Eddy Simulation (DES) model at high Reynolds number flow. The first stage has obtained the grid with better flow visualization that captures a well developed vortex shedding in the wake of rod and airfoil. However, this result is the early onset to a further grid refinement assessment in the next stage.

The next stage of grid refinement has provided good insights on the grid independency of especially the finest grid proposed. The GCI inspection of the flow variables showed that there was a gradual reduction when the grid system was refined. Also, the when the extrapolated value from Richardson Extrapolation are compared with current results, the results of finer grid (case E) showed good performance. Hence, the grid from case E is appropriate to be used further in the next analysis of rod airfoil due to the GCI values are all less than 0.5%.

Acknowledgments

The authors would like to wish and acknowledge University of Technology Malaysia for the grantment of Ainuddin Wahid Scholarship that has support this study financially.

References

[1] Amiet, RK. 1975 Acoustic radiation from an airfoil in a turbulent stream. *J. Sound and Vib.* **41** 407-420
[2] Arbey, H and Bataille, J. 1983 Noise generated by airfoil profiles placed in a uniform flow. *J. Fluid Mech.* **134** 33-47
[3] Paterson, RW. and Amiet, RK. 1976 Acoustic radiation and surface pressure characteristics of an airfoil due to incidence turbulence. NASA CR-2733
[4] Blake, W.K. 1986 Mechanics of Flow-induced Sound and Vibration. 2 *Vols. Academic Press.* New York
[5] Oerleman, S. and Migliore, P. 2004 Aeroacoustic Wind Tunnel Tests of Wind Turbine Airfoils. *National Aerospace Laboratory NLR,* Emmeloord, The Netherlands 319
[6] Jackson, R., Graham, JMR. and Maull. DJ. The lit on a wing in a turbulent flow. *Aeronautical Quarterly*
[7] Jurdic, V. 2005 Effect of an angle of attack and airfoil shape on turbulence-interaction noise. *AIAA/CEAS Aeroacoustic Conference Meeting and Exhibit*, 22-25 May
[8] Bachmann, T., Klan. S., Baumgartner, W., Klaas, M., Schroder, W. and Wagner, H. 2007 Morphometric characterization of wing feathers of the barn owl Tyto alba pratincola and pigeon Columbia livia, *Front Zoo* **4** 1-15.
[9] Jacob, MC., Boudet, J., Casalino, D., and Michard, M. 2005 A rod-airfol experiment as benchmark for broadband noise modelling. Theorotical and Computational Fluid Dynamics. **10** 171-196

[10] Casalino, D., JHacob, M. and Roger, M. 2003 Prediction of Rod-Airfoil Interaction Noise using Ffowccs-William Hawkings Analogy. *AIAA Journal.* **41** 2.

[11] Sorguven, E., Magagnato, F., and Gabi, M. 2003 Acoustic prediction of a cylinder and airfoil configuration at high Reynolds numbers with LES and FWH. *ERCOFTAC* Bulletic **58** 47-50.

[12] Boudet, J., Grosjan, N. and Jacob, M. 2005 Wake-airfoil interactions as broadbandnoise source: A large eddy simulation study. *International Journal of Aeroacoustic* 4 **1** 93-115.

[13] Magaganato, F., Sorguven, E. and Gabi, M. 2003 Far-field noise prediction by large eddy simulation and Ffowccs-Williams and Hawkings Analogy. *AIAA* 3206

[14] Grecshner, B., Thiele, F., Jacob, M. and Casalino, D. 2008 Prediction of sound generated by a rod- airfoil configuration using EASMDES and the generalized Lighthill/FWH Analogy. *Computers and Fluids* 37 **4** 402-413.

[15] Berland, J., Lafon, P., Crouzet, F., Daude, F. and Baily, C. 2010 Numerical insight into sound sources of a rod-airfoil flow configuration using direct noise calculation. *AIAA* 3705

[16] Yong Li, Wang, X., Chen, Z. and Li, Z., 2014 Experimental study of vortex-structure interaction noise radiated from rod-airfoil configurations. *Journal of Fluids and Structures.* **51** 313-325

[17] Fluent 6.2, 2006 Documentation File, ANSYS Manual

[18] Salim M. Salim and Cheah SC. March 18-20 2009 Wall y+ strategy for dealing with wall-bounded turbulent flows. *Proceedings of Int. Multiconference of Engineers and Computer Scientists.* **2**

[19] Richardson, LF and Gaunt JA, 1927. The deferred approach to the limit. Part I. Single lattice. Part II. Interpenetrating lattices, *Philosophical Transactions of the Royal Society of London. Series A. Containing Papers of A Mathematical or Physical Character.* **226** 299-361

[20] Stern, F., Wilson, RV, Coleman, WH. And Patterson EG , .2001. Comprehensive approach to verification and validation of CFD simulations Part I. Methodology and Procedures. *Journal of Fluids Engineering* 123 **4** 793-802

[21] Roache, PJ. 1994. Perspective: A method for uniform reporting of grid refinement studies. *Journal of Fluid Engineering.* 116 **3** 405-411

[22] Wilcox, DC. 2006. The effect of circular cylinders between crossflow. *Journal of Fluids and Structures.* 1 **2** 239-261

Seismic Simulation Test of a Fire Protection Riser Pipe with Fixed Groove Joints

S W Kim[1,*], B G Jeon[1], S W Ahn[2] and S W Wi[2]

[1] Seismic Research and Test Center, Pusan National University, 49 Busandaehak-ro, Mulgeum, Yangsan, Kyungnam 50612, Republic of Korea
[2] Korea Testing Certification, 27 Heungan-daero, Gunpo, Gyeonggi, 15809, Republic of Korea

*Corresponding author: swkim09@pusan.ac.kr

Abstract. In this study, seismic simulation tests were conducted on a stainless steel pipe with fixed groove joints for fire protection systems using cyclic loading. When a seismic load occurred, the deformation of the specimen and the response by the relative displacement between the elements were analyzed. To examine the behavior of the tee joints and elbow (vulnerable elements of the specimen) in the range higher than the ±1° minimum deformation angle in the event of an earthquake, the deformation angle was measured using an image processing method. The von Mises stress was also measured at the tee joint and elbow using a 3-axis strain gauge, and the result was compared with the allowable stress criterion to confirm the seismic performance.

1. Introduction

With the improvement of living standards and the continued industrial development, the importance of air-conditioning machinery, industrial machinery, and piping materials has been recognized. Pipes are used as applications in all industrial fields, such as machinery, electricity, electronics, fibers, refrigeration air conditioning, and plants, and they are also widely utilized in areas related to fire protection, chemistry, and safety, thereby increasing the importance of the piping area by day. Although pipes are connected to major structural sections, they do not support external loads and perform their own given functions. Thus, they can be classified as non-structural elements.

The trend of earthquake occurrence, among natural disasters, shows that earthquakes with magnitudes between 5.0 and 5.9 are increasing worldwide. Owing to this trend, studies for securing the seismic safety of non-structural elements are being conducted mainly in countries with frequent earthquakes [1,2]. Verification criteria for evaluating the mechanical performances of piping-related products, such as vibration, bending, and internal pressure, have been presented. Studies on seismic performance evaluation that consider the characteristics that may occur due to the behavioral difference between two supports in the event of a seismic load as well as the characteristics of pipes that exhibit the damage trend of low-frequency fatigue failure due to repeated displacement, however, are required. In the U.S., the National Fire Protection Association (NFPA) suggests the seismic design criteria of fire protection facilities, including piping [3].

In this study, a piping system with fixed groove joints was constructed by referring to NFPA13, and seismic simulation tests were conducted using cyclic loading. The behavior of the specimen due to the displacement-dominant behavior caused by the deformation of the structure or the relative displacement between the structural elements in the event of a seismic load was analyzed.

2. Test Conditions and Method

In this study, a riser pipe system was selected as the test facility, and a steel frame was designed and fabricated to implement side sway using the relative displacement between the structural elements. The test facility was a riser pipe system with two tee joints and one 90° elbow. It was constructed to have pipe connectors within 305 mm from the floor and 610 mm from the ceiling, in accordance with NFPA13.

To consider the actual operating conditions, the pipe was filled with water, and a 2 MPa internal pressure was applied using a booster. For the allowable side sway suggested by the International Building Code (IBC) as shown in Table 1, ten static cyclic loading tests were conducted [4]. h_{xx} represents the floor height of the corresponding floor.

Table 1. Allowable side sway of IBC

Risk category		
I or II	III	IV
0.020 h_{xx}	0.015 h_{xx}	0.010 h_{xx}

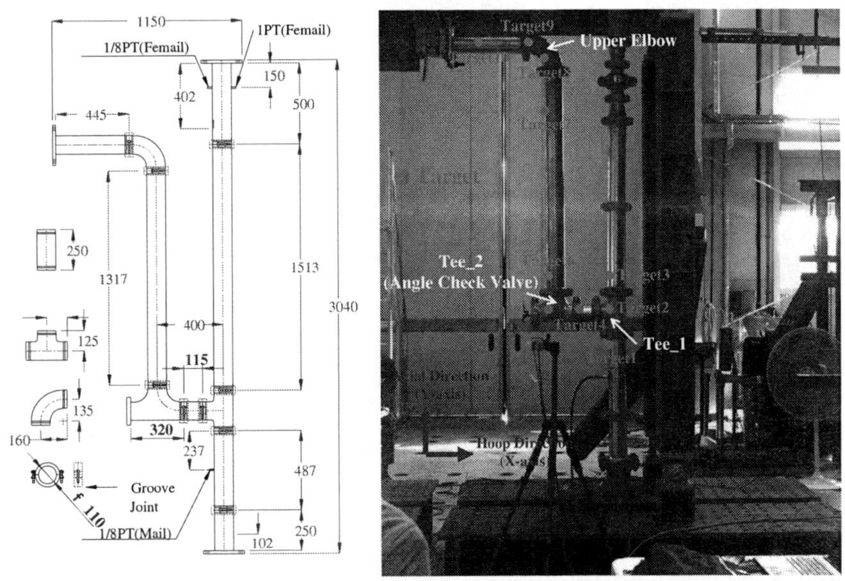

Figure 1. Riser pipe system with fixed groove joint

Figure 1 shows the riser pipe with fixed groove joints obtained using the image measurement system for measuring the deformation angle [5,6]. Table 2 shows the major components and subassembly lists of the specimen. The height of the specimen was 3 m (3.04 m including the plate for connecting the load cell).

Figure 2 shows the specimen installed for the side sway test of the riser pipe system as well as the installation locations of the sensors. As for the sensors, a 3-axis load cell (MC63-3A, Dacell) was fastened to all the ends of the riser pipe with bolts to measure the drawing force. Previous studies reported that the damage to piping caused by earthquakes is low-frequency fatigue failure, and that damage caused by displacement-dominant behavior may occur at the fittings, such as the tee joints and elbows, where the nonlinear behavior is concentrated [6]. Therefore, a 3-axis strain gauge (FRA-5-11-5L, Tokyo Sokki) was attached to the center of the tee_2 joint (3SG-1) and to the crown location of the upper elbow (3SG-2). For cyclic loading, two actuators were connected to the upper slab of the steel frame. In addition, a frame connected with hinges was used so that the load could be applied only to the pipe system when cyclic loading was performed using the actuators. The loading displacement and the load were measured using the load cells of the actuators and an LVDT (linear variable differential transformer; SDP-300D, Tokyo Sokki).

Table 2. Major components and subassembly lists

Part name	Specifications	Quantity
Pipe	KS D 3507 SPP, 100A 4.5T	-
Elbow	KS D 3507 SPP, 100A 4.5T, 90° elbow	1
Tee	KS D 3507 SPP, 100A 4.5T, 100A	1
Angle check valve	L-type	1
Hammerless valve	Hammerless valve	1
Joint	Groove joint (fixed), steel, manufacturer C	11

Figure 2. Side sway test of the riser pipe system

The fixed groove joints must endure an at least ±1° deformation in the event of an earthquake. Moreover, according to the results of previous studies, leakage occurs in tee joints with groove joints due to the gradually amplified cyclic loading at the lower than 1.15° deformation angle [7]. For this reason, for the seismic safety evaluation of pipe connectors, which are major piping elements vulnerable to earthquakes, it is necessary to consider the deformation angles of major piping elements. Therefore, the displacement responses were measured at each point of Figure 1 using the image processing method, and the deformation angles were measured using the measured displacement responses.

3. Test Results and Analysis

Ten cyclic loading tests were conducted for the allowable side sway (0.02 h_{xx}, ±60 mm) suggested by IBC. After the seismic simulation tests, test facility deformation, leakage, cracking, parts separation, and damage did not occur. In this study, the deformation angles were measured using the image processing method, and the displacements were measured at the tee joints and elbow, the riser pipe elements vulnerable to earthquakes. Figure 3 shows the deformation angles measured at the tee_2 and elbow points, and Table 3 shows the maximum and minimum deformation angles measured from the cyclic loading test. As can be seen from Table 3, the maximum deformation angle range of the specimen was 3.83° for the upper elbow, 1.38° for the tee_1 joint, and 1.68° for the tee_2 joint (angle check valve). Therefore, when a structure with a 3 m floor height is loaded with a 0.02 h_{xx} magnitude, the maximum deformation angle range applied to the elbow of the riser pipe is more than 1°, resulting in a higher than 1° deformation angle, the minimum deformation criterion of fixed groove joints, in the event of an earthquake.

Table 3. Deformation angle at vulnerable components of the riser pipe

Position	Target	Maximum deformation angle (Degree)		
		+	-	Max. range
Tee_1	1~4	0.55	-0.83	1.38
Tee_2 (angle check valve)	4~6	0.57	-1.11	1.68
Upper elbow	7~10	1.88	-1.95	3.83

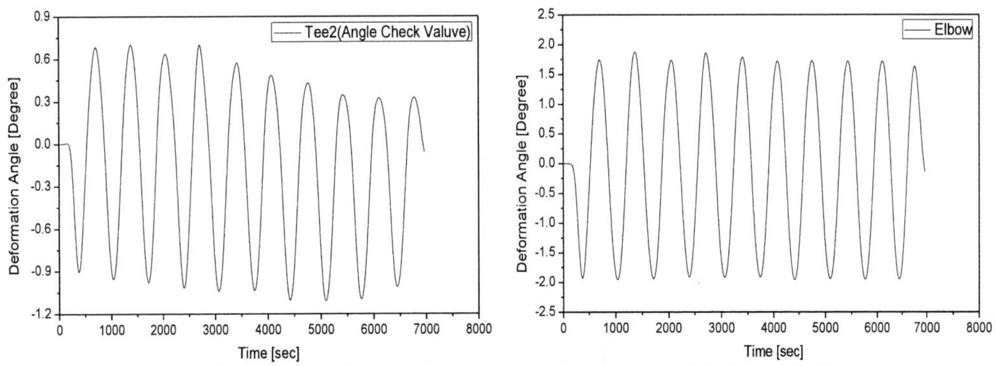

Figure 3. Deformation angles measured at the tee_2 joint and elbow

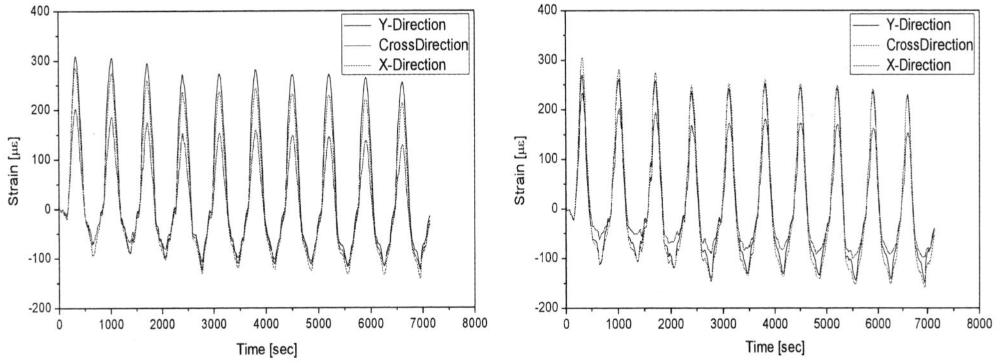

Figure 4. Strain responses measured at the tee_2 joint and elbow

Figure 4 shows the responses measured at the tee_2 joint and elbow of the specimen using the 3-axis strain gauges. Table 4 shows the measured maximum strain values and confirms that the strain values measured at the tee_2 joint and elbow were relatively small.

Table 4. Maximum strain responses

Location name	Orientation	Measured strain (μm/m)	
		Max.	Min.
SG3-1	X	309.0	-124.2
	45°	286.3	-142.2
	Y	201.9	-124.2
SG3-2	X	270.1	-147.9
	45°	306.2	-158.3
	Y	233.2	-100.5

The minimum tensile strength (0.2% offset) of SUS 304 stainless steel, the piping material used in this study, is 205 MPa, and its allowable stress is 137 MPa because it accounted for 2/3 of the tensile strength [8]. The stress criterion of the Class 1 piping of a nuclear power plant under the Level D service loading condition suggested by ASME Boiler & Pressure Vessel Code, SEC III, DIVISION I, Subsection NB, NB-3656 is less than a third of the allowable stress. The Level D service loading condition is a load combination that occurs with an extremely low probability, and includes seismic loads [9]. Therefore, if the von Mises stress measured from the specimen is lower than 410 MPa, the specimen is considered safe against earthquakes. Figure 5 shows the von Mises stress values measured at the tee_2 joint and upper elbow of the specimen using the 3-axis strain gauges. In Table 5, the measured von Mises stress is compared with the allowable stress, and based on this, it was found that the measured stress was lower than the allowable stress criterion. When construction was performed according to the method suggested by NFPA13, as shown in Figure 5 and Table 5, the strain and von Mises stress that occurred to the pipe with fixed groove joints were low, and the margin was high compared to the allowable stress criterion.

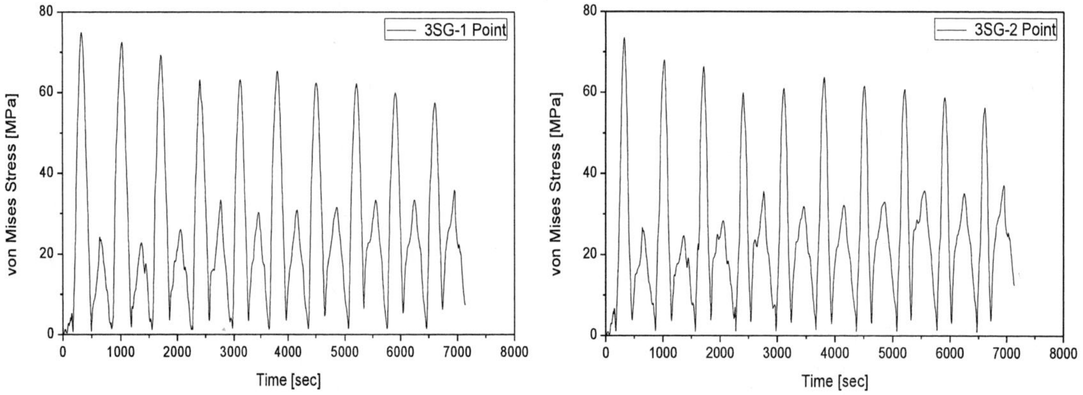

Figure 5. von Mises stress measured at the tee_2 joint and elbow

Table 5. Comparison of the measured and allowable stress values

Location	Measured max. stress (S, MPa)	3Sm (MPa)	S/3Sm
SG3-1	74.82	410	5.35
SG3-2	73.34	410	5.45

4. Conclusion

In this study, for the evaluation of the seismic performance of a riser pipe with fixed groove joints, cyclic loading tests were conducted using the pipe installation method suggested by NFPA13 and the allowable side sway of IBC. After the seismic simulation tests, test facility external deformation, leakage, cracking, parts separation, and damage were not observed through visual inspection, and infinitesimal deformation of the components was not found as well. Moreover, the von Mises stress values of the tee joints and elbow, the vulnerable elements of the riser pipe, were lower than the allowable stress, and it was confirmed that the margin of the riser pipe with fixed groove joints for the seismic performance was high.

Acknowledgments

This work was supported by the Technology Innovation Program (or Industrial Strategic Technology Development Program) (10078266, National Standard Technology Improvement program) funded By the Ministry of Trade, Industry & Energy (MOTIE, Korea). Moreover, the authors would like to thank the KOCED Seismic Research and Test Center for their assistance with the test equipment.

References

[1] Oh SH and Shin SH 2016 Correlation analysis of Gyeongju Earthquake waveform and structural damage scale *Journal of the Architectural Institute of Korea Structure & Construction* **32(12)** pp. 33-44

[2] Yoon BI 2018 Nonstructural elements resulting from Pohang Earthquake and direction of future seismic design nonstructural elements *Review of Architecture and Building Science* **62(4)** pp. 23-28

[3] NFPA 13 standard for the installation of sprinkler systems

[4] IBC 2012 International Building Code International Code Council

[5] Kim SW, Choi HS, Jeon BG, Hahm DG, Kim MG 2018 Strain and deformation angle for a steel pipe elbow using image measurement system under in-plane cyclic loading *Nuclear Engineering and Technology* **50(1)** pp. 183-195

[6] Kim SW, Choi HS, Park DU, Baek ER, Kim JM 2018 Water level response measurement in a steel cylindrical liquid storage tank using image filter processing under seismic excitation *Mechanical Systems and Signal Processing* **101** pp. 274-291

[7] Bursi OS, Reza MS, Abbiati G, Paolacci F 2015 Performance-based earthquake evaluation of a full-scale petrochemical piping system *Journal of Loss Prevention in the Process Industries* **33** pp. 10-22

[8] ASME Section II Part D

[9] ASME 2007 Boiler & Pressure Vessel Code, SEC III Division I Subsection NB NB-3656

IOP Conf. Series: Materials Science and Engineering **491** (2019) 012025 doi:10.1088/1757-899X/491/1/012025

Numerical Study on the Flow Characteristics of 2-D Free Jet at a Low Reynolds Number

B Shin[1,*] and S Tashiro[2]

[1] Department of Mechanical Design Systems Engineering, University of Miyazaki, Miyazaki 889-2192, Japan

[2] Graduate School of Engineering, University of Miyazaki, Miyazaki 889-2192, Japan

[*] Corresponding Author: brshin@cc.miyazaki-u.ac.jp

Abstract. Two-dimensional free jets have been numerically studied in the laminar flow regime. The numerical solution was obtained by solving the incompressible Navier-Stokes equations with SMAC finite-difference method. The variation of the axial velocity, potential core and jet spreading width were investigated to make clear the characteristics of free jet at a low Reynolds number. It was found that the fully developed region where the self-similarity is satisfied exists in the downstream of the collapsing point of the potential core. The variation of the core is depended on the Reynolds number. It showed that the half-width of geometry is useful in identifying the similarity. The predicted velocity distributions were in good agreement with those of experimental and theoretical results.

1. Introduction

Jet flow is one of the fundamental and important topics in fluid mechanics, and due to its geometric simplicity it has been studied in various fields such as fluids and combustion engineering, and can be seen in a variety of fields around us. In the jet flow, the boundary of the jet, spread, flow distance, shear structure, and heat transfer effects are sensitive to the structure of the initial stages of the jet. Especially in this regime, the intermixing of complex flow elements such as instability, coherent structure, and entrainment flow due to the development of free shear flow exist, therefore it is difficult to analyze. Many theoretical and experimental researches on free jet have been studied up to now since free jet flow is known to have a self-preservation and similarity [1]-[5]. In particular, theoretical analysis on the characteristics of laminar jet by Schlichting [1] has been well known but is based on boundary layer theory, and thus the velocity distribution at the outer edge of the jet flow do not match those from experimental results even in regimes where similarity laws are valid. In order to solve this problem, various theoretical studies have been conducted for decades [6]-[11]. Unfortunately, however, almost of them introduce the concept of boundary layer, there is naturally a limit to which it could be applied to various flow regions including the initial region of the jet flow. On the other hand, in recent years, due to the improvement in computational environment and rapid propagation of computational fluid dynamics (CFD), numerical analysis of complex flow phenomenon has been conducted. Research done with CFD is advantageous when analyzing jet flow accompanying the flow phenomena change due to flow conditions such as Reynolds number or the shape of the nozzle, but are mostly focused on turbulent flow at the high Reynolds number, and very few studies focus on the low Reynolds number regime.

In this study, a computational code based on SMAC finite difference method [12] was developed in order to investigate the jet flow characteristics at the low Reynolds number (*Re*) regimes from the initial jet flow region to the fully developed region, and the two-dimensional (2-D) incompressible free jet in the range of Re = 30 - 300 is numerically analyzed. Also, from the analysis, characteristics of free jet such as variation of the axial velocity, similarity, potential core and jet spread of the jet flow are clarified.

2. Numerical method

2.1. Fundamental equations and SMAC scheme

The fundamental equations for the numerical simulation are the continuity equation and the incompressible Navier-Stokes equations.

$$\nabla \cdot \boldsymbol{u} = 0 \tag{1}$$

$$\frac{\partial \boldsymbol{u}}{\Delta t} + \nabla \cdot \boldsymbol{uu} = -\nabla p + \frac{1}{Re} \nabla^2 \boldsymbol{u} \tag{2}$$

In this study, the fundamental equations (1) and (2) can be written by applying the Simplified MAC (SMAC) scheme as [13]

$$\frac{\boldsymbol{u}^* - \boldsymbol{u}^n}{\Delta t} = -\nabla p^n + (\frac{1}{Re} \nabla^2 \boldsymbol{u} - \nabla \cdot \boldsymbol{uu})^n \tag{3}$$

$$\frac{\boldsymbol{u}^{n+1} - \boldsymbol{u}^*}{\Delta t} = -\nabla \phi \tag{4}$$

$$\nabla^2 \phi = \frac{\nabla \cdot \boldsymbol{u}^*}{\Delta t} \tag{5}$$

$$p^{n+1} - p^n = \phi \tag{6}$$

where $\boldsymbol{u}$, and p denote the velocity and pressure, and ϕ represents the pressure increments, superscript n is the time level, asterisk * denotes the intermediate time level between n and n+1. In the computation of SMAC scheme, the dependent variables $\boldsymbol{u}$ and p are defined at different points of the staggered grid. The use of such a staggered grid is advantageous in that spurious errors are completely removed as in the MAC scheme. The Euler forward scheme was used to discretize the time derivative term, and for the spatial derivative, the 2nd order central difference method was basically applied. However, in the computation of high Reynolds numbers flow, a QUICK scheme [14] in the convection term was used in order to obtain accuracy and stability of the computation. Also, the boundary value problems in equation (5) were determined by using the SOR method with a maximum iteration number of 20.

2.2. Computational domain and boundary conditions

As can be seen from figure 1, the computational domain was set by 100d in the x direction, $\pm 30d$ in the y direction from the center of the nozzle with 2000×1200 square grids, where d represents the nozzle width. The boundary conditions were imposed by the initial velocity distribution at the nozzle exit, the Dirichlet conditions for pressure at the downstream, and the Neumann conditions for velocity and pressure in all other regions. Also, the Reynolds number was defined by the average velocity at the nozzle exit and the nozzle width d.

3. Numerical results

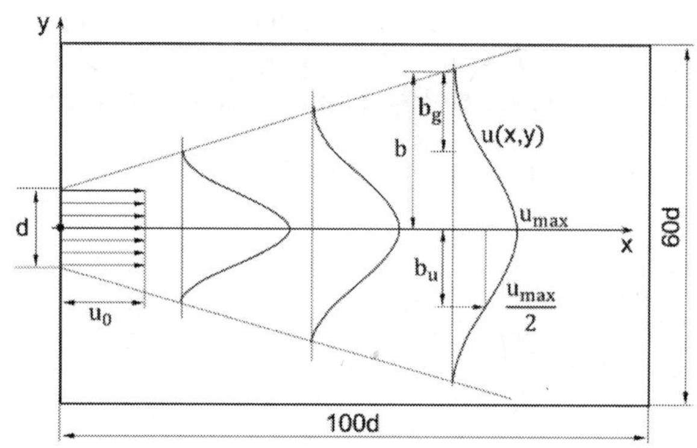

Figure 1. Jet flow geometry and computational domain.

3.1. Comparison with experimental and theoretical results

First, in order to check the accuracy and reliability of the present jet flow code by the SMAC scheme, predicted results were compared with exiting results. In this computation, iterative computations of the explicit time-marching method with Δt of 0.02 were conducted. The converged solution was obtained when the continuity conditions were lower than 1×10^{-5}. The time average values were taken from the solutions for 200 non-dimensional times.

Figure 2 shows time averaged axial velocity distributions at several downstream sections with non-dimensional distance y/d, when the velocity distribution (u_0) of the Poiseuille flow with $Re = 457$ is set as the initial condition at the nozzle exit for comparison with experiments. In this Reynolds number regime, the jet flow basically showed an asymmetric unsteady flow behaviour [15], but the time averaged velocities are in good agreement with experimental results [2] from the initial region of the jet to the fully developed regions, for instance, at $x/d = 3$ and 10.

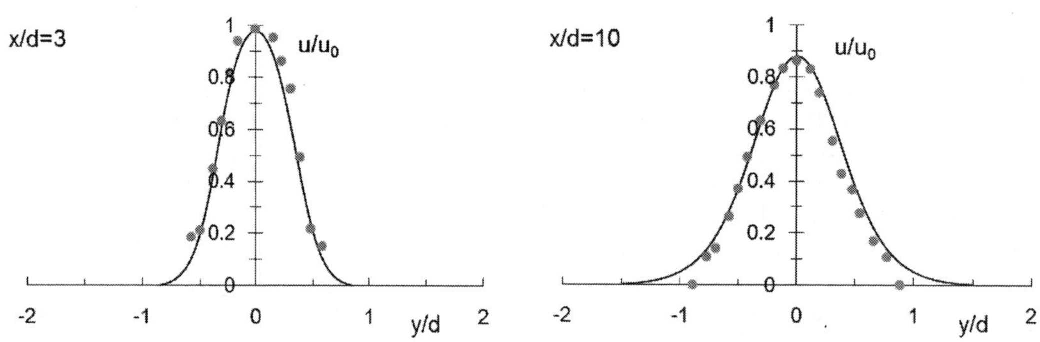

Figure 2. Velocity distribution of Poiseuille flow jet in two different sections of x/d at $Re = 457$.

Figure 3 shows the axial direction velocity distribution at $Re = 200$ with the uniform flow boundary condition at the nozzle exit. Here, axial velocity u is normalized by the maximum velocity u_{max} on the centre line at given x/d. In general the numerical results well predicted compared with the experimental results [7] as in the case of Poiseuille flow conditions, especially at regions extremely close to the nozzle exit. Figure 4 depicts a comparison of the present prediction to Schlichting's theoretical solutions [1]. These results were obtained from the computations with the initial conditions of the Poiseuille flow and the uniform flow at the nozzle exit. In this figure, y is normalized by the half width of velocity b_u which is the distance from the centre line to the point where u/u_{max}=0.5. At fully developed flow regions (for example x/d = 15), the present predictions are in good agreement to the theoretical solutions even if the inflow conditions are different from each other. There is a somewhat discrepancy near the outer edge, but it is due to the fact that at theoretical solutions based on the theory of boundary layer: the solution is determined by $u = 0$ at $y \to \infty$. As investigated so far, the present method well simulated the 2-D jet flow velocity filed from the nozzle exit to the fully developed region where the self-similarity is satisfied with high accuracy, resulting that the present numerical method is considered to be sufficiently reliable.

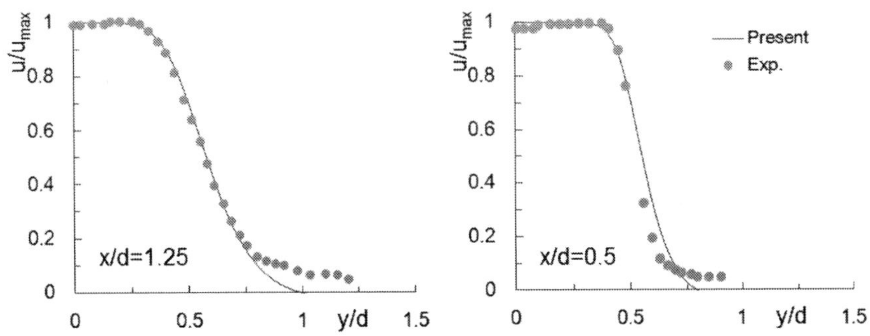

Figure 3. Velocity distribution of uniform flow jet in two different sections near the nozzle exit at Re = 200.

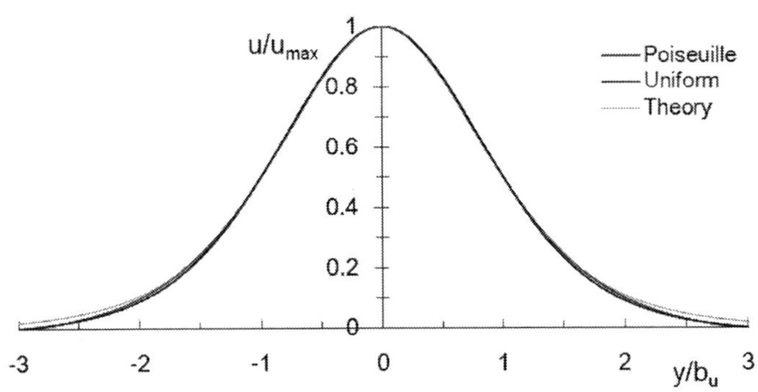

Figure 4. Velocity distribution of Poiseuille flow jet and uniform flow jet in the section of x/d = 15 at Re = 100.

3.2. Characteristics of velocity distribution

In order to investigate the jet velocity at an arbitrary cross section in the axial direction, figure 5 shows the axial velocity profiles along the y/b_u from the initial region to the developing region. This result was obtained from the uniform flow boundary condition with Re of 100. At initial regions close to the nozzle, the distribution is far from a Gaussian curve, shows large deviation from the similarity laws of velocity, but as the flow reaches downstream the jet flow develops, and approximately at regions after $x > 6d$, the similarity in velocity profiles begin to appear. Also, since these profiles are in good agreement with Sato and Sakao's experimental results [15] as well as with the analytical solution by Schlichting [1], it could be known that these distributions, just as results in figure 4, are nearly velocity profiles at fully developed flow regions which satisfy the self-similarity of jet.

On the other hand, there are several ways to estimate the characteristics of the jet depending on the choice of non-dimensional parameters [16]. Figure 6 indicates axial velocity profiles with respect to non-dimensional distance by half width of geometry b_g(=b/2, see figure 1) instead of b_u. As in figure 5, it could be seen that the jet will develop toward the downstream and the velocities satisfy the similarity in the fully developed regions. In figure 5, however, velocity curves by the half width of velocity are difficult to distinguish among them since the curves have two fixed points at y/b_u of 0.0 and 1.0, but for ones with half width of geometry in figure 6, the differences in velocity distribution in the developing and developed regions are easy to know. Therefore, the evaluation of velocity distribution using the half width of geometry is useful for observing jet flow development.

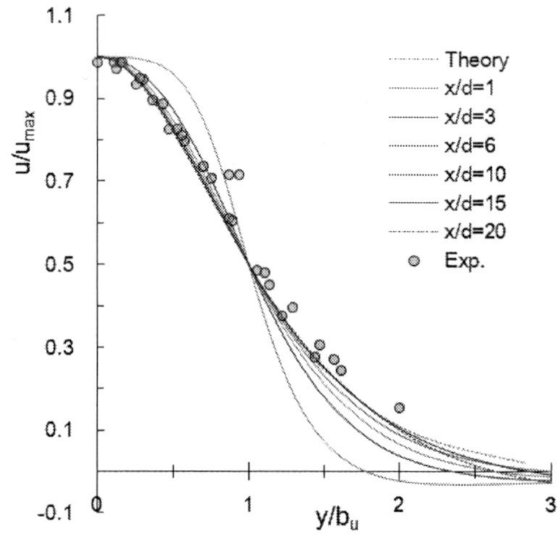

Figure 5. Axial velocity profiles with respect to the half-width of velocity in several downstream sections of *x*/*d* at *Re* = 100.

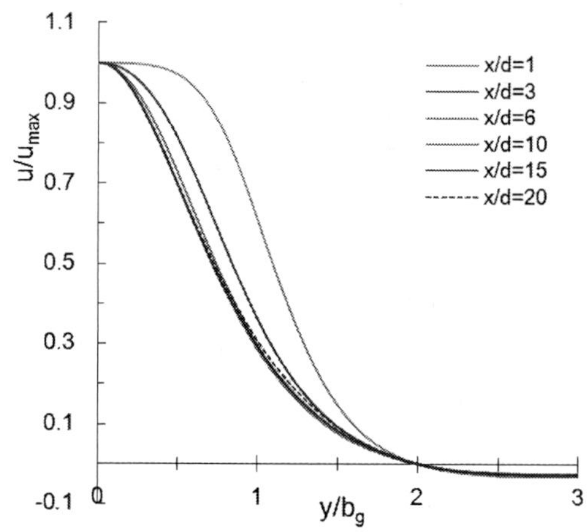

Figure 6. Axial velocity profiles with respect to the half-width of geometry in several downstream sections of *x*/*d* at *Re* = 100.

Figure 7 shows the change in velocity distribution with respect to non-dimensional distance [6] using half width of velocity and distance (x) in the axial direction. The variation of velocity due to this non-dimensional parameter shows the distinction of boundary between initial regions and regions satisfying similarity more vividly compared to figures 5 and 6. Figure 8 depicts the variation of velocity with respect to the non-dimensional parameter $(y - b_g)/x$, which uses the half width of geometry to imitate figure 7. As discussed in figure 6, the evaluation using the half width of geometry shows the transition of the velocity change clearer than that of the half width of velocity, and even at developed regions with $x/d \geq 6$, shows less overlap than compared to figures 6 and 7. From this, it could be inferred that at Re = 100, the fully developed region which satisfy similarity locates on the downstream after following transient process after breakdown of the potential core (refer to figure 9).

3.3. Potential core

Figure 9 plots the distance to the position where the potential core collapses. Here, the potential core length, x_p, was defined as the distance from the nozzle exit to the point where $u_c \approx 0.95u_0$. u_c and u_0 represent axial velocities at the jet centre and the nozzle exit, respectively. When $Re \leq 300$, x_p extends toward downstream with almost linear gradient of $x_p/d \approx 0.04Re$, but as a result of additional investigation for regions of $Re > 300$, the gradient was decreased to around $14 \leq x_p/d \leq 15$. These results are somewhat different from the case of turbulence flow known as $x_p/d = 12$ [17]. Figure 10 illustrates the change in the width of potential core (y_p/y_0) along the dimensionless distance $\xi(= (x/d)/Re)$ in the axial direction [8],[18] with the half width of nozzle y_0. It could be seen that the core region decreases and eventually disappears as x moves downstream from the nozzle exit. In the range of $100 \leq Re \leq 300$, the relationship between change in y_p/y_0 and ξ is in approximately similarity, and as seen in figure 9, the distance when the potential core breaks down is around $\xi = 0.04$.

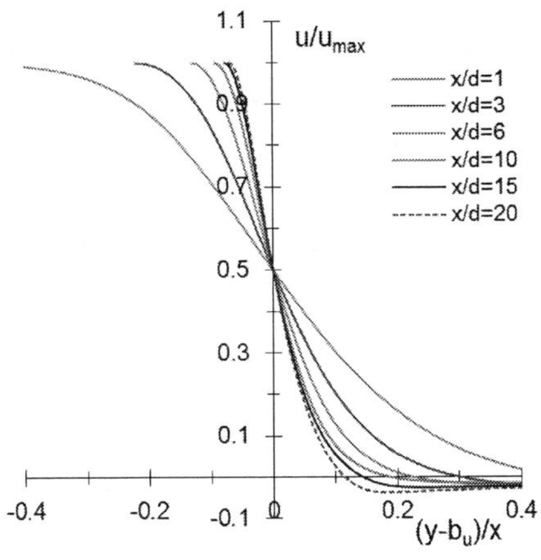

Figure 7. Axial velocity profiles with respect to the $(y - b_u)/x$ in several downstream sections of *x*/*d* at *Re*=100.

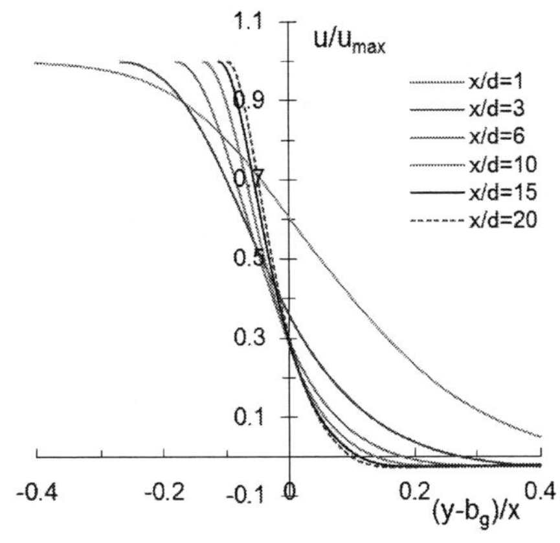

Figure 8. Axial velocity profiles with respect to the $(y - b_g)/x$ in several downstream sections of *x*/*d* at *Re*=100.

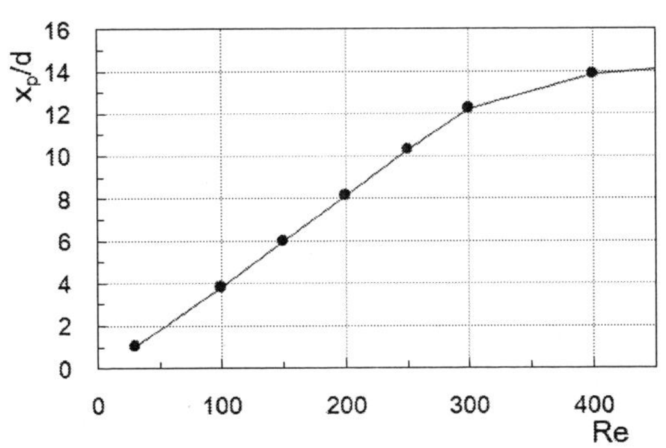

Figure 9. Potential core length with Reynolds number.

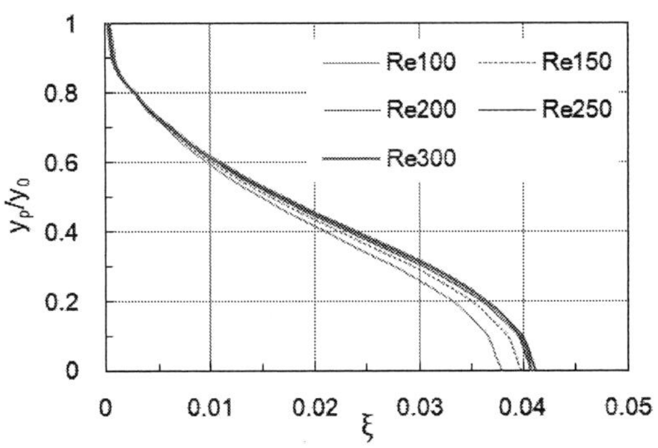

Figure 10. Variation of potential core width with $(x/d)/Re$.

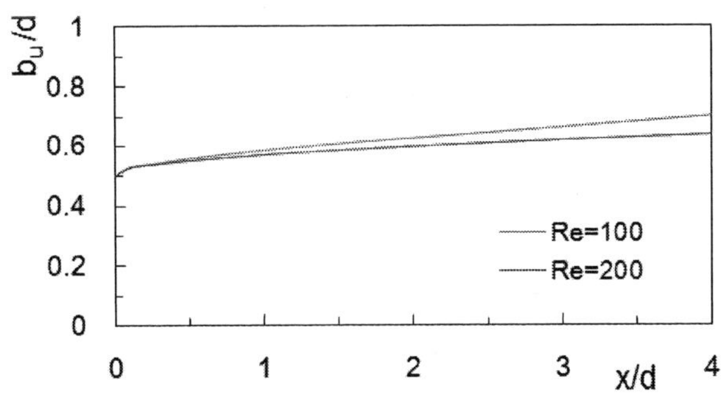

Figure 11. Variation of jet spread with half-width of velocity along the jet axis at Reynolds numbers of 100 and 200.

3.4. Jet spread

Figure 11 depicts the variation of jet spread. Jet spread is one of the most important fundamental characteristics in the analysis of jet flow characteristics. In this study, therefore, the jet width and its change over the axial direction were investigated by using the half width of velocity. The jet width spreads as the flow flows downstream from the nozzle exit, and in contrast to figure 10, as the Reynolds number increases the spread gradient tends to decrease. Also, it is confirmed that the tendency of jet spread obtained from this simulation is almost same as the experimental investigation [7].

4. Conclusions

In order to investigate the jet flow characteristics at the low Reynolds number regimes, two-dimensional incompressible free jets were computed by using SMAC finite difference method. By comparing the numerical results with experimental and theoretical results, it was confirmed that the present numerical method is sufficiently reliable to predict the jet flow with accuracy. From the investigation of the axial velocity distribution, similarity, potential core and jet spread of free jet, it was clarified the 2-D jet flow characteristics in the wide range of jet from the initial jet flow region to the fully developed region, which was difficult with the conventional analytical method. Also, it showed that the evaluation of velocity distribution by the half width of geometry is useful for observing jet flow development.

Acknowledgments
This work was partly supported by JSPS KAKENHI Grant Number JP17K06164.

References

[1] Schlichting H 1979 *Boundary Layer Theory* (New York: McGraw-Hill)
[2] Eugene P S and Thomas L L 1971 *Experimental Investigation of an Axisymmetric Fully Developed Laminar Free Jet* (NASA TN D-6304)
[3] Thomas L L and Eugene P S 1972 *Experimental Investigation of an Axisymmetric Free Jet with an Initially Uniform Velocity Profile* (NASA TN D-6783)
[4] Deo R C, Mi J and Nathan G J 2008 *Phys. Fluids* **20-**075108 p 075108-01
[5] Grandchamp X, Fujiso Y, Wu B and Hirtum A Van 2012 Steady *ASME J. Fluids Eng.* **134** p 011203
[6] Wygnanski I and Fiedler H 1969 *J. Fluid Mech.* **38** pp 577-612
[7] Greene G C 1974 *Comparison of measured and calculated velocity profiles of a laminar incompressible free jet at low Reynolds numbers* (NASA TN D-7510)
[8] Hatta K and Nozaki T 1975 Two-Dimensional and Axisymmetric Jet Flows with Finite Initial Cross Sections *Bulletin JSME* **18** pp 349-357
[9] Akaike S and Nemoto M 1988 *ASME J. Fluids Eng.* **110** pp 392-398
[10] Lee D D, Kihm K D and Chung S H 1997 *ASME J. Fluids Eng.* **119** pp 716-718
[11] Pani B S, Lee Joseph H W and Lai Adrian C H 2009 *J. Hydro-environment Research* **3** pp 121-128
[12] Amsden A A and Hallow F H 1970 *J. Comput. Phys.* **6** pp 322-325
[13] Shin B R and Ikohagi T 1991 *Computers & Fluids* **19** pp 479-488
[14] Leonard B P 1979 *Computer Methods in Appl. Mech. and Eng.* **19** pp 59-98
[15] Sato H and Sakao F 1964 *J. Fluid Mech.* **20** pp 337-352
[16] Abramovich G N 1963 *The Theory of Turbulent Jets* (Cambridge: MIT Press)
[17] Rajaratnam N 1976 *Turbulent Jets* (Amsterdam: Elsevier Sci. Pub. Co)
[18] Rankin G W, Sridhar K, Arulraja M and Kumar K R 1983 *J. Fluid Mech.* **133** pp 217-231

IOP Conf. Series: Materials Science and Engineering **491** (2019) 012026 doi:10.1088/1757-899X/491/1/012026

Investigation on the Effect of Process Parameters on Impact Strength of Fused Deposition Modelling Specimens

PL. Ramkumar

Mechanical department, Institute of Infrastructure, Technology, Research and Management (IITRAM), Ahmedabad, Gujarat, India – 380026

plramkumarno1@gmail.com

Abstract. Fused deposition modelling is one of the most prevalent rapid prototyping technology which is used to construct working parts of difficult geometry. With more development in the technology, it is now widely used for manufacturing quality parts. To get more realistic data from prototype testing, it is essential that the prototypes manufactured from FDM have properties similar to production quality parts. There is thus a need to study the mechanical properties of parts fabricated from FDM and how different process parameters affect these properties. The objective of this research is to focus on a comparative study on the effect of process parameters like infill density, layer thickness and β-angle on impact strength of FDM Specimens printed in honeycomb and rectilinear infill patterns with acrylonitrile butadiene styrene (ABS) as print material. From the experiments we could conclude that increase in infill density and layer thickness results in higher impact strength. Also, honeycomb pattern exhibit better impact strength than rectilinear pattern.

1. Introduction

Rapid Prototyping has become the focus of today's manufacturing industry, enabling engineers and designer to quickly manufacture prototypes of parts or assemblies using the CAD files. The main advantage of rapid prototyping is as it follows additive manufacturing method unlike the traditional subtractive manufacturing methods like machining [1]. Fused Deposition Modelling, more commonly known as 3D printing is the most common technique employed among all the Rapid Prototyping technologies available [2]. In FDM process material (generally thermoplastic) is selectively dispensed through a heated nozzle, whose motion is controlled in 3D space using CNC technology [3-4]. FDM technology, developed by S. Scott Crump in late 1980s, was initially used for modelling and prototyping parts. With more development in the technology, it is now widely used for manufacturing production-quality parts. In the present economy, most products have short life cycles. FDM thus is a boon for engineers in quickly fabricating prototypes, which enables them to test and reiterate on their design, saving valuable time and money [5]. Process parameters in FDM process like infill density, raster width, pattern of printing, raster angle etc. greatly influences the quality of the constructed parts [6-8].

Recently, mechanical properties of parts produced using various rapid prototyping technology has been investigated and it is found that tensile strength of the specimen loaded parallel to build direction exhibited higher strength than the specimen loaded in perpendicular direction [9]. The variation of tensile strength and young's modulus with respect to change in process parameters like raster angle, raster width and air gap on tensile strength and young's modulus has been investigated, it is found that optimum laser width along with raster orientation results in higher yield stress [10]. The effect of different process parameters like layer thickness, feed rate and build direction on tensile and bending strength of PLA samples printed using FDM technology has been analysed [11]. The author observed an increase in ductility as feed rate and layer thickness is increased. The most contentious parameter in FDM process is layer thickness. As many investigation has been performed on the knowing the effect of layer thickness still there is an inconsistencies in result obtained by different researcher [12].

In this paper an effort has been made to study the effect of three process parameters like Infill Density, Layer Thickness and β-angle (explained in section 2) on Impact strength of FDM specimens. Also, a comparative study has been performed to know the effect these process parameters on impact strength for two different infill patterns (i.e., Honeycomb and Rectilinear pattern)

2. Experimental Hypothesis

In this paper a comparative study between honeycomb and rectilinear infill patterns has been performed and also the effect of different process parameters on specimens printed with above mentioned infill patterns are discussed. The process parameter studied in this paper are infill density, layer thickness and build direction. Infill density corresponds to the air gaps in the parts printed. It is expressed in percentage of ratio of the weight of the part printed to weight of completely solid part of equal volume. Since FDM is an additive manufacturing (AM) technology where parts are built layer by layer, layer thickness influences the overall properties of the part. Layer thickness determines the degree of bonding between adjacent layers, and thus strength of the part in the build direction. Build direction is the direction of layer build-up during the printing process. The β-angle defined in this paper is the angle between build direction and direction of impact loading. As the orientation of the part can be varied, different β-angle values are possible.

It is expected that with increasing infill density, the impact strength will increase as more rasters have to be broken. It is also expected that increasing the layer height would decrease the impact strength as the effective bonding between layers will reduce. Rectilinear is the default infill pattern employed in most 3D printer interfacing software's. Honeycomb sandwich structures are known to have better strength to weight ratio when loaded with compressive load in the direction normal to hexagons.

Thus, an impact load in the same direction should yield higher strength. Detailed description of design of experiment and the levels of parameters used, is provided in section 3.

3. Methodology

The material used in all the specimens in this study was ABS (Acrylonitrile butadiene styrene). The ABS offers better chemical resistance and heat stability. All the specimens were built using Mendel Prusa i3 RepRap 3D printer. The CAD model of the specimen were designed using PTC Creo© software. The software used to interface with 3D printer Hardware Repetier Host. Slic3r© program was used to slice the CAD in into layers and G-code path for the extruder to follow.

The ASTM D256-10 (Standard methods for determining the IZOD pendulum impact resistance of plastics) was used for testing of all specimens. In this method the specimen is held vertical and is broken by a single swing of the pendulum. Irrespective of the infill pattern, the specimen printed have the first and the last three layers of solid. Also, the specimen has a three-contour perimeter for all the layers.

A level two full factorial experiment design was used where all combination of levels of all the parameters are run. Thus, the total number of combinations possible is 2^k. The number of parameters is three and two levels for each parameter were selected as shown in table 1, hence the total number of different specimens becomes 2^3 i.e. 8. Two samples per combination were printed, hence a total of 32 specimens were printed and tested with 16 of each infill patterns. As a 100% infill density specimen could closely represent a solid isotropic ABS specimen, two more specimens with 100% infill density and β-angles of 0° and 90° were tested. This was done to determine the difference in impact strength of FDM ABS specimen and Solid Isotropic.

Table 1: Process parameters and their levels.

Parameters	Level 0	Level 1
Layer thickness (mm)	0.25	0.35
Infill density (%)	40	60
Beta (degrees)	0	90

The parameters, other than the ones selected in this study, were kept constant.
Raster width: 0.4 mm
Extrusion Temperature: 230℃
Bed Temperature: 105℃
Raster orientation: [0, 90] for Rectilinear infill pattern and [0, 60,120] for Honeycomb infill pattern.

4. Statistical Analysis

Table 2 and table 3 show the impact resistance and impact strength values from the experiments as per the design scheme considered for both honeycomb and rectilinear pattern. Statistical analysis using ANOVA has been performed to know the factors which are significantly effecting the response as presented in the table 4. 5% level of significance ($p=0.05$) has been considered in this analysis. From the analysis we notice that beta angle is significantly effecting the response rather than the other two parameters.

Table 2: Impact resistance and impact strength value for honeycomb pattern.

Honeycomb Pattern				
Layer height (mm)	In fill density (%)	Beta angle in (degree)	Impact Resistance (J/m)	Impact Strength (kJ/m^2)
0.250	40	0	126.13	18.01
0.350	40	0	119.04	17.00
0.250	60	0	139.96	19.99
0.350	60	0	150.98	21.56
0.250	40	90	67.71	9.67
0.350	40	90	80.90	11.55
0.250	60	90	77.46	11.06
0.350	60	90	84.25	12.03

Table 3: Impact resistance and impact strength value for rectilinear pattern.

Rectilinear Pattern				
Layer height (mm)	In fill density (%)	Beta angle in (degree)	Impact Resistance (J/m)	Impact Strength (kJ/m^2)
0.250	40	0	127.85	18.26
0.350	40	0	108.16	15.45
0.250	60	0	98.22	14.03
0.350	60	0	136.81	19.54
0.250	40	90	61.02	8.71
0.350	40	90	70.76	10.11
0.250	60	90	64.07	9.15
0.350	60	90	70.76	10.11

Table 4: Results of ANOVA for honeycomb & rectilinear pattern

Sourse	DF	Seq SS	Adj SS	Adj MS	F	P
Layer Height	1	71.5	71.5	71.5	1.11	0.352
Infill Density	1	433.2	433.2	433.2	6.72	0.061
Beta Angle	1	6372.6	6372.6	6372.6	98.86	0.003
Error	4	257.8	257.8	64.5		
Total	7	7135.2				

5. Results and discussion:

From the ANOVA analysis as shown in the table 4, we notice that beta angle is significantly effecting the response rather than the other two parameters. But still there will be some variation in the impact resistance of printed material due to infill density and layer thickness. These two parameters can be predominantly neglected. The effect of process parameters on the impact resistance of both honeycomb and rectilinear print patterns are as shown in figure 1. If we observe the chart for rectilinear and honeycomb pattern, it can be clearly noted that honeycomb pattern specimen has predominately good impact resistance and impact strength compared to rectilinear pattern. The reason behind this is as the probocation of crack in case of honey comb pattern is quite difficult compared to rectilinear pattern. As this pattern has interlocking layers in the form of honeycomb, the crack cannot easily penetrate.

It is also seen that with increasing infill density and layer height, the net impact energy of the specimen increases for both the infill patterns as raster packing is denser, thus more rasters are to be broken. Whereas, the net impact energy drops significantly when the β-angle is changed from 0 to 90 degrees. The specific reason behind the decrease in impact energy is the orientation of build layer of the specimen. An inter-layer shear force component comes into play, offering more resistance to breaking by impact. As the impact blow is perpendicular to the direction of build layer, the layer is not possessing sufficient strength to with stand the blow. Therefore, the impact energy in parallel direction (0 degree) is better than the specimens printed in the perpendicular direction (90 degree) when both the patterns are compared, we can see that the honeycomb patterns have better impact energy than rectilinear patterns.

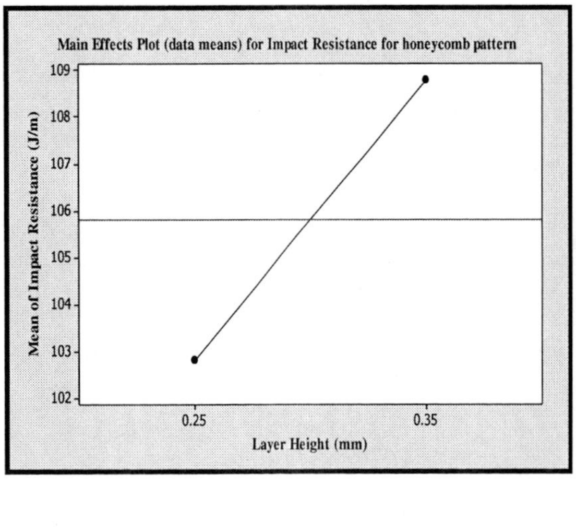

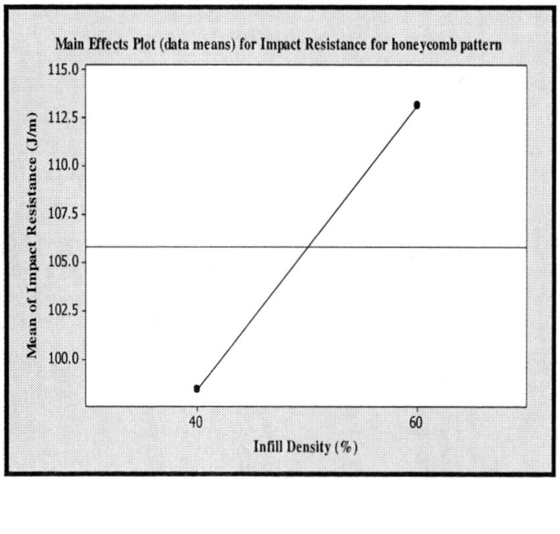

IOP Conf. Series: Materials Science and Engineering **491** (2019) 012026 doi:10.1088/1757-899X/491/1/012026

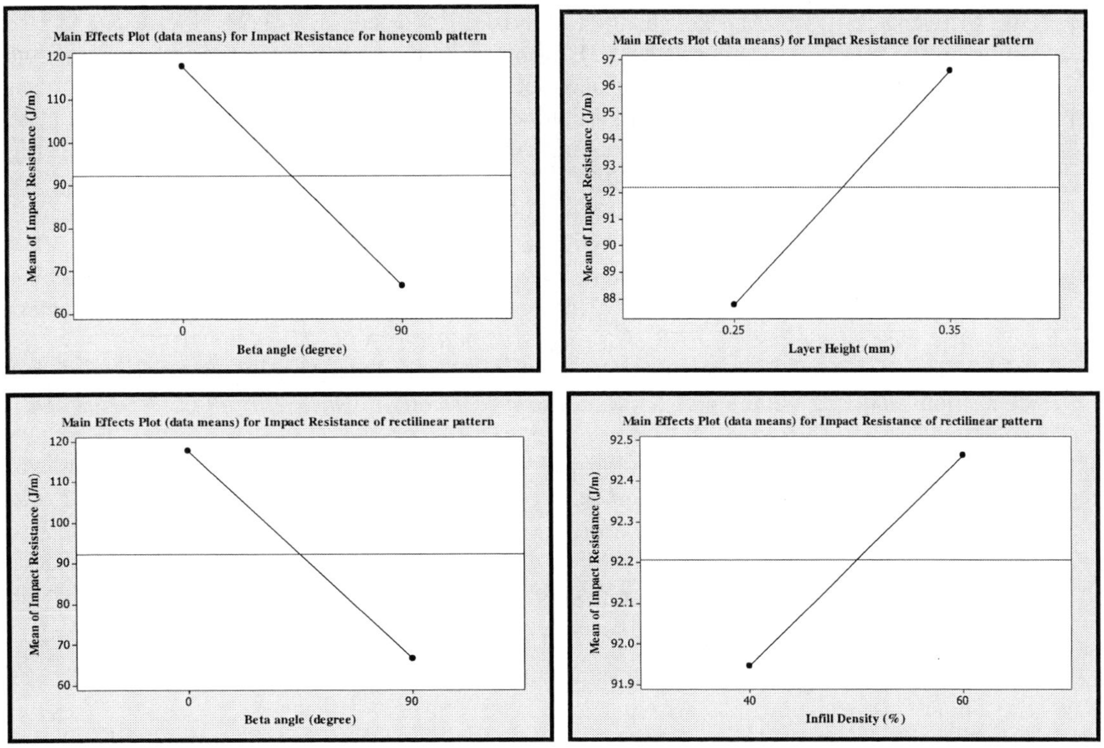

Figure 1. Effect of process parameters on impact resistance of both honeycomb and rectilinear pattern

6. Conclusion

In this work, the effect of process parameter on impact resistance and impact strength of the FDM specimen printed with both Honeycomb Infill pattern and Rectilinear Infill pattern has been compared and presented. The material used throughout the study was ABS. All the specimens were built using in-house built DIY Rep Rap 3D printer. Flexural testing was performed as per the ASTM standard D 790. From the investigation we could observe that increasing infill density and layer thickness increases the impact strength of specimen as raster packing is denser, thus more rasters are to be broken. When direction of impact loading and build direction coincide (β-angle = 0^0), impact energy increase. Honeycomb infill pattern exhibits more impact strength than rectilinear, this is mainly because of the crisscross nature of the pattern due to raster orientation of [0°, 60°, 120°].

References

[1] Kay RN 2014 Effect of Raster Orientation on the Structural Properties of Components Fabricated by Fused Deposition Modeling

[2] Mohamed OA, Masood SH, Bhowmik JL, Nikzad M and Azadmanjiri J 2016 Effect of process parameters on dynamic mechanical performance of FDM PC / ABS printed parts through design of experiment. *J Mater Eng Perform*; 2922–35

[3] Tingting H, Shanggang W and Ketai H 2015 Quality control for fused deposition modeling based additive manufacturing: Current research and future trends *1st Int. Conf. on reliability Sys. Engg Ieee xplore* (Beiging) p 1-6

[4] Kim C, Espalin D and Cuaron A 2015 A study to detect a material deposition status in fused deposition modeling technology *Ieee Int. Conf. on Advanced Intelligent Mechatronics* (South Korea) p 779-83

[5] Sood AK, Ohdar RK and Mahapatra SS 2009 Parametric appraisal of mechanical property of fused deposition modelling parts *Materials and Design* (India) pp 287-95

[6] Wu W, Geng P, Li G, Zhao D, Zhang H, Zhao J 2015 Influence of layer thickness and raster angle on the mechanical properties of 3D-printed PEEK and a comparative mechanical study between PEEK and ABS, *Materials* pp 5834–5846.

[7] Ning F, Cong W, Hu Y, Wang H 2016 Additive manufacturing of carbon fiberreinforced plastic composites using fused deposition modeling: effects of process parameters on tensile properties, *Journal of Composites Materials* pp1–12.

[8] Vaezi M, Chua CK 2011Effects of layer thickness and binder saturation level parameters on 3D printing process, International journal Advanced Manufacturing Technology pp 275–284.

[9] Kim GD and Oh YT 2008 A benchmark study on rapid prototyping processes and machines: quantitative comparisons of mechanical properties, accuracy, roughness, speed, and material cost *Proc. Inst. Mech Engg. Part B: J of Engg. Manufacture* Vol 222(2), pp 201-15

[10] Fodran E, Koch M and Menon U 1996 Mechanical and dimensional characteristics of fused deposition modeling build styles *Solid freeform fabrication Symp, University of Texas, Austin* pp 419-42

[11] Chacóna JM, Caminero MA, García-Plaza E, Núñez PJ 2017 Additive manufacturing of PLA structures using fused deposition modelling: Effect of process parameters on mechanical properties and their optimal selection *Materials and Design* p 143-157.

[12] Rankouhi B, Javadpour S, Delfanian F, Letcher T 2016 Failure analysis and mechanical characterization of 3D printed ABS respect to later thickness and orientation, Journal of Failure Analalysis and Prevention pp 467–481.

IOP Conf. Series: Materials Science and Engineering **491** (2019) 012027 doi:10.1088/1757-899X/491/1/012027

Perforation Tests of Aluminum Alloy Specimens for a Wide Range of Temperatures Using High-Performance Thermal Chamber - Experimental and Numerical Analysis

A. Bendarma [1, 2, *], T. Jankowiak [1], A. Rusinek [3] and T. Lodygowski [1] and M. Klosak [2]

[1]Poznan University of Technology, Institute of Structural Engineering, Piotrowo 5, 60-965 Poznan, Poland
[2]Universiapolis, Ecole Polytechnique d'Agadir Bab Al Madina, Qr Tilila, Agadir, Morocco
[3]National Engineering School of Metz, Laboratory of Mechanics, Biomechanics, Polymers and Structures, 1 route d'Ars Laquenexy, 57000 Metz, France

[*]E-mail address: b.amine@e-polytechnique.ma, Tel: +212-6-61-09-88-72

Abstract. The paper describes a work focused on the process of perforation of aluminum alloy sheet at a wide range of specimen temperatures. This has been obtained by using the specially designed thermal chamber to heat specimens before impact. Based on this experimental series. Experimental, analytical and numerical investigations have been carried out analyze the perforation process, the ballistic properties of the aluminum alloy impacted by a conical nose shape projectile are studied. The experimental investigations have been extended by numerical simulations using a general purpose software Abaqus/Explicit. A good correlation has been obtained. To study the temperature effects. The range of available temperatures is from the room ambient temperature up to 300 °C. The time needed to heat the specimen and to stabilize its temperature is 20 minutes. This study has allowed to discuss the ballistic behavior and resistance of sheet plates of the aluminum alloy under dynamic loading and different temperatures.

1. Introduction

In the international literature has been a few experimental data available in which specimens are subjected to impact loading at elevated temperatures. The usual approach is to carry out perforation tests at room temperature and to extrapolate results using numerical simulations at high temperatures.

Many authors have deal with perforation analysis, from theoretical approaches such as discussed in [1-2] to more practical considerations as reported in [3-5]. [6-8] studied the influence of the projectile nose shape (conical, blunt and hemispherical) and the projectile diameter on ballistic properties and failure modes of thin steel targets. In order to simulate the behavior of impacted and perforated structures, the finite element method with an explicit time integration procedure was an effective technique. [6-10] Numerical simulations, in particular by the FE method, are also effective supplements for theoretical and experimental investigations which were carried out to analyze the dynamic behavior of impacted materials. However, no direct data concerning perforation failure modes at high temperatures are reported or published. The thermal softening of the material is usually tested using quasi-static experiments and its extrapolation to high strain rates is often a rough simplification. The idea of carrying out dynamic tests using thermal chamber is a solution of the problem. In this work, a new highly-performant thermal chamber is used to heat up metallic thin plate specimens and to impose an initial temperature higher than the room temperature. The maxi-mum temperature reached with this original device is 300 ˚C. Aluminium alloy has been selected in this work due to it no or insignificant strain rate sensitivity [11]. Experimental results obtained have been compared with available literature data.

2. Equipment for specimens heating

To reach a perforation tests at high impact velocity, a pneumatic gas gun has been used (Figs 1 and 2). It permits to reach a maximum initial impact velocity V_0 of 125 m/s for a projectile mass equal to m_p = 28 g. During this work, the present device is equipped with a thermal chamber. The temperature may vary from room temperature to 300 ˚C. A waiting time of about t≈20 min was estimated to heat up the specimen and to reach a uniform temperature distribution in the specimen. For it, a thermocouple was fixed in the middle of the reference plate to calibrate the thermal chamber. Another thermocouple was fixed inside the thermal chamber to enable a homogenous temperature control. Using the previous set-up described, Aluminum alloy plates have been subjected to perforation. The plate dimensions were 130 mm ∗ 130 mm for thicknesses equal to t=1 mm. Which is fixed along its perimeter to provide a perfect fixation and to avoid sliding. The gas gun set-up is shown in Figure 1.

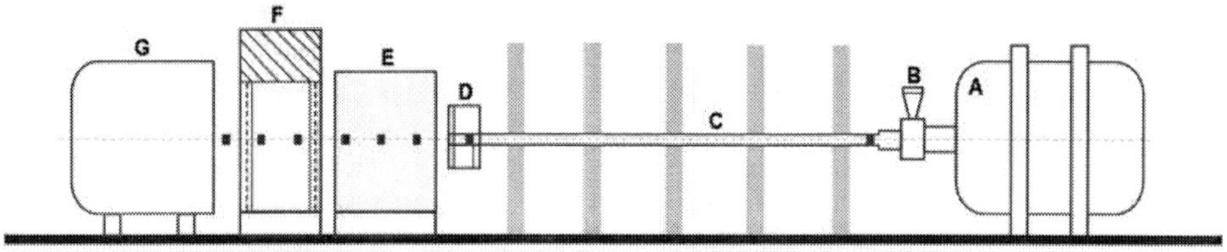

Figure 1. Gas gun set-up used for perforation tests at high impact velocities and temperatures [12]

A wide range of impact velocities from 40 to 125 m/s has been covered during the tests. The oven used for the test is presented Figure 2. A sarcophagus is used around the plate specimen to have a uniform temperature distribution after a certain waiting time. Therefore, the two sides of the specimen are heat up in the same time. After it, due to conductivity the entire specimen is heated to reach the initial temperature imposed.

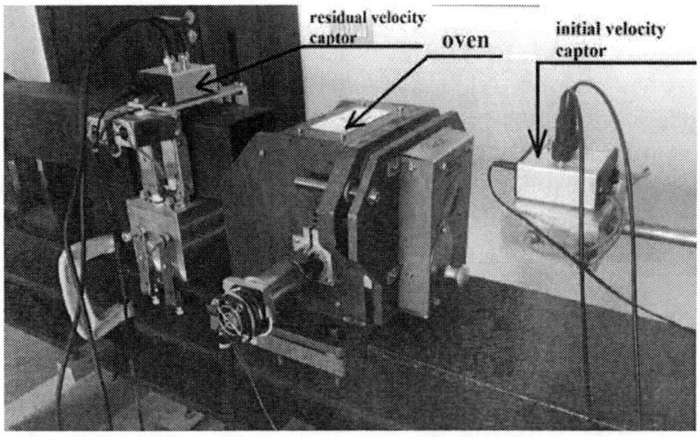

Figure 2. Oven used for perforation tests at high impact velocity and high temperature.

3. Experimental results

3.1. Perforation test at room temperature

In this section the mechanical behavior of aluminum sheets under impact loading has been described. Experimental, analytical and numerical investigations have been carried out to analyze the perforation process [6]. A wide range of impact velocities from 40 to 180m/s and temperature from 21 to 300 ˚C has been covered during the tests. The residual velocity of the projectile can be calculated using the equation proposed by [13] Eq.1,

$$V_R = (V_0{}^{\kappa} - V_B{}^{\kappa})^{1/\kappa}, \tag{1}$$

Where V_0 is the initial velocity, the constant V_B is equal to 40 m/s, and the ballistic curve shape parameter k is equal to 1.65, the absorbed energy by the plate E_d can be calculated using Eq.2

$$E_d = \frac{m_P}{2}\left(V_0{}^2 - V_R{}^2\right). \tag{2}$$

The results in terms of the ballistic curve V_R-V_0 are reported in Fig. 3a.

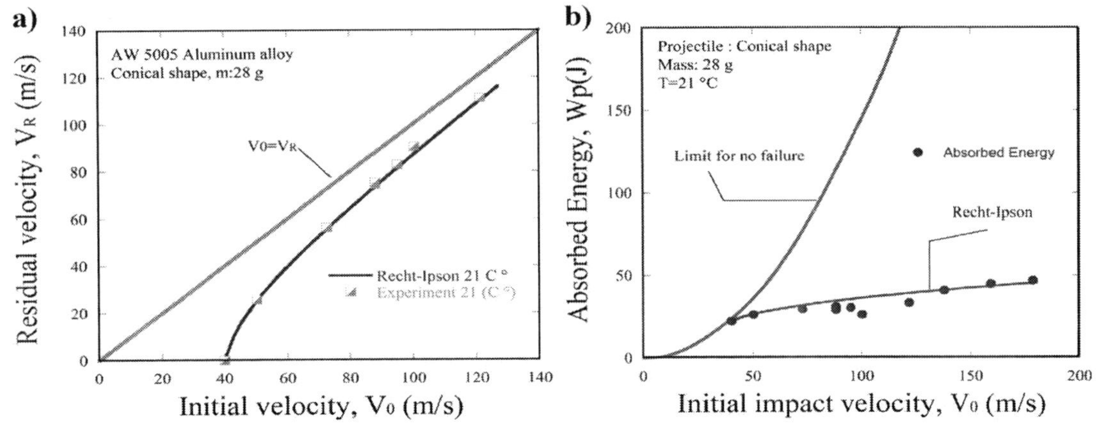

Figure 3. a) Ballistic curve obtained during perforation and determination of the ballistic limit, b) Energy absorbed by the plate during impact test, determination of the failure energy,[12]

The difference of the initial and residual kinetic energy can be calculated using the experimental data, then based on the Recht-Ipson approximation, the energy absorbed by the plate can be calculated, see Fig. 3b. Using Eq.2 the minimum energy to perforate is 28 J ($m_P = 28\ g$ and $V_0 = V_B = 40\ m/s$).

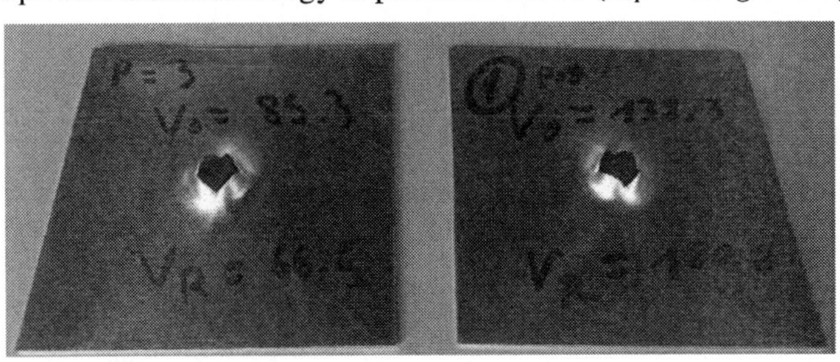

Figure 4. Experimental observation of failure patterns, V_0= 85.3 m/s and 132.3 m/s, [12]

As showed in Figure 4 for the initial impact velocity of V_0 = 85 m/s the failure pattern has been obtained with four petals with the residual velocity of V_R = 66.5 m/s. The same failure is observed for the V_0 = 132.3 m/s with V_R = 120.2 m/s, respectively, the number of petals is equal to 4 from 40 to 125 m/s. [14] contain a complete description concerning the number of petals depending on the projectile shape and the failure mode.

IOP Conf. Series: Materials Science and Engineering **491** (2019) 012027 doi:10.1088/1757-899X/491/1/012027

3.2. Perforation test at elevated temperature:

To complete this work a based ballistic studies have been performed. Experimental results in term of the failure mode are reported. A typical failure behavior are observed, the failure modes observed depends on both impact velocity and initial temperature value. Using a conical projectile, a failure mode by petaling occurs inducing radial necking due to a process of piercing. The conical projectile perforates the target plate and the plastic strain is localized at the extremities of the petals. Analytical predictions discussed in [15] are fully confirmed at room temperature, whereas more discrepancy in petals number is reported at higher temperatures. Figure 5 present a comparison between experimental results at room temperature and for a temperature of 100 and 300 ˚C for 1 mm plate thickness. It can be observed that increasing the initial temperature of the specimen changes the ballistic limit (no perforation state) to lower values: the ballistic limit obtained is approximately 40 m/s (20 ˚C), 35 m/s (100 ˚C), 34 m/s (200 ˚C) and 32 m/s (300 ˚C) using the conical projectile. Whereas the shape parameter k is 2.1 (100 ˚C), 1.83 (200 ˚C), 1.9 (300 ˚C) respectively. The other measured points are also shifted when higher residual velocities V_R are measured for elevated temperatures. The energy absorbed during the impact does not change considerably with the impact velocity. Experimental results stay in accordance with analytical predictions using Eq. 2 [13].

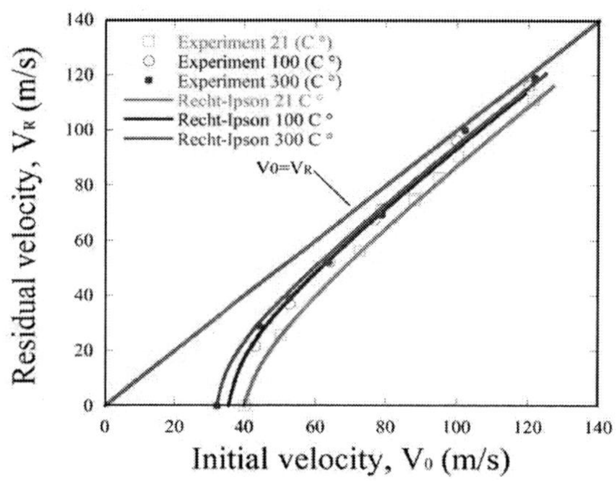

Figure 5. Impact velocity V_0 vs residual velocity V_R - experimental results for T=21˚C, T=100˚C and T=300˚C; using conical projectiles

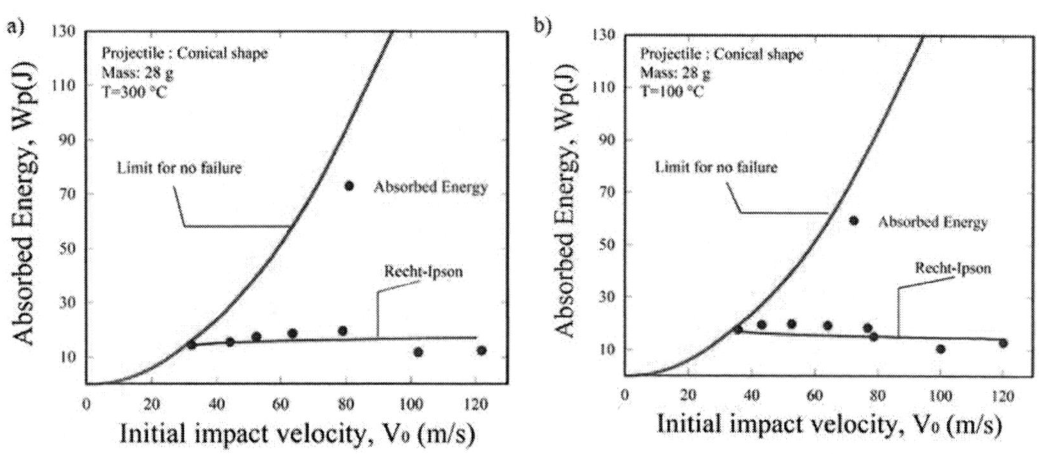

Figure 6. Energy absorbed by the plate during impact test, a/ T=100 ˚C and b/ T=300 ˚C; using conical projectile

Perforation process causes an instant increase of temperature localized in the perforated zone. Plastic deformation energy dissipated during the perforation process is transformed into thermal energy and provokes a considerable increase of temperature in petals. As shown in figure 6 the energy is quite constant at 100 °C and 300 °C, in the contrary at room temperature the absorbed energy increase when the initial impact velocity increase.

4. Numerical simulation

4.1. Material parameters

In the context of this work's applications, a model has been chosen, it is an empirical model from Johnson-Cook to describe the behavior of the studied material. Johnson and Cook propose an empirical law [16] based on experimental results and designed for rapid implementation in calculation codes. This model includes the influences of deformation velocity; strain hardening and temperature:

$$\sigma = \left(A + B\varepsilon_{pl}^{n}\right)\left(1 + C\, ln\frac{\dot{\varepsilon}^{p}}{\dot{\varepsilon}_{0}}\right)\left(1 - T^{*m}\right) \tag{3}$$

Where A is the yield stress, B and n are the strain hardening coefficients, C is the strain rate sensitivity coefficient, $\dot{\varepsilon}_0$ is strain rate reference value and m is the temperature sensitivity parameter. In this work, isothermal conditions are assumed. Therefore, the last term of the JC model related to the non-dimensional temperature where T^* is the homologous temperature.

The material constants are obtained from experimental tests. The Parameter C has been calculated using presented experimental tests for quasi-static loading. These constants are shown in Table. 1 [12].

Table 1. Material parameters for Johnson-Cook model, [12]

A (MPa)	*B* (MPa)	*n* (-)	*C* (-)
147	60	0.9	0.003

4.2. Failure criterion

During our analysis an analytical and numerical approach for the failure criterion was also considered as recently proposed in [13] for this aluminium alloy. Two functions with four constants H, I, J and K are determined using an optimization method (based on algorithm using MATLAB software).

$$\varepsilon_Y\left(\dot{\varepsilon}\right) = \begin{cases} He^{(I\ log_{10}(\dot{\varepsilon}))} & \dot{\varepsilon} \le \dot{\varepsilon}_{Transmition} \\ J - (K\ e^{\ log_{10}(\dot{\varepsilon})}) & \dot{\varepsilon} \ge \dot{\varepsilon}_{Transmition} \end{cases} \tag{4}$$

Where $\dot{\varepsilon}_{Transmition}$= 1 1/s.

The exemplar solution of this system of equations for another metal (Aluminum) reported in [12] is presented in Figure 7.

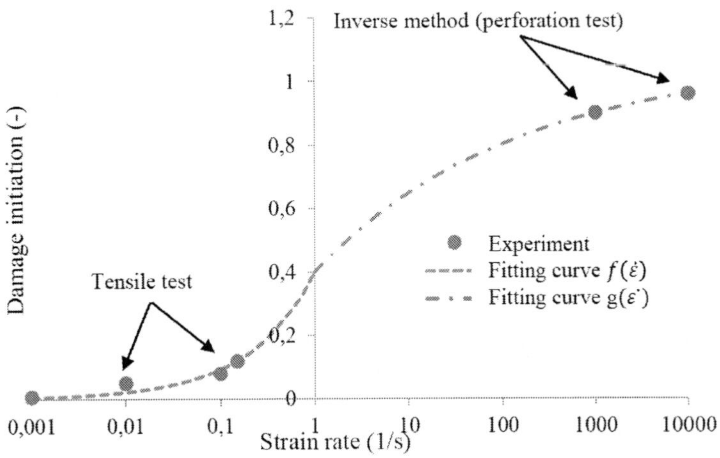

Figure 7. Plot of failure strain versus plastic strain rate using optimized Model III, [12]

4.3. Numerical simulation

Using finite element code Abaqus software the perforation test has been simulated, the mesh is denser in the projectile-plate contact zone, the thickness of the plate in this area is 1.0 mm, and the velocity is defined in the predefined fields with impact velocities range from 40 to 125 m /s as obtained in experiment. The projectile behavior has been defined as rigid body. The friction coefficient is assumed to be equal to 0.2 [12]. When the value of failure strain is changed, a decrease in petals numbers with a 72° nose angle has been observed. An analytical model for the number of petals prediction proposed by [14] has been used and confirmed by FE simulations.

5. Results and discussion

In this section a comparison between experimental and numerical results has been done. As it might be seen in Figure 8, there is a good correlation between the experimental and numerical results, the numerical model has demonstrated more ductile behavior at the failure start, it has revealed ductility at the terminal phase of failure.

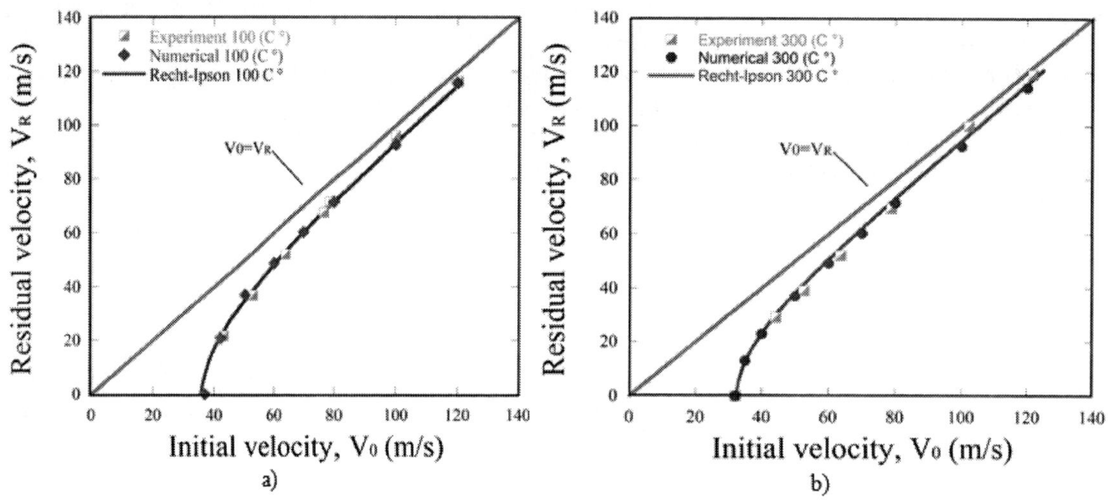

Figure 8. The ballistic curve in experiment and in simulation using thermal condition a/ T=100 °C, b/ T=200 °C and b/ T=300 °C; using conical projectile

It can be concluded that the thermal chamber set-up proved its designed performances and opened new perspectives to carry out perforation tests in a wide range of temperatures. Important results were obtained for aluminum alloy AW 5005 heated to elevated temperatures. The failure mode in form of petals confirmed analytical considerations. The studies on initial impact velocity and residual impact velocities reproduced typical behavior of material. The energy absorbed during perforation is quasi-constant for the studied range of velocities (up to 121 m/s). Using conical shaped projectile, the average value was 26 J at room temperature and decreased to the average of 18 J at 300 °C.

6. Concluding remarks

The analysis of the experimental results shows that the material is very strain rate and temperature sensitive. It has been shown in the stress range ($10^{-4}\ s^{-1} \leq \dot{\varepsilon} \leq 10^{4}\ s^{-1}$). The experimental study also showed that temperature has a significant influence on the mechanical behavior. Experiments have shown that the ballistic limit and the failure mode of the structure are strictly related to the projectile's nose shape. A specific analysis shown that the number of petals after perforation increases with the conical angle of the projectile. In the end, a simulation and a numerical analysis to reproduce the experimentally observed dynamic mechanical behavior for the studied structure, has been achieved using the Abaqus finite element code.

References

[1] Børvik, T., Hopperstad, O.S., Langseth, M., Malo, K.A., (2003), Effect of target thickness in blunt projectile penetration of Weldox 460 E steel plates, *International Journal of Impact Engineering*, 28 (4), pp. 413-464.

[2] Recht, R.F., Ipson, T.W. (1963), Ballistic perforation dynamics, *Journal of Applied Mechanics*, 30 (3), pp. 384-390.

[3] Abed, F. H., Voyiadjis, G. (2007), Adiabatic Shear Band Localizations in BCC Metals at High Strain Rates and Various Initial Temperatures, *Int. J. Multiscale Comput. Eng*., 5 (3–4), pp. 325–349.

[4] Jankowiak, T., Rusinek, A., K.M. Kpenyigba, Pesci, R. (2014), Ballistic behavior of steel sheet subjected to impact and perforation. Steel and Composite Structures, 16 (6), pp. 595-609.

[5] Jankowiak, T, Rusinek, A, Wood.P, (2015), A numerical analysis of the dynamic behaviour of sheet steel perforated by a conical projectile under ballistic conditions. *Finite Elements in Analysis and Design*, 65, pp. 39-49.

[6] Kpenyigba, K. M., Jankowiak, T., Rusinek, A., & Pesci, R. (2013). Influence of projectile shape on dynamic behavior of steel sheet subjected to impact and perforation. *Thin-Walled Structures*, *65*, 93-104.

[7] Rusinek, A., Rodríguez-Martínez, J. A., Arias, A., Klepaczko, J. R., & López-Puente, J. (2008). Influence of conical projectile diameter on perpendicular impact of thin steel plate. Engineering Fracture Mechanics, 75(10), 2946-2967.

[8] Rusinek, A., Rodriguez-Martinez, J. A., Klepaczko, J. R., & Pęcherski, R. B. (2009). Analysis of thermo-visco-plastic behaviour of six high strength steels. *Materials & Design*, *30*(5), 1748-1761.

[9] Quinney, H., & Taylor, G. I. (1937). The emission of the latent energy due to previous cold working when a metal is heated. *Proceedings of the Royal Society of London. Series A, Mathematical and Physical Sciences*, 157-181.

[10] Xue, L., Mock Jr, W., & Belytschko, T. (2010). Penetration of DH-36 steel plates with and without polyurea coating. *Mechanics of materials*, *42*(11), 981-1003.

[11] R. Julien, T. Jankowiak, A. Rusinek, P. Woo, P., (2016) Taylor's Test Technique for Dynamic Characteri-zation of Materials: Application to Brass, Exp. Tech., 40, pp. 347-355.

[12] Bendarma, A., Jankowiak, T., Łodygowski, T., Rusinek, A., & Klósak, M. (2017). Experimental and numerical analysis of the aluminum alloy AW5005 behavior subjected to tension and perforation under dynamic loading. *Journal of Theoretical and Applied Mechanics*, *55*(4), 1219-1233.

[13] Recht, R., & Ipson, T. W. (1963). Ballistic perforation dynamics. *Journal of Applied Mechanics*, *30*(3), 384-390.

[14] Atkins, A. G., & Liu, J. H. (1998). Necking and radial cracking around perforations in thin sheets at normal incidence. *International journal of impact engineering*, *21*(7), 521-539.

[15] Landkof, B., & Goldsmith, W. (1985). Petalling of thin, metallic plates during penetration by cylindro-conical projectiles. *International Journal of Solids and Structures*, *21*(3), 245-266.

[16] Johnson, G. R., Cook, W. H., (1985), Fracture characteristics of three metals subjected to various strains, strain rates, temperatures and pressures. *Engineering Fracture Mechanics*, 21(1), pp. 31-48.

IOP Conf. Series: Materials Science and Engineering **491** (2019) 012028 doi:10.1088/1757-899X/491/1/012028

A Study on Fatigue Strength Improvement for Tension Clamp of Railway Using Work Hardening

Jeong-Gil Cho[1], Jun-Woo Kim[2] and Jeong-Seo Koo[3,*]

[1]Complex Research Center for Materials & Components of Railway, Seoul National University of Science and Technology.
[2]Department of Rolling Stock System, Seoul National University of Science and Technology.
[3]Department of Railway Safety Engineering, Seoul National University of Science and Technology.

[*]koojs@seoultech.ac.kr

Abstract. Since the tension clamp is essential for the safe operation of trains, the problem of the broken tension clamp must be carefully managed. The damage of the tension clamp can be caused by various reasons such as track conditions and dynamic behaviours of trains. In this paper, the improvement of durability through the improvement of fatigue strength was studied by considering the experiment results and the FE analyses. First, the analytical model was verified by comparing the results of the stress and strain distributions with the initial clamping force and the results of the FE analyses. And then, the method to improve the fatigue strength characteristics was analytically evaluated by considering the behaviour of tension clamp (middle band and spring arm) and work hardening characteristics.

1. Introduction

The tension clamp is a key device that can withstand the various loads acting on the rail, fixes the gauge, reduces the vibration and sinks of the track bed, and plays a very important role in railway derailment safety. Especially, considering the reduction of the maintenance cost of the track, the maintenance of the tension clamp becomes a very important problem in the recent situation where the installation of the concrete bed track with the large demand of the tension clamp increases when compared to the ballast bed track.

Through many researches and developments, the advanced railway countries of the Europe such as England and Germany have supplied various types of tension clamps to meet the track characteristics around the world. The tension clamps are widely used not only in intercity railways including high-speed railways but also in urban railways in Korea.

In the previous study on tension clamp, Lee et al. [1] verified that it is important to secure proper initial clamping force for improvement of durability through experiments and numerical analyses. Choi et al. [2] analysed the behaviour of the tension clamps according to initial clamping condition, load changes due to train operation, linear/curved tracks, and inner/outer gauge differences using experiments and numerical analyses. In addition, Bae et al. [3] compared the stress state of the tension clamp with the initial clamping force by an experiment and a numerical analysis. However, despite many of the above studies, tension clamps sometimes fail due to operating conditions of various trains,

IOP Conf. Series: Materials Science and Engineering **491** (2019) 012028 doi:10.1088/1757-899X/491/1/012028

differences in installation and maintenance status of the tracks, so they tend to fail in securing adequate service life as suggested by the manufacturer.

In this study, we tried to find the analytical model verified by comparing the experiment result of the stress and strain distributions with the clamping force (torque) and the finite element analysis result, and then have tried to find a way to improve the durability considering behaviour of the tension clamp and work hardening characteristics.

2. Characteristics of the tension clamp

2.1 Clamping condition of the tension clamp
The model used in this study is the SKL 15 developed by Vossloh, as shown in Figure 1[4].

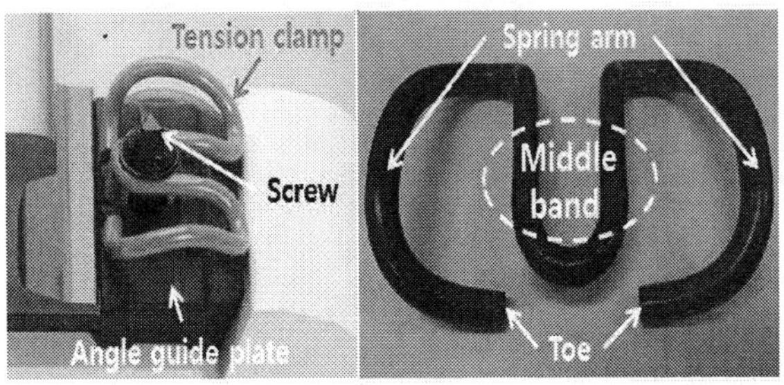

Figure 1. Rail fastening system

When clamping the clamp screw with a proper tightening torque, the middle band of the tension clamp should be in contact with the frame part of the angle guide plate, and the maximum allowable gap is 0.5(mm).

2.2 Failure of the tension clamp
The tension clamp, which must have long-term durability, is the most important component of the rail fastening system. However, in the case of the real field, fractures frequently occur due to various unknown factors. In fact, most of these showed a fatigue fracture behaviour that propagated from the point near the angle guide plate to the inside, as shown in Figure 2 ~ 3.

Figure 2. Fractured tension clamp

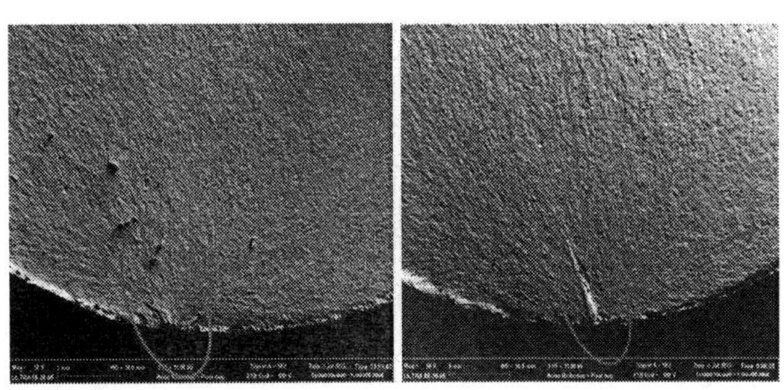

Figure 3. Fracture plane of the tension clamp

3. Experiment of the tension clamp

3.1 Strain measurement location and experiment method
In order to confirm the deformation of the mid-band and the strain at the beginning point of the fracture according to the clamping force, the displacement and the strain were measured by using LVDT and a 3-axis strain gauge, as shown in Figure 4.

In the case of the strain gauge, it was attached to the portion nearest to the point where the fracture started because it was in contact with the angle guide plate. After the comparison with the experiment results of the strain gauge through the FE analyses, the strain at the crack initiation point was predicted.

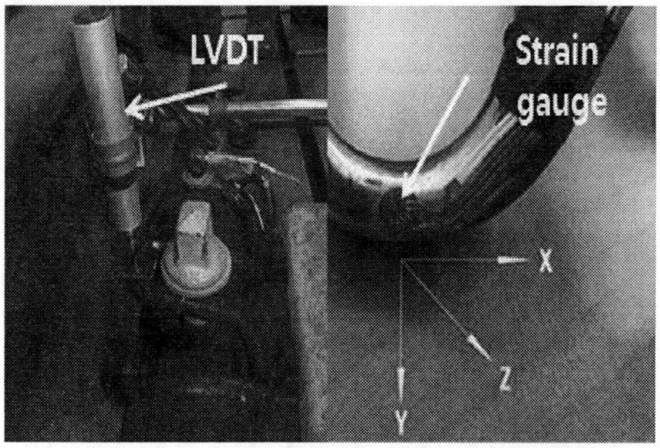

Figure 4. Experiment of the tension clamp

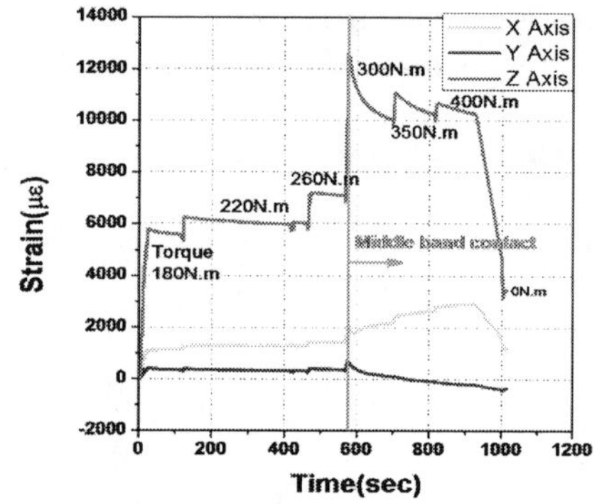

Figure 5. Strain results of tension clamp according to clamping forces

3.2 Experiment result according to clamping force
As shown in Figure 5 and Table 1, when the clamping force is 300(N·m) or less, no contact occurs, and when the clamping force is removed, the original shape can be restored. But, in the case of over 300(N·m), the bottom of the middle band contacts the angle guide plate, and after the clamping force is removed, the plastic deformation occurs and the original shape can't be restored.

Table 1. Experiment results of the tension clamp according to clamping forces

Torque (N·m)	Vertical middle-band deflection(mm)	Natural frequency(Hz)		Contact
		1st	2nd	
0	0	-	-	
180	11.96	445	565	X
220	12.93	463	581	
260	14.52	468	582	
300	19.52	489	625	O
350	20.00	513	635	
400	20.14	527	631	

4. FE analysis

4.1 FE modeling and analysis condition
In order to precisely analyse of the tension clamp, a FE model is needed to consider various curvatures. So, we constructed a FE model for the numerical analysis of tension clamps using 3D SCAN of the product, as shown in Figure 6.

Modelling is performed by using Hyperworks[5], a commercial software, and it is meshed using a tetra mesh, which is a general solid element. The applied properties are shown in Table 2.[1,2,6] And the analytical conditions were the same as the experiment conditions of the tension clamp, using the displacement equal to the deflection of the middle band in the vertical direction according to the clamping force.

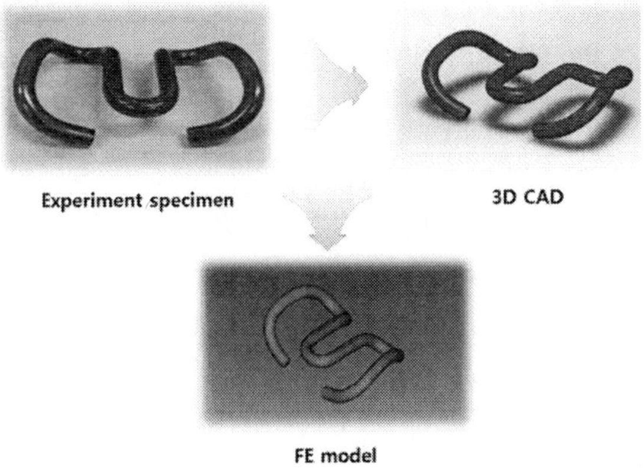

Figure 6. Tension clamp(SKL 15)

Table 2. Mechanical properties of materials

Properties	Value
Density	7.85(kg/m^3)
Tensile strength	1,300(MPa)
Yeild strength	1,000(MPa)
Young's modulus	200(GPa)
Poisson's ratio	0.27

4.2 Comparison of the experiment result and FE analysis result

To obtain the principle strain according to the clamping force of the experiment result, we applied the each direction result of the 3-axis strain gauge to the equation (1).

$$\varepsilon_{\max} = \frac{\varepsilon_x+\varepsilon_y}{2} \pm \sqrt{\left(\frac{\varepsilon_x-\varepsilon_y}{2}\right)^2 + \frac{\gamma_{xy}}{2^2}} \tag{1}$$

When comparing the principle strain of the measurement point with the finite element analysis results according to the clamping force, the error rates were within the acceptable range (within 7.7%) as a whole except for the clamping force of 300(N·m) since the contact with the angle guide plate occurred. Based on this, it was confirmed that the ratio between the measurement point and the maximum strain point was about 1.83 times, as shown in Table 3 and Figure 7.

Table 3. Comparison of the FEM and experiment result

Torque (N·m)	Vertical middle-band deflection(mm)	Measure point						Max. point	B/A
		X-strain (με)	Y-strain (με)	Z-strain (με)	Theoretical max. principle strain (με)	FEM max. principal strain (με)[A]	Error (%)	FEM max. principal strain (με)[B]	
180	11.96	1100	350	5600	3550	3575	0.7	6535	1.83
220	12.93	1250	370	6000	3842	3895	1.4	7120	1.83
300	19.52	2074	173	10334	6377	5922	7.7	10826	1.83

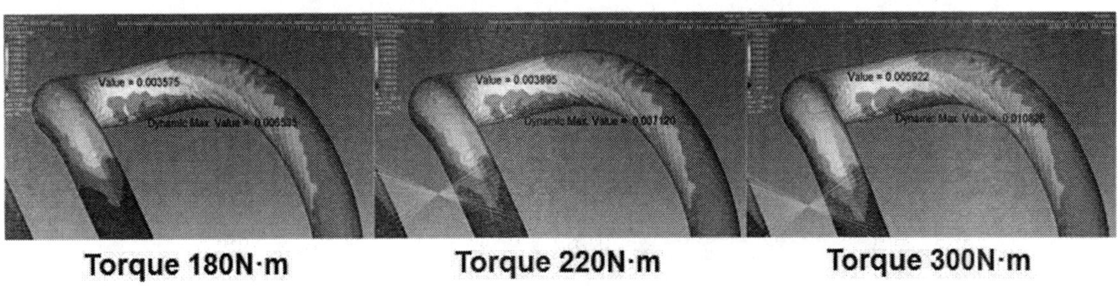

Figure 7. FEM results

4.3 FE analysis considering operation condition

In order to check the behavior of the tension clamp during the operation of the train, we measured the wheel load, lateral force, displacement, the strain of the tension clamp, and the vertical displacement of the rail.

Among these, the vertical displacement of the rail, which directly affects the tension clamp when the train passes through, is shown in Figure 8, and it is confirmed that the behaviour is within a range of 1(mm).

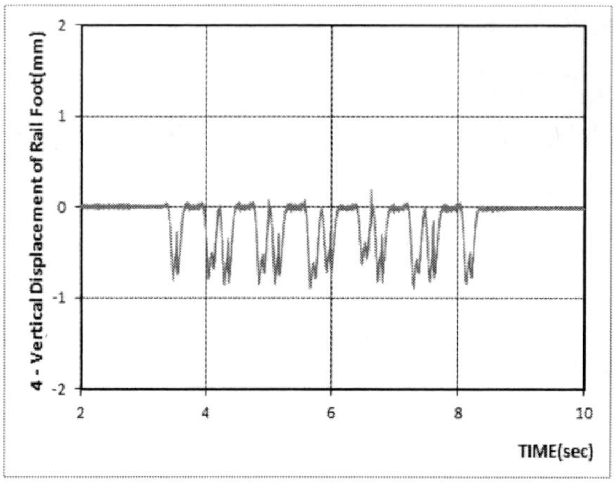

Figure 8. Vertical displacement of rail due to running of train

It can be checked that the bottom of the rail moves to downward direction within a range of 1(mm) due to the running of the train while supporting the rail by using the initial clamping force of the tension clamp. And it can also be understood that the stress-strain acting on the tension clamp is lessened by the running of the train after applying the initial clamping force. So, we constructed a FE analysis model using LS-dyna[7], a commercial software, to confirm the equivalent stress range during train running.

The results of the analysis of the von Mises stress range via the displacement control of the spring arm after applying various initial clamping forces to the tension clamp are shown in Figure 9.

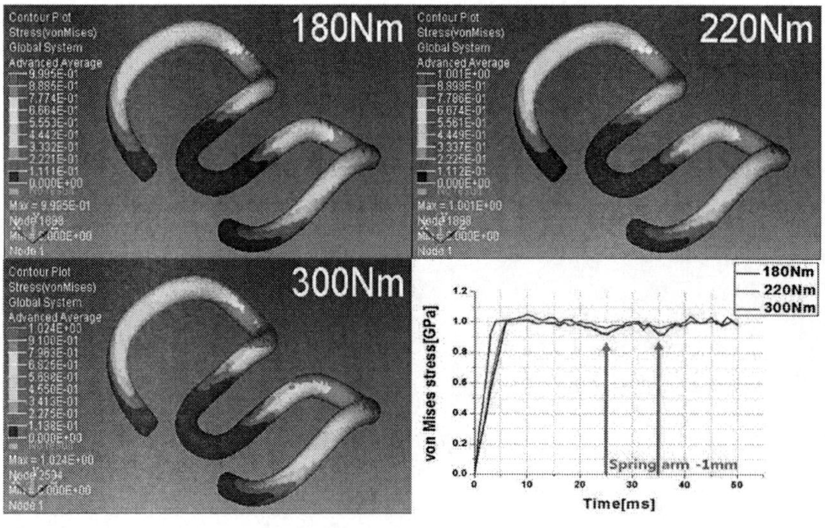

Figure 9. von Mises stress distribution due to initial clamping force and train running

In the case of initial clamping force of 180(N·m) and 220(N·m), von Mises stress distribution can be confirmed up to the yield strength by the initial clamping force, and we can confirm that the von Mises stress decreases when the displacement of the spring arm is moved down by 1(mm). However, in the case of the initial clamping force of 300(N·m), in which the middle band contacts the angle guide plate, it is already exceeded yield strength by initial clamping force and it can be confirmed that the change of the von Mises stress is not too significant even if the displacement of the spring arm is moved down by 1(mm). Therefore, it is considered that work hardening exceeding the elastic range of the material has already been considered as an initial clamping force.

5. Improvement of fatigue strength by clamping method

5.1 Work hardening

In the initial clamping force condition of 300(N·m) in which the middle band contacts the angle guide plate, plastic deformation exceeding the yield strength occurs and a varying dynamic stress is applied to it by the running of the train. In this case, since it behaves near the yield strength of the material, it is sensitive to fatigue and is very vulnerable to small surface defects.

Therefore, fatigue strength is expected to be improved by using work hardening [8] and additional analysis is performed to confirm it.

5.2 Surface plasticity model and elastic recovery model after surface plasticity

For comparison of the surface plasticity model and the elastic recovery model after surface plasticity, we have performed FE analyses according to two scenarios, as shown in Figure 10.

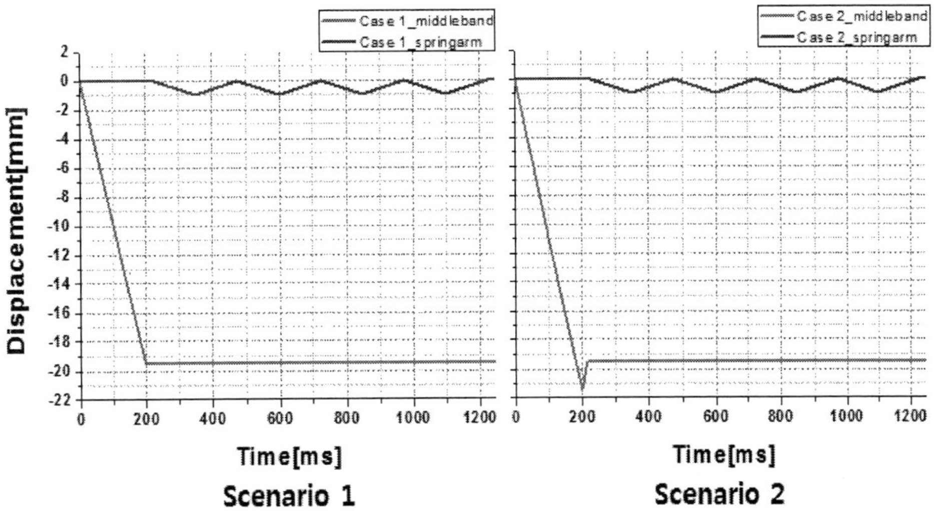

Figure 10. Displacement control condition according to two scenarios

For the scenario 1 (surface plasticity model), the displacement of middle band was given as 19.52(mm) using the experiment results of 300(N·m), and then the displacement of the spring arm was reduced by 1(mm) considering the running speed of the train.

But, in the case of Scenario 2 (elastic recovery model after surface plasticity), the displacement of the middle band was increased by 21.47(mm), which is 10% more than Scenario 1, and after restoring elasticity to 19.52(mm) and the displacement of the spring arm is the same as the scenario 1.

IOP Conf. Series: Materials Science and Engineering **491** (2019) 012028 doi:10.1088/1757-899X/491/1/012028

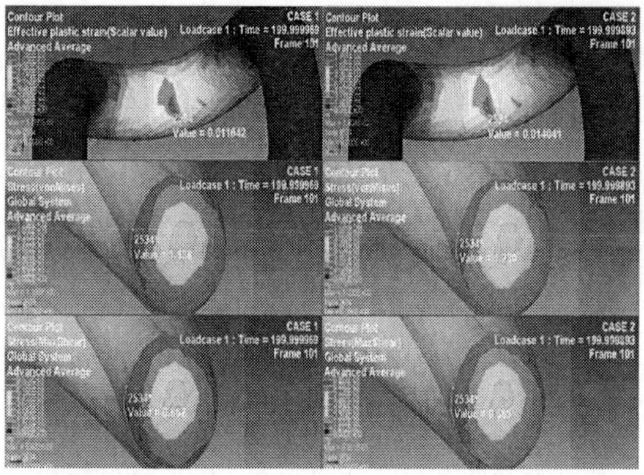

Scenario 1 **Scenario 2**

Figure 11. Comparison of the scenario 1 and 2 results - 200(ms)

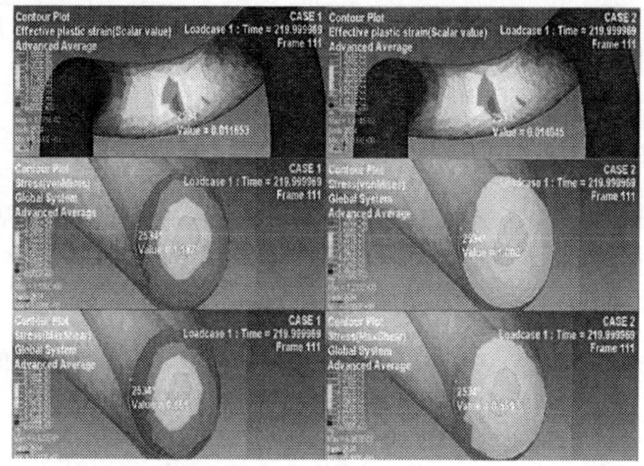

Scenario 1 **Scenario 2**

Figure 12. Comparison of the scenario 1 and 2 results - 220(ms)

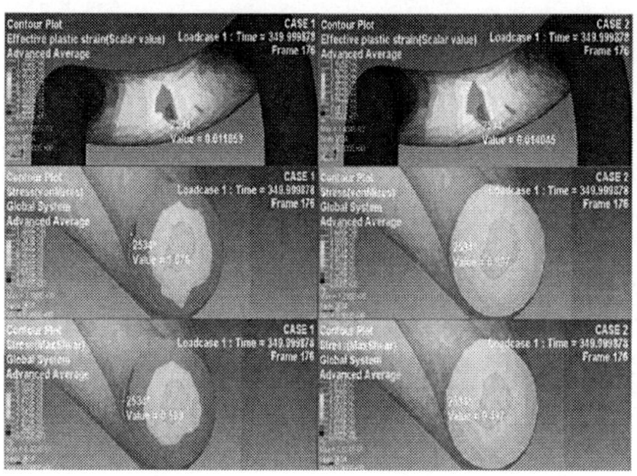

Scenario 1 **Scenario 2**

Figure 13. Comparison of the scenario 1 and 2 results - 350(ms)

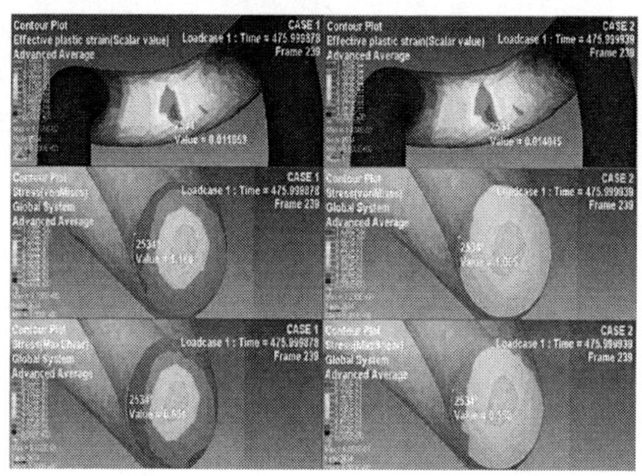

Scenario 1 **Scenario 2**

Figure 14. Comparison of the scenario 1 and 2 results - 475(ms)

Figures 11 ~ 14 are the FE analysis results showing the time-dependent cross-sectional distribution of the major part where the fracture occurred. These are comparison of plastic strain, von Mises stress, and maximum shear stress according to the two scenarios. And Figure 15 shows the variation of the von Mises stresses of the main regions where the fracture occurred.

In scenario 2, compared to scenario 1, the yield strength increased due to the work hardening effect and then, it can be confirmed that the range of stress variation due to the displacement of the spring arm is the same, but the variation is relatively lower portion when compared with the scenario 1.

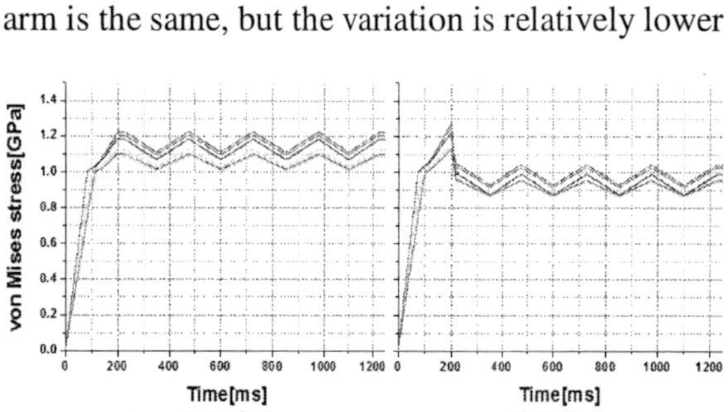

Figure 15. Comparison of von Mises stress distribution of major parts by two scenarios

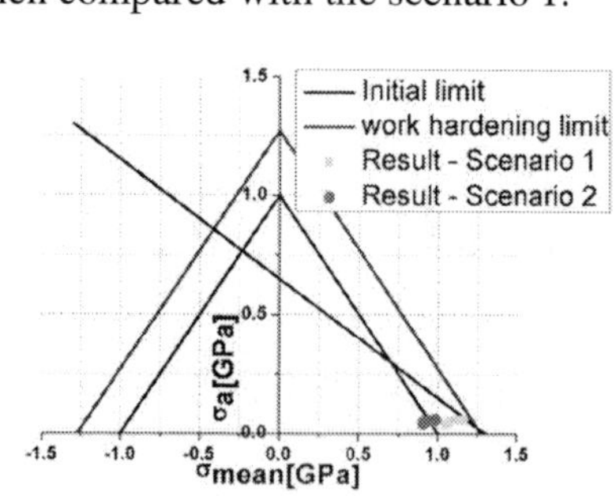

Figure 16. Comparison of Goodman diagram results of major parts by scenario

The fatigue strength is calculated by using the Goodman diagram, as shown in Figure 16, and it can be confirmed that the scenario 2 shows a relatively safe range because its yield strength is increased when compared with the scenario 1.

6. Conclusion

In this paper, the fatigue strength improvement considering experiments of the tensile clamp, FE analyses, and work hardening effect was studied. The results are as follows.

(1) The experiment was performed to confirm the strain distribution according to the clamping force of the tension clamp. At the clamping force of over 300(N·m), the bottom of the middle band was contacted to the angle guide plate, and when the clamping force was removed, it was confirmed that plastic deformation occurred.

(2) The FE analyses were performed using the accurate shape information of the tension clamp via 3D SCAN to verify the analytical model by confirming the similarity between the FE analysis results and the experiment results. And then the strain distribution of the part where the maximum strain occurred is predicted.

(3) As a result of the analysis of the initial clamping force condition of 300(N·m) proposed by the manufacturer, due to the variation of stress activated from the train, it behaves near the yield strength of the material, so it is sensitive to fatigue and is very vulnerable to small defects on the surface.

(4) In order to improve the fatigue weakness, analytical evaluation was performed considering the effect of work hardening.

It is expected that the durability can be improved if the tension clamp is fastened to use the work hardening effect, as suggested in this study.

Acknowledgments

This study was supported by the Research Program funded by the SeoulTech (Seoul National University of Science and Technology).

References

[1] Lee D. W., Choi J. Y., Baik C. H. and Park Y. G. 2013, "Influence of Initial Clamping Force of Tension Clamp on Performance of Elastic Rail Fastening System", *Journal of the Korean Society of Civil Engineers*, 33(3), pp. 1243-1251.

[2] Choi S. H., Park B. H., Yun K. M., Bae H. U. and Lim N. H. 2015, "Behavior of Tension Clamp in Rail Fastening System", *Korea Academy Industrial Cooperation Society*, 16(12), pp. 8812-8819.

[3] Bae S. B., Choi S. H., Son B. J. and Lim N. H. 2013, "A Study on the Behavior of the tention clamp in the fastener", *Journal of Conference & Annual Meeting of the Korean Society for Railway*, pp. 1067-1070.

[4] Vossloh, Rail fastening systems for slab tracks –System 300, Publications.

[5] "Hyperworks user manual", 2013, Altair Engineering Inc.

[6] DIN EN 10089, Hot rolled steels for quenched and tempered springs technical delivery conditions, 2003

[7] "LS-dyna user manual", Version 971 R4, 2009, Livermore Soft Technology Corp.

[8] D. T. Llewellyn. 1997, "Work hardening effects in austenitic stainless steels", *Materials Science and Technology*, 13(5), pp 389-400.

IOP Conf. Series: Materials Science and Engineering **491** (2019) 012029 doi:10.1088/1757-899X/491/1/012029

Some Aspects of Influence of Turbopump Unit Fabrication and Assembly Method on Seal Clearances

A.V. Ivanov

Deputy Chief Designer for new technologies of NPO Energomash Named after Academician V.Glushko, JSC, bldg. 1, Burdenko Str., Khimki, 141400, Russian Federation

iav308@inbox.ru

Abstract. Each element of turbo-pump unit has its own dimensions, tolerances of form and surface arrangements that differ from the rated ones. When assembling a turbo-pump unit they are located with some skew due to the mounting offsets of the rotor and stator sealing surfaces. Such skew arise due to a combination of hard-to-predict factors, such as the inaccuracy of the machine, the clearance of the centering surfaces, the rigidity in machining, the reliability of the fixing of the component, the state of the component, the balance of the system, and so on. The main factors affecting the installation bias of the rotor and stator sealing elements include the structural diagram of the unit, the workflow for manufacturing component parts, the workflow of assembling and controlling quality. The present article considers the dependencies for calculating the values of the mounting bias. It is necessary to reduce the effect of the installation bias on the size of the sealing gap to reduce leakage in the seals of pumps and turbines.

1. Introduction

Turbo-pump units of liquid rocket engines represent a complex structure having numerous parts and interfaces included. Both the turbo-pump unit stator and rotor are a complex spatial structure representing a package of parts tied together with fixing elements. For instance, the oxidizer pump rotor of RD-170 engine turbo-pump unit is a complex structure where nine parts and four ball bearing inner rings are mounted and tightened with the help of three fastening nuts and four screws. Each of these elements has its own dimensions, tolerances of form and surface arrangements that differ from the rated ones. As a result, when assembling a turbo-pump unit they are located out-of-alignment, with some skew [1, 2, 3], due to the mounting offsets of the rotor and stator sealing surfaces.

2. Factors affecting the value of the mounting offsets

Usually, non-contact seals [4] are utilized in the liquid rocket engine turbo-pump units, the working clearance (δRw), in which is equal to the difference of the mounting clearance (ΔRm) and the total strain and temperature deformations of the rotor and stator sealing elements during its operation, and the local minimum clearance is equal to the difference of the radial working clearance and the sum of all mounting offsets during fabrication and assembly:

$$\delta R_w = \Delta R_m - \int_l (\delta R_T + \delta R_s) dl \quad (1)$$

$$\delta R_{min} = \Delta R_w - \int_l (\delta R_E + \delta R_\delta) dl = \Delta R_m - \int_l (\delta R_T + \delta R_s) dl - \int_l (\delta R_E + \delta R_\delta) dl \qquad (2)$$

Generally, three basic groups of factors affect the value of the mounting offsets (offsets of the sealing surfaces axes relative to the centerline):

- Unit structural arrangement determining the quantity and relative position of the rotor and stator elements;
- Component parts' fabrication processes determining the accuracy and actual errors of the parts fabrication;
- Assembly and its quality control processes determining the relative positioning accuracy, as well as the relative offset of the rotor and stator sealing parts.

In order to accurately determine the mounting offset values, it is required to carry out highly laborious calculations considering the influence of each error, stiffness of each part and value of the axial drawing-up force for each fastening nut or other fixing elements.

The following maximum allowable errors of the rotor and stator elements fabrication shall be known and set in the course of fabrication:

- Misalignment of seat and seal diameters;
- Non-parallelism of thrust faces;
- Non-perpendicularity of thrust faces relative to seat diameters;
- Seat surfaces form deviation;
- Gaps in bearings and seatings.

The influence of some distinctive features of the high-speed turbo-machines' parts and assembly units' fabrication on the surfaces location tolerances affecting the seal clearances is given below.

Figure 1a shows the result of the shaft machining upon the misalignment of centers due to the misalignment of holes for the centers. As shown on the figure, the misalignment of the centers results in the surfaces form deviations (tapering). The machined surfaces are coaxial, but such machining produces deviations from the specified seating clearances, the component parts joint stiffness and setting strength change leading to the seal clearance change.

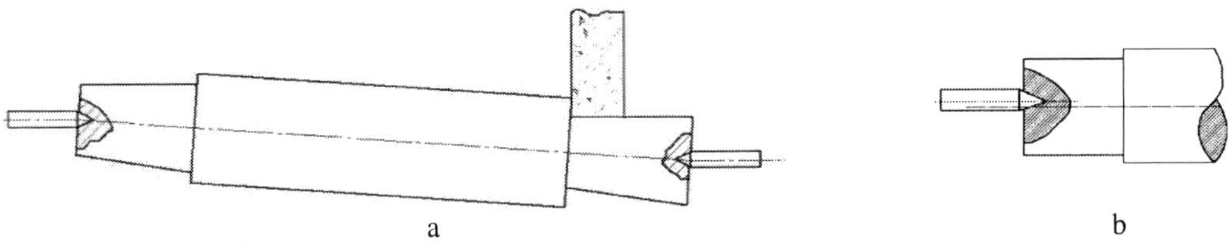

Figure 1. The form and position deviations during the shaft machining in the centers: a - misalignment of centers; b - radial motion variation of the rotating center

When the machine is in a continuous operation, the regular adjustment of the centers' similar height and parallelism of their joint axis shall be carried out. If such activities are not carried out, the misalignment of the machined surfaces will appear (Figure 1b).

Figures 2, 3 show, in a quite extreme form, the location deviations that may occur due to machining process imperfection. If the force of the shaft contraction with the tailstock center FA is much less than the radial force FR, the alignment hole may not be fully adjacent to the center surface (Figure 2). The same misalignment of the alignment hole with respect to the center may be present if the alignment surface is damaged. In this case, the axis of the shaft centers will be out-of-alignment relative to the part rotation axis that will result in the machined surfaces misalignment unequal with respect to the centers axis. These surfaces will remain coaxial with respect to each other.

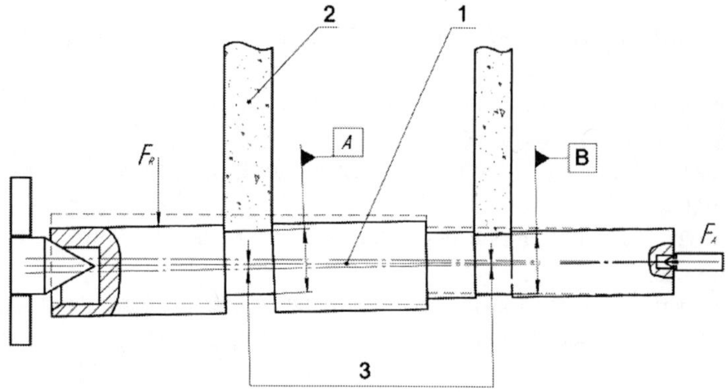

Figure 2. The surfaces misalignment due to the insufficient contraction force F_A
1 - axis of rotation; 2 - grinding wheel (knife); 3 - misalignment

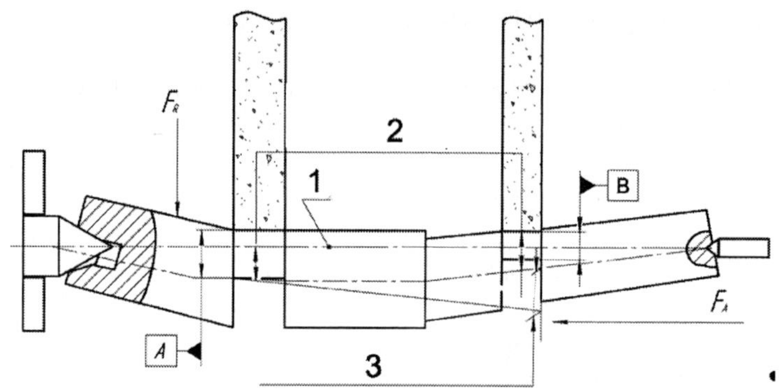

Figure 3. Misalignment at high values of F_R and F_A contraction forces
1 - axis of rotation; 2 - misalignment relative to the centers axis; 3 - misalignment between A and B surfaces

At high values of FA and FR forces, a shaft elastic bending is observed (Figure 3). In this case, the shaft bent axis turns around the part rotation axis during machining. After removing the shaft from the machine, its axis draws straight, the machined cylindrical surfaces become out-of-alignment and the end surfaces become non-parallel.

The misalignment of the seat surfaces for bearings and seals in the rotor is usually significantly less than in the stator, since these surfaces are often made in one part; the quantity of the rotor parts is usually less than that of the stator. However, due to the greater amount of skew addends in the unit, the shaft misalignment shall be set quite accurately. The non-parallelism of the parts' faces attached to the shaft results in considerable skew (offset) values of the sealing surfaces. The rotor part face non-perpendicularity with respect to the seat surface upon the considerable length and absence of clearance for the seating on the shaft also results in the skew increase.

The location deviations of the sealing surfaces of the stator sealing elements, especially housings, result from a combination of various, often hard-to-predict, factors such as the actual inaccuracy of the machine and machining device, clearance for the alignment surfaces, runout variation error, stiffness of the machine - device - part - cutting tool system, possible contamination of mounting seats, part fixing reliability, device condition, system balance, etc. It is extremely difficult to determine by calculation the possible fabrication errors due to the reasons specified [1].

The vast majority of housings and other large stator elements are machined in two settings. When machining in the second setting, the minimum diameter surface machined together with the surfaces available in the first setting (D in Figure 4) is used as the reference one. The reference surface runout is controlled in the second setting and if it does not exceed the value specified in the design documentation then the machining is allowed. The surfaces machined in the second setting may have rough location deviations with respect to the surfaces machined in the first setting.

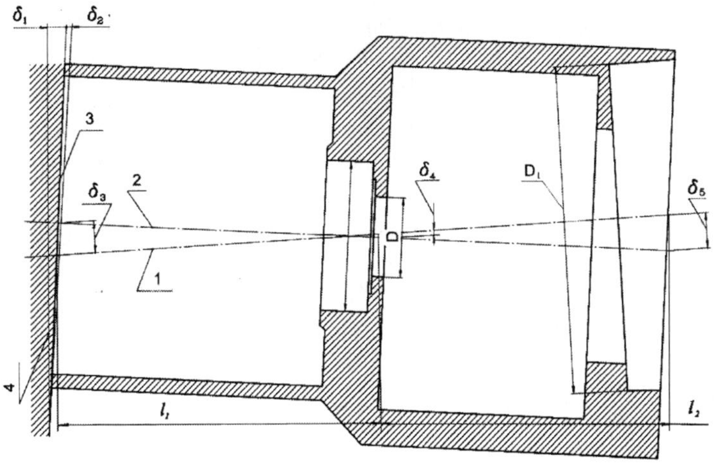

Figure 4. Diagram for possible surfaces location deviations during the housing machining
1 - axis of rotation; 2 - reference axis; 3 - device plane; 4 - plane perpendicular to the spindle axis

Figure 4 shows a possible option of the housing machining errors at the device face runout δ1, housing face runout δ2 and housing axis offset by δ3 relative to the axis of rotation (this offset can reach the values comparable to the value of the radial sealing clearance). In this case, the surface offset D up to the value δ4 allowable in accordance with the design documentation results in the misalignment of the machined surface relative to the seat surface for the bearing reaching the value of $\delta_5 = \frac{l_1 \cdot \delta_4 + l_2 \cdot (\delta_3 + \delta_4)}{l_1}$ that will exceed the value specified in the documentation, usually equal to δ4. The incorrect application of the device may entail even more rough fabrication.

The assembly is the final, labor-intensive and costly process that defines the performance characteristics, cost of production and competitiveness of any product. The relative labor intensity of the assembly works over the past 40-50 years has been continuously increasing due to the fact that the workpiece preforming and machining methods are advancing faster than the assembly method [5].

When assembling the unit, the offset value is affected by the offsets of the rotor and stator parts within the mounting clearances, rotation of the rotor parts with respect to each other, tightening forces of the rotor fastening nuts and coupling screws, as well as the method for the housings mutual alignments and connection. The correctly selected method to control the assembly process and quality of the complete assemblies allows to identify the parts and assembly units with unacceptable combinations of fabrication errors affecting the offset values.

The process of the turbo-pump unit assembly and control may be carried out according to one of the three options.

Option I. The parts having errors within the design documentation requirements are submitted for assembly with no restrictions. The errors are summarized in the assembly units in a random manner. In this case, the probability of assembly with both minimum and maximum sum of errors exist. With this method, the maximum possible sum of errors shall not exceed the level allowable for the normal operation. This option with the large number of parts may require an improved accuracy of fabrication [6].

The second process option is based on the fact that the probability of assembly with the worst combination of errors is low. As in the first case, the assembly is carried out using any set of parts, but the complete assemblies undergo special inspection and if the assemblies with the unacceptable level of overall errors are found they are sent for rearrangement and reassembly. The rotor surfaces' runout inspection may be an example of the above. The rotor sealing surfaces are, as a rule, used as the reference surfaces. The inspected runout values include:

- Runout due to the rotor mounting deformations;
- Runout due to the misalignment of the rotor elements' interfacing surfaces;
- Runout due to the offset of the rotor elements within the seating clearances between the rotor elements;
- Runout due to the non-parallelism of the interfacing faces and non-perpendicularity of the thrust faces and the seat diameters;
- Runout due to the accuracy of the reference surfaces fabrication.

The third process option provides for the use of special fabrication, assembly and inspection operations ensuring the errors minimization to the required level, for example, the sealing surfaces machining in the assembled rotor or stator. Such operations are widely used in the aircraft industry when manufacturing gas turbine engines.

As a rule, the second process option is used. In this case, an accurate and careful calculation of the mounting offsets of the rotor and stator sealing elements is required. In turbo-pump units the inaccuracies (dimensional tolerances, surfaces location tolerances) of the large amount of the rotor and stator parts and assembly units have an influence on the offsets. The degree of influence depends on both the actual deviation values and the relative position of the parts with respect to each other as per the angle of turn in the particular unit example. The deviations summation may be performed using two methods: Probabilistic method and minimum-maximum method. The probabilistic method considers the phenomenon of scattering and probability of various combinations of the constituent elements' deviations. The minimum-maximum calculation method considers only the extreme deviations of the constituent elements.

The practice of testing the turbo-pump units fabricated in small series has demonstrated that it is possible to securely protect the unit against the adverse influence of fabrication errors if the overall deviations calculated as per the minimum-maximum method do not exceed the allowable values. Since the turbo-pump units are produced in small series or in single quantities and the issue of reliability of the unit and its elements is of paramount importance, the calculation as per the probabilistic method is irrational. It should be also taken into account that the minimum-maximum method calculations in the course of the unit design allow to identify the error combinations that are dangerous for the operation and take the required measures in advance, during the design stage. When the investigated parameter maximum values are unacceptable, the measures excluding the possibility of such values occurrence in the actual units shall be provided. Definitely, the fact that the actual values of skews in the unit, at least the mounting ones, are usually less than the maximum values determined as per the maximum-minimum method is considered in this case.

The turbo-machines' rotors represent a structure designed as a package of individually fabricated parts centered relative to each other or to the shaft, tightened with a common nut or other fastening elements. The misalignment of any sealing surface is the resulting error of all the parts included in the rotor, that significantly complicates the calculation of the mounting offset values. The tightening, as a rule, is carried out using significant axial force in order to eliminate the joint clearances. The shaft shoulders' non-perpendicularity with respect to the threads may result in the rotor bend. When being installed on the shaft, the nut, within the thread clearance, may radially offset and rotate to eliminate the skew ensuring the firm adherence of the nut face against the rotor mating part, as well as circumferentially equalize the tightening force. In this case, the nut face runout relative to the pitch diameter not exceeding the half minimum thread clearance may be allowed. If the threaded connection is coated, the thread clearance decreases and the same shall be considered when specifying the tolerances. In order to achieve the steady-state and axially-symmetrical rotor stiffness, it is required to provide for the tight zero-clearance adherence of all faces of its parts due to the tightening force. The sufficient tightening of the both rotor and stator parts shall be provided at deformations resulting from the internal and external loads, for instance, the hydro-gas-dynamic and unbalanced loads. The required tightening force value depends on the parts dimensions, their fabrication error values, their stiffness, the value and nature of external and internal loadings. It should be noted that when attaching the parts included in the rotor or stator packages with the help of a group threaded connection, an additional error occurs due to the unequal tightening force associated with the material elastic deformations and contact deformations on the joint surfaces. These additional errors shall be considered when calculating the angular deviation of the rotor or stator sealing surfaces.

Another special feature of the rotors is that their constituent parts: Bushings, turbine wheels, screws, impellers may be installed in various relative positions equivalent from the structural point of view. However, the presence of macro deviations and deviations of the contacting surfaces location results in the fact that by changing the relative angular position of the parts it is possible to obtain the versions of the same rotor with different characteristics including those not satisfying the specified accuracy requirements. The rotor assembly is the process of forming its assembly axis by selecting the optimal relative position of the parts. The rotor axis formed in such a manner may be only rectified by disassembling the rotor with subsequent turn or replacement of the parts. This process is highly labor-intensive and requires the rotor parts modifications in some cases [7].

3. Calculation the value of the mounting offsets

In the result of generalizations based on the application of the dimension chain theory with inherent assumptions, we receive a dependence to determine the mounting offset of the seal diameter in the housing with respect to the diameter for ball bearings due to the misalignments:

$$\delta R_{E,\delta}^{st} = \frac{l}{L} \cdot \int_{l} \left(E_{st} + \frac{\delta_{st}}{2} \right) dl = \frac{l}{L} \cdot \left(\sum_{m} (E_{st})_{m} + \sum_{n} \left(\frac{\delta_{st}}{2} \right)_{n} \right) \tag{3}$$

l – distance between the sealing and the thrust ball bearing;

L – distance between the ball bearings.

Est – misalignment between the elements forming the stator part of the sealing from the bearing to the sealing surface, in terms of radiality;

δst – diametrical fit clearance between the elements forming the stator part of the sealing from the bearing to the sealing surface, in case of the interference fit the value in the calculation shall be assumed equal to 0.

The mounting offset of the rotor sealing element with respect to the rotor centerline shall be determined as per the following dependence:

$$\delta R_{E,\delta}^{rot} = \frac{l}{L} \cdot \int_{l} \left(E_{rot} + \frac{\delta_{rot}}{2} \right) dl = \frac{l}{L} \cdot \left(\sum_{i} (E_{rot})_{i} + \sum_{j} \left(\frac{\delta_{rot}}{2} \right)_{j} \right) \tag{4}$$

Erot – misalignment between the elements forming the rotor part of the sealing from the bearing to the sealing surface, in terms of radiality;

δrot – diametrical fit clearance between the elements forming the rotor part of the sealing from the bearing to the sealing surface.

The clearance in the seals with fixed wall (groove and labyrinth seals) and the clearance setting are influenced by the mounting offset due to the both stator and rotor elements misalignments and fit clearances. The clearance in the floating and semi-flexible ring seals is influenced only by the mounting offset of the rotor elements, i.e $\delta R_{E,\delta}^{st} = 0$.

The mounting offsets reach significant values even for small-sized units. To provide for the non-contacting operation of the seals, the radial clearance value may be considered as sufficient if it exceeds the sum of the half of the rotor sealing surface mounting offset value and the mounting offset value of the shaft axis passing through the axes of the mounting surfaces of the ball bearings' inner rings.

In order to reduce the minimum clearances in the pumps and turbines, it is required to reduce the mounting offset effect. Measures to reduce these values may be divided into four directions, all related to improving the fabrication accuracy:

- Reduction of the bore diameters misalignments;
- Reduction of the fit clearances influence due to the strengthening of tolerances and change-over to tight fits (interference fits);
- Limiting the sealing diameter misalignment during the assembly of housings, pumps and turbines;
- Co-machining - grooving of the housings seal diameter or the rotor assembly.

These measures are quite sophisticated and costly. They shall be implemented only after a careful analysis of the requirement and feasibility. If possible, the desired result shall be achieved in a different way, since the improving of accuracy, as a rule, results in the improved quality, but also increases the fabrication cost.

4. Conclusions

The following may be concluded based on the above considered influence of the turbo-pump unit fabrication and assembly method on the seal clearances:

- Mounting offsets have a significant effect on ensuring the noncontacting operation of the seals;
- Mounting offset calculation requires the consideration of the large number of factors;
- it is reasonable to calculate the mounting offsets of the rotor and stator sealing as per the minimum-maximum method.

Reference

[1] Ivanov A. V. Vliyanie dopuskov formi i raspolozheniya povekhnostei uplotnitelnogo koltsa yf zazori v uplotneniyah TNA // Vestnik Voronezhskogo Gosudarstvennogo Tehnicheskogo Universiteta, Seriya “Energetika”. (Influence of Form and Location Tolerances of the Sealing Ring Surfaces on the Turbo-pump Unit Seal Clearances // Bulletin of the Voronezh State Technical University, "Energy Production" series.) Tom 3. – 2007 – no. 6 – 71–74. (in Russ.) (Vol.3. - 2007 - no. 6 – 71–74. (in Russian))

[2] NASA CR-174886. Seal Technology For Liquid Oxygen (Lox) Turbopumps, 1985 – 390 p.

[3] Satoshi Takada, Masataka Kjkuchi, Takayuki Sudou, Fumiya Iwasaki, Yoshiaki Watanabe, Makoto Yoshjda. Critical Performance of Turbopump Mechanical Elements for Rocket Engine // Trans. JSASS Space Tech. Japan Vol. 7, No. 26, 2009 – pp. 65–70.

[4] Yong-Bok Lee, Kyoung-Wook Kim, Sol-Ji Ryu, Jin Taek Chung. Leakage performance and rotordynamic characteristics of pump floating ring seals for turbopump // Proceedings of ASME Turbo Expo 2014: Turbine Technical Conference and Exposition, June 16 - 20, 2014, Düsseldorf, Germany, 2017. – 8 p.

[5] Bezyazichnyi V.F. Tehnologiya sborki naukoemkih izdelyiy // Naukoemkie tehnologii v mashiostroenii. – 2012. – no. 4 – pp. 28–37. (in Russ.) (Scientific Products Assembly Method // High Technology in Machine Industry – 2012. – no. 4 – pp. 28–37. (in Russian))

[6] Zaharov V.A. Tehnologiya sborki I kontrolya podshipnikovyh opor gazoturbinnyh dvigatelei – M.: (Method for Assembly and Inspection of Gas Turbine Engine Bearing Assemblies – M.:) Mashinostroenie, 1985 – 128 p. (in Russ.) (Mashinostroenie Publishing Company, 1985 – 128 p. (in Russian)

[7] Ilyina M.E. Metod upravlrniya tehnologicheskim protsessom sborki rotora GTD diskovogo tipa // Izvestiya Volgogradskogo gosudarstvennogo universiteta. – 2006 – no. 2 – pp. 25–27. (in Russ.) (Method of Controlling the Process of Assembling the Disk-type Gas Turbine Engine Rotor // Izvestiya of Volgograd State University – 2006 – no. 2 – pp. 25–27. (in Russian))

IOP Conf. Series: Materials Science and Engineering **491** (2019) 012030 doi:10.1088/1757-899X/491/1/012030

Factors Determining the Pipelines Installation Features during the Liquid Rocket Engine Assembly

V.V. Charykov[1,*] and A.V. Ivanov[2]

[1] Post-graduate student, Deputy Technical Director, NPO Energomash Named after Academician V.Glushko, JSC, bldg. 1, Burdenko street, Khimki, 141400, Russian Federation
[2] Deputy Chief Designer for new technologies of NPO Energomash Named after Academician V.Glushko, JSC, bldg. 1, Burdenko Str., Khimki, 141400, Russian Federation

*charykov_vv@npoem.ru

Abstract. A liquid rocket engine contains a vast number of pipelines. The present article describes some of the features with respect to the pipelines fabrication, introduces the main factors influencing the mismatches during the engine assembly, proposes an approach to the pipelines fabrication based on the application of the advanced design, calculation, fabrication and inspection methods based on the 3D geometrical model of an object. The existing technology of the assembly provides for the manufacture of the engine layout, the reference design of the templates. Manufactured pipeline in accordance with the standard for each specific instance of the engine requires adjustment. The exclusion of modifications and adjustments during assembly requires the use of modern methods of design, production and control based on the use of the model-oriented approach. Including calculation mathematical modeling of all processes occurring during the manufacture and assembly of pipelines.

1. Introduction

The liquid rocket engine consists of a large number of high-load units. The units are connected by means of the pipelines (Figure 1) operating under high pressure, temperature and vibration loads. It is required to ensure the maximum strength and tightness of all connections at the minimum weight and installation loads in order to provide the liquid rocket engine reliable operation.

Figure 1. Layout of the liquid rocket engine general assembly components

The majority of pipelines in the modern engines are installed "in-situ" in the course of the assembly. The existing assembly method provides for the engine mockup making, pipelines standardization based on the mockup and making of patterns [1, 2]. Since the engine mockup contains units having unique dimensions that differ from the nominal ones, the reference pipelines are fabricated for these particular unique dimensions [3]. So the pipeline fabricated in accordance with the reference standard requires adjustment for each particular engine. In addition, the development of rocket engineering requires a constant improvement of strength characteristics of the materials applied that creates additional difficulties in the pipelines fabrication.

2. Features the pipelines fabrication

The pipelines fabrication method depends on their structural design [4]. The following may be related to the main features of the liquid rocket engine pipelines design (Figure 2).

- spatial configuration;
- fabication tolerances for the engine structural elements;
- application of high-strength steels and nickel alloys;
- difference of geometrical dimensions for each particular engine.

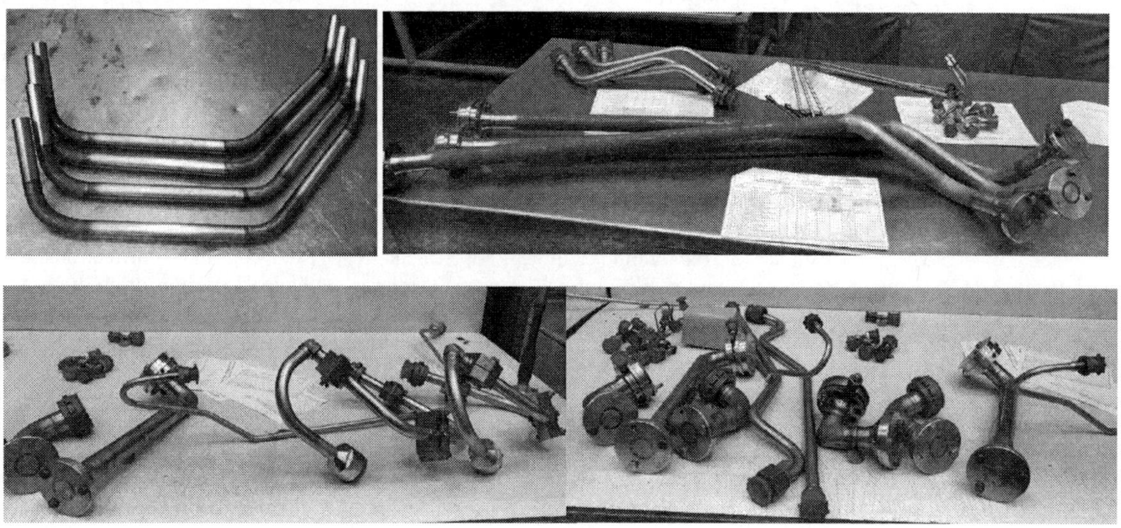

Figure 2. Examples of small and mid-diameter pipelines of the liquid rocket engine

The following issues shall be solved in order to provide for the assembly without prior adjustment at multi-knee spatial bending of pipes:

- determination of the extreme curvature of bending when no defects occur;
- determination of the actual coordinates of pipelines junction points for each particular engine;
- formation of the pipeline 3D design model considering the actual dimensions;
- development of a control program for a pipe-bending machine with numerical program control for each particular pipeline of each particular engine;
- fabrication of pipelines using pipe-bending machines with numerical program control.

Thus, the exclusion of modifications and adjustments during assembly requires application of state-of-the-art design, fabrication and inspection methods based on the use of model-centric technologies (technologies based on the use of geometrical 3D models).

Such an approach corresponds to the current engineering development trends, i.e. moving away from traditional approaches and methods based on the assembly accompanied by numerous metalwork operations: adjustments and modifications. Replacement of such operations with a "virtual" adjustment - the development of the geometrical 3D model of the particular pipeline for the particular engine.

At the same time, it shall be noted that the above organizational and technical measures may solve the issue of pipelines assembly with no modification only upon consideration of a sufficiently wide range of design and process factors. The basic of the same are as follows:

- pipelines junctions point location tolerances;
- pipeline configuration;
- pipeline minimum bend radii depending on its material and diameter;
- presence and length of pipeline linear portions;
- accuracy of spatial coordinates determination;
- actual mechanical properties of the pipeline material;
- springback of the pipeline material depending on its dimensions during fabrication;
- pipeline bending method;
- residual stresses acting in the structure;
- pipeline material relaxation.

3. The value of mismatches in the course of the pipelines installation

The value of mismatches in the course of the pipelines installation may be significant, it determines the spatial position of the mating surfaces and is defined as the vector sum of a large number of values. In general, the value of mismatches in the course of the pipelines installation may be determined as follows:

$$\Delta\vec{A} = \delta\vec{A}_l + \delta\vec{A}_a + \delta\vec{A}_p + \delta\vec{A}_s + \delta\vec{A}_r + \delta\vec{A}_{is} + \delta\vec{A}_t + \delta\vec{A}_f, \qquad (1)$$

$\delta\vec{A}_l = \delta\vec{A}_{lx} + \delta\vec{A}_{ly} + \delta\vec{A}_{lz}$ - deviation from the true position due to the fabrication tolerances;

$\delta\vec{A}_{lx} = \int_x (l_x - l_{xn})\mathrm{d}x$ – x offset;

$\delta\vec{A}_{ly} = \int_y (l_y - l_{yn})\mathrm{d}y$ – y offset;

$\delta\vec{A}_{lz} = \int_z (l_z - l_{zn})\mathrm{d}z$ – z offset;

l_x, l_y, l_z – actual coordinates along axes x, y, z correspondingly;

l_{xn}, l_{yn}, l_{zn} – nominal values of coordinates along axes x, y, z correspondingly;

$\delta\vec{A}_a$ - deviations due to the inspection accuracy when determining the spatial coordinates while inspecting the actual dimensions of junction points;

$\delta\vec{A}_p$ - deviations due to the difference between the actual mechanical properties and nominal values;

$\delta\vec{A}_s$ - deviations due to the springback of the spatial pipeline in the course of its fabrication;

$\delta\vec{A}_r$ - deviations due to the pipeline material relaxation;

$\delta\vec{A}_{is}$ - deviations due to the pipeline internal stresses;

$\delta\vec{A}_t$ - deviations due to the temperature deformations resulted from the temperature difference in the course of the pipeline fabrication and engine assembly;

$\delta\vec{A}_f$ - deviations due to the cross-section deformation in the course of the pipelines fabrication.

In this case, the largest contribution to the mismatches is made by the springback that depends on such factors during bending as the pipeline mechanical properties and wall thickness, bending radius, pipeline shape, bending methods, etc. The uniformity of the pipeline fabrication material mechanical properties significantly influences the springback. Currently, the springback value is usually determined based on the experimental data obtained during the process development.

Moreover, minimum plastic deformations resulting in the pipeline cross-section deformation shall be ensured during the pipelines fabrication.

Accounting of the specified deviations is possible when applying advanced numerical methods of 3D mathematical modeling.

It shall be noted that one of the cornerstones of the analytical modeling successful implementation is the adequacy of the applied model that means the mathematical qualitative and quantitative description of a real-life physical object with all the constraints, assumptions and boundary conditions. Traditionally, within the framework of the flight-type engines' common design and manufacture practice, the analytical studies were carried out mainly to analyze working processes in the liquid rocket engine units and elements. The design analysis of processes was not carried out, the primary focus was on the process development with registration and subsequent repetition of the achieved parameters and performance. Nowadays, the application of such approach leads to significant loss of time for the process mastering of production and can make the product range non-competitive.

4. Approach to the fabrication and assembly of pipelines applying modern technologies

The modern engineering development stage requires to calculate and model both work and manufacturing processes. As to the pipelines, it is required to model the entire process of their fabrication including the workpiece heat treatment, bending with account for mechanical elastic and plastic deformations of local heatings, warping and buckling failure during fabrication, elastic and residual stresses and deformations, cold-work strengthening, non-uniformity of mechanical properties, etc.

Thus, sophisticated multidisciplinary analytical mathematical models having a high level of compliance with the real-life physical and mechanical properties of materials, product design, running technical and production processes are applied in the course of the pipeline fabrication analytical modeling that take into account the entire scope of affecting factors and loads.

The use of the geometrical model maximally similar to the real-life object is an important requirement. For instance, it is desirable to use the geometrical model developed with consideration of the actual part/assembly unit dimensions in order to analyze the strength characteristics of a particular example.

At the same time, it shall be noted that modern design and production engineering approaches basically handle quite simple hydraulic, thermal, kinematic, static and dynamic (vibration) calculations performed based on simplified geometrical models using averaged (rated) material characteristics. The results are sometimes quite different from the full-scale test results and require the introduction of corrective empirical coefficients that is near impossible when designing new products. Consequently, such analytical models do not allow predicting the products behavior during the entire life cycle with the accuracy required to exclude deviations during the fabrication and testing, and the attempt to do so at the early stages of development, even before the models have become more equivalent to reality, turns into an actual "false start" that hands over the baton of problems/incompliances to the next stages of the product life cycle increasing the costs and expenditures, as well as the overall time of the product launch [5].

A generalized algorithm for applying the modern approaches to the design, fabrication and assembly of pipelines including the application of the object geometric model in the course of the design, fabrication and inspection is shown on Figure 3. This algorithm is rather simplified and may be amended for each specific case taking into account the fabrication and operation conditions.

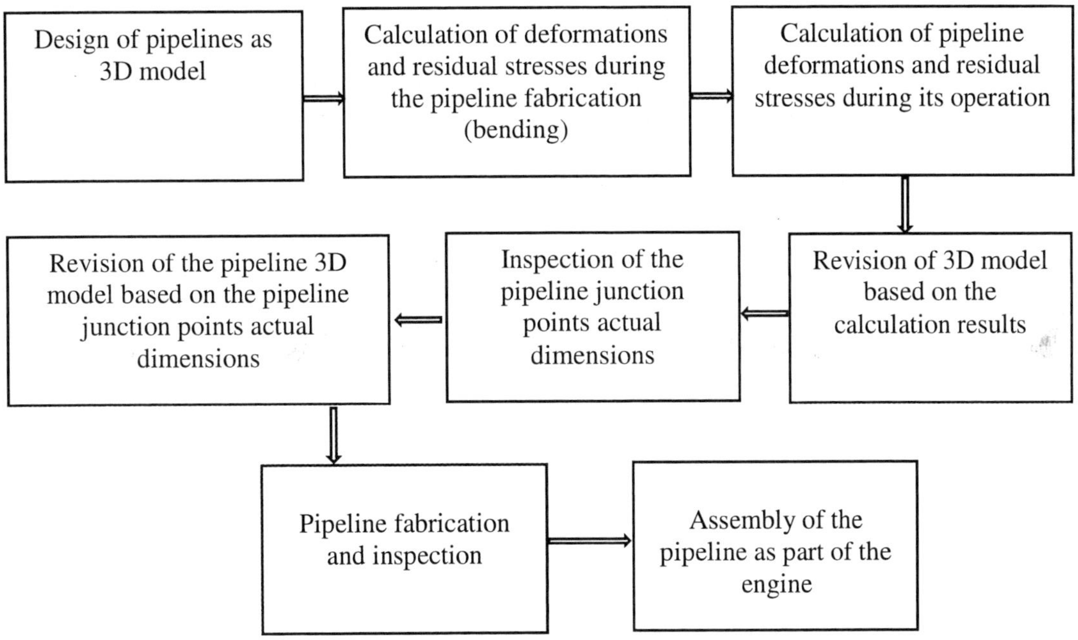

Figure 3. Pipeline fabrication algorithm using the model-centric approach

The application of the specified approach, except for the exclusion of the pipelines adjustment and pre-bending during assembly, will allow to move beyond the typical todays situation when the number of design variations caused by the introduced errors or new data is evenly distributed over the entire life cycle starting from the design and to the production and fabrication preparation. It is a common knowledge that the later the variations are introduced, the higher is the cost of their implementation. With the model-centric approach it becomes possible to transfer the majority of variations to the design stage (in this case, the "design" stage includes the 3D model obtained based on the results of measuring the actual spatial geometry of the junction points) and minimize the costs of the variations introduction and implementation. In addition, the use of such method will allow to eliminate the "human factor" in the course of the pipelines fabrication and engine assembly.

It should be noted that the pipelines design in the form of the electronic design documentation included in the engine electronic model is of crucial significance for the application of the advanced pipeline fabrication and inspection methods and may be implemented in the two following ways:

- development of the electronic model by reproducing the reference pipeline;
- pipeline development based on the product electronic model.

The first method may be used for the engines produced in series or the engines that already have reference pipelines. Such a model for ensuring the assembly without modifications is required to be adjusted to the engine digital model developed as per the nominal dimension values. With this method the reference pipeline is digitized with the help of a 3D scanner receiving a point cloud. Then, a route is constructed based on the received data along which the pipeline cross-section is constructed. The main disadvantage of this method is that the obtained pipeline model will correspond not to the nominal dimensions, but to the dimensions of the particular engine with all its deviations and mismatches. The main advantage lies in the fact that it is not required to develop an electronic model of the entire engine to start the work , i.e. one can get to work immediately upon taking the decision on digitalization.

The second method is applicable for the products being designed. The route is constructed based on the engine electronic model using the basic connecting points and the influence of all elements included in the electronic model. The main advantage of this method is the obtaining of the pipeline model developed as per the nominal dimensions taking into account all the constraints imposed by the designer requirements, as well as the possibility of optimization due to the analytical modeling in parallel with the design. The main disadvantage is that this method requires the development of the complete electronic model of the entire engine.

Both the first and the second methods of obtaining the pipeline electronic models may be used depending on the engine development stage, requirements to its elements, availability of instrumentation equipment.

5. Conclusions

The above approach represents one of the steps in utilizing the product "digital twins" - a complete mathematical description of not only the generalized, "ideal" product, but also of each particular example with its inherent geometrical, hydraulic, strength, thermal and other parameters and performance.

Thus, the task of installing the liquid rocket engine pipelines with no adjustments and modifications is an integrated task combining the design, inspection and fabrication issues. The application of state-of-the-art design, inspection and fabrication tools based on the use of the model-centric 3D computer modeling allows to address the solution of the issue of installing the pipelines with no adjustments.

References

[1] Technologia proizvodstva zhidkostnyh raketnyh dvigateleiy / V.A. (Liquid rocket engine manufacturing technology / V.A. Moiseev, V.A. Tarasov, V.A. Kolmykov, A.S. Filimonov. – M.: Bauman Moscow State Technical University, 2008. – 381 pp. (in Russian)

[2] Vorobeiy V.V., Loginov V.E. Technologia proizvodstva zhidkostnyh raketnyh dvigateleiy – M.: (Liquid rocket engine manufacturing technology – M.: MAI, 2001. – 496 pp. (in Russian)

[3] Satiukov V.A., Sokolkin Y.V. Tehnologicheskaiya mehanika toplivnyh magistraleiy zhidkostnyh raketnyh dvigateleiy. (Process mechanics of liquid rocket engine feed lines.) M.: FIZMATLIT. 2009. – 144 pp. (in Russian)

[4] McAllister E. W. Pipeline rules of thumb handbook: a manual of quick, accurate solution to every day pipeline engineering problem. Burlington, Ma.: Gulf professional publ., 2009. 747 p.

[5] Tsifrovoe proizvodstvo. (Digital production.) Metody, ekosistemy, tehnologii. (Methods, ecosystems, technologies.) Rabochiy doklad departamenta korporativnogo obucheniya shkoly upravleniya Skolkovo. 2018, - 110 pp. (in Russ.) (Working paper of the corporate training department of Skolkovo school of management. 2018, - 110 pp. (in Russian))

IOP Conf. Series: Materials Science and Engineering **491** (2019) 012031 doi:10.1088/1757-899X/491/1/012031

On Unification of Used Materials and Processes at the Enterprise

V V Zakharov

JSC "NPO Energomash named after V.P. Glushko", 1st Burdenko Street, Khimky, Moscow Region, 141400, Russian Federation

Email: zakharov_vv@cb.npo

Abstract: The unification is one of possible ways to reduce costs and labor intensity. When unique technologies are used, the unification of technological processes can not be massive, but it can greatly simplify the manufacture of assemblies and units by reducing the amount of plating coating and finding the best coating to provide reliable material prevention. A liquid rocket engine is a subject to enormous loads during its operation, and the loss of tightness in the fuel and oxidizer channels can lead to irreversible consequences. The unification of the materials of the sealing elements is one of the necessary directions to ensure the safety of the engine. Thus, by unifying the engineering process of plating coating a stable tightness of detachable joints can be achieved, the production cycle may be optimized by reducing the number of toolings. By the end the unique technologies ensuring the tightness of components and assemblies will not be lost.

1. Introduction

The succession of technical solutions in the form of technological unification reduces the variety of processes and technical equipment, eliminates duplication of works during the technological preparation of production, reduces its labour intensity and duration, expands the application scope of progressive means and processes, implements the resource saving policy and provides production flexibility.

For the purposes of technological process unification, one should elaborate production processes for duplicate parts and tooling, as well as develop single processes, such as of plating coating application, and further module technology of coating application in multiproduct manufacture.

Sealing elements, which ensure tightness, are essential structural components of liquid-propellant rockets. Effective solutions for sealing element design significantly influence on profitability, reliability and energy conversion efficiency of a power unit.

Complete tightness often remains an unfeasible wish not only because of fundamental physical problems, but also due to the required solution cost effectiveness [1]. One of the key factors of a successful performance of works is to solve the problems of a design and technological unification and standardization of products and technological processes.

A variety of service conditions and tightness requirements led to the creation of a variety of sealing elements, mainly based on mechanical sealings [2]. Quite often repair costs exceed the price of destroyed assemblies or even the whole power unit in hundred or thousand times.

The problem is now becoming ever more relevant with lack of state funding and unwillingness of business to invest into long-term projects. Optimization of production processes by means of unification will provide us with a certain positive economical effect and minimize labour intensity of process flows. Which, in its turn, will decrease production costs. Whereby the positive economic effect spreads on both technological processes (main indicators of unified process quality are productivity, service life, reliability and other indicators reducing operational costs) and personnel, for example, reducing training time.

2. Main Part

Being a scientific production association, "NPO Energomash" aims at unification mainly through the material selection, optimization of the number of items and used materials, which all together will reduce production and labour costs, e.g. for sealing element manufacture or plating coating application.

One of the major requirements considered during unification and standartization is design for manufacturability. It should be noted that a technologist and designer should closely cooperated during designing works in order to manage design for manufacturability.

Design for manufacturability means the combination of its properties that in desired conditions ensure optimal labour costs, finances, materials, and time for technical preparation of production, operation, repair and pursuant to established quality indicators are observed.

Production manufacturability should ensure reduction of time and money for design and technological preparation of production and for manufacturing of the product. Production manufacturability should provide reduction in labour intensity and production costs, including costs of technical preparation for production. The main problems, arising during creation of a new rocket and space technology, are primary associated with the demands to improve efficiency and increase service life. Operational parameters depend on permissible levels of pressure, temperatures and speeds, the limits of which are often determined by the capabilities of used sealings [3].

JSC "NPO Energomash" is a principal designer of liquid-propellant rockets with large power and high energy characteristic.

Large powers lead to high voltages in the assemblies and units of the engine. Respectively, it is required a high tightness of numerous detachable joints and the use of elastic metal sealings, which provide the necessary contact with sealing surfaces, sufficient to ensure tightness. Reliability and tightness of detachable joint depend on the initial load on the sealing surfaces, the value of which should be sufficient to provide the required contact pressure.

Processes, for sealing purposes, are determined by liquid and gas properties, heat transfer and heat loss, phase changes, wear and corrosion, balance of forces and torques applied on moving parts, vibration state, etc. Rather frequently these determining parameters are so closely interconnected, that it is hard to calculate precisely the sealing system behaviour in service [4]. In view of this, it is necessary to find the solution or to prove that it is impossible to create a definite seal with the requirements stipulated by the existing state of technology.

Strict requirements for operation reliability of units and component supply lines with detachable joints are applied to the sealing elements of modern re-usable liquid-propellant rockets.

Sealing elements work at heavy deformations and frictions of the joint contacting surfaces caused by high power, dynamic and thermal loadings, which leads to rivets and fretting wear of these surfaces.

To ensure a longer service life, the sealing elements must be elastic and made from high-strength alloys with good ductility and fatigue strength.

Elastic sealing elements should provide the specified tightness and maintain serviceability within larger temperature and pressure ranges.

Consequently, our goal is to improve sealing reliability and decrease labour intensity. For this, we elaborate a unified technology for sealing manufacture, including: 1. Selection of base material; 2. Maturization of surface preparation for coating technology; 3. Optimization of coating layer thickness.

In order to fulfill these goals, the first stage of works in "NPO Energomash" Labs included necessary researches of defects and reasons of their occurrence on sealing surfaces of parts. These works are required to achieve required parameters of used materials.

Years of experience and collected statistics show that when assembling an engine for qualification tests, and often after them, various connections are formed with leackages. In the course of our experimental works, we investigated a number of different sealing elements installed in especially critical assemblies and units.

The main material for the sealing elements is alloy steel, which has high strength characteristics while maintaining high ductility when the temperature goes down to -253°C. The steel is not sensitive to stress raisers along the whole range of cryogenic temperatures [5]. Tests of detachable connectors simulators with sealings made from the selected steel in liquid hydrogen at 35 *MPa* under conditions close to real showed that the sealing maintains high tightness, and leakage does not exceed 7.5 × 10-11g/s, which is significantly lower than acceptable values. Thus, in order to ensure protective characteristics and, consequently, to increase tightness, we estimated unified Cu+*Ag* plating coating.

During the complex inspection of sealing elements after tests, we face the following difficulties:

1. Defects in the form of through deepenings, which are located across the contact track, were identified by macroanalysis of the working sealing surfaces (Figure 1)
2. There are numerous relatively deep stripes oriented in the circumferential direction on the sealing surface (Figure 1)
3. Electron microscopical study showed presence of silver in the form of "knobby" bulbs on the surface of through deepenings
4. Using Surftest *SJ-400* profilograph, the profilogram of the sealing surface in the circumferential direction (Figure 1) was made and the surface roughness coefficient was determined (roughness coefficient of the sealing surface is *Ra* = 1.76 (should be within 1.25 as per the design documentation))

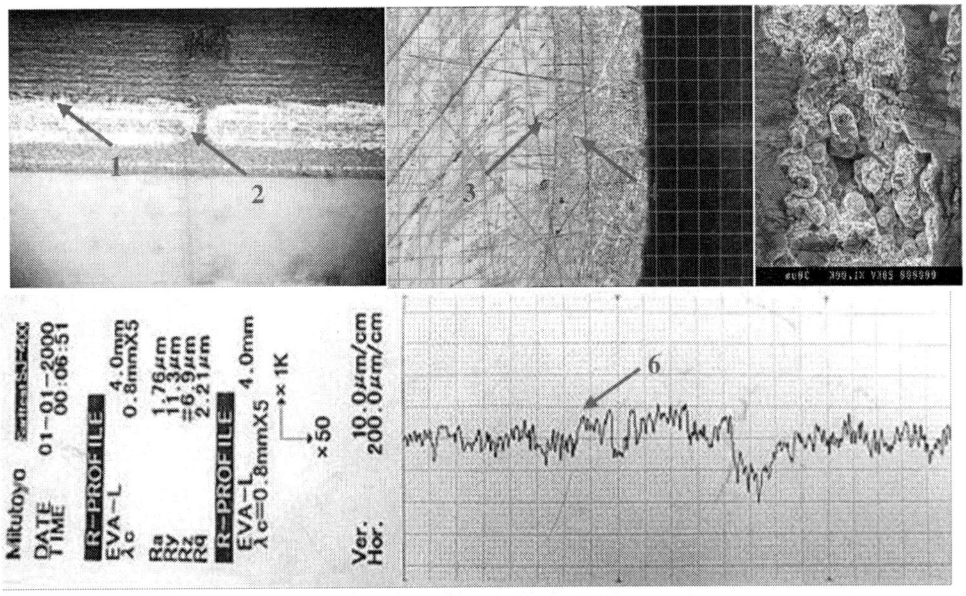

Figure 1. Deviation visualisation
(1–deformation trace; 2–through deepening;
3,4–demarcation of copper and silver layers;
5– "knobby" bulbs on silver coating; 6–roughness).

5. A defective area of wear with significant plastic deformation in the radial direction (from inner to outer diameter) is present. The sealing surface has a circular deformation trace due to the contact with the sealing surface.

6. Electron probe microanalyzer studied the areas of cracking and particles pitting and showed that the X-ray spectrum contains both silver lines and copper lines, which indicate complete wear of the silver coating in these local areas.

7. Peeling of the silver coating along the outer surface of the first layer of silver coating is presented. The peeling cause was revealed to be the low adhesion of the silver outer layers to the first layer surface.

8. The microstructure of the coatings was studied on transverse microsections (Figure 2).

At the transition corner between the sealing surface and inner diameter of the part, the copper coating was "polished" down to zero, and the first silver layer is absent (mechanically removed). Silver in this area has a two-layer structure and ~ 30 μm thickness (Figure 4).

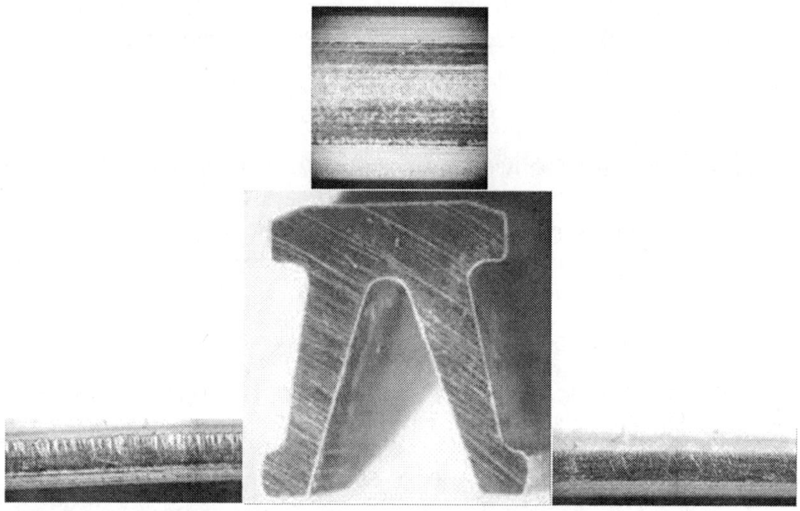

Figure 2. Macroimages of sealing section, sealing surfaces and outer diameter surface.

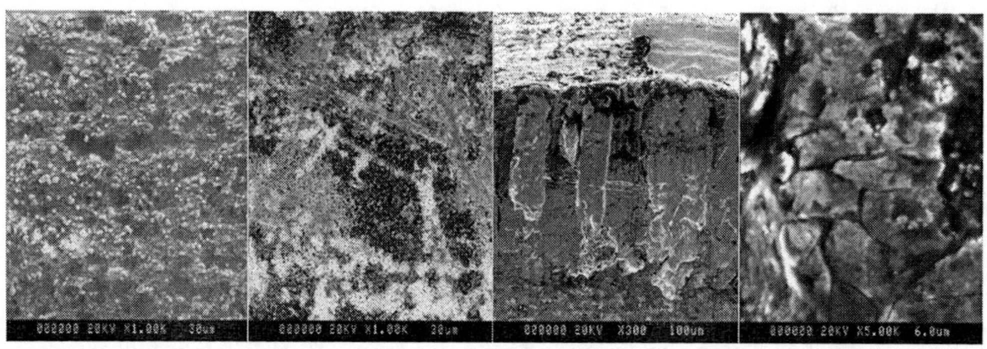

a b c d

Figure 3. Surface of undamaged silver coating (a); surface of deformation trace (b); radial grooves (c); cracking of silver coating and pitting of scales on fretting surface (d).

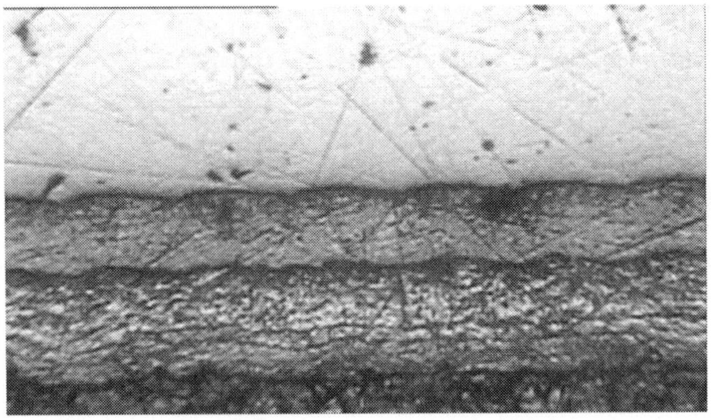

Figure 4. Copper and silver coating microstructure
Stripping of the silver coating goes along the boundary between the first and second layer

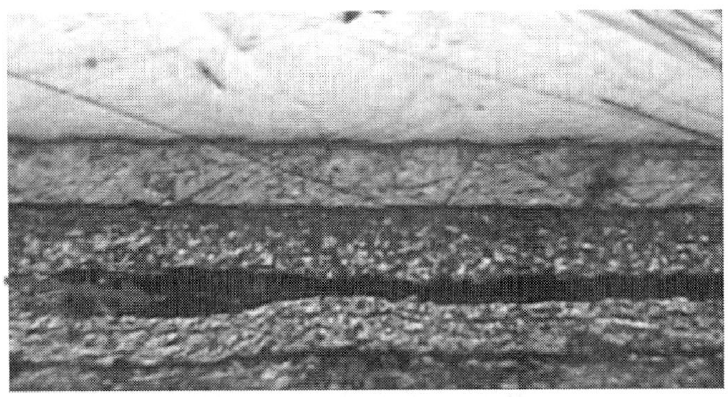

Figure 5. Coating microstructure.
Stripping of the silver coating goes along the boundary between the first and second layer is shown.

All these deviations lead to excess of maximum permissible tolerances for tightness and leakages.

Conducted researches show that the identified fractographic indicators of wear on sealing surfaces, such as multiple surface (fatigue) cracking, formation of fragmented particles, their pitting, subsequent movement and deformation between the contacting surfaces, present the traces of fretting wear, which intensity depends mostly on normal pressure and amplitudes of the reciprocal deflections of the contacting surfaces (Figure 3).

Circular deformation traces on the sealing surface, as well as on the surface of the sealing outer diameter, indicate that they are pressed under the pressure to the corresponding surfaces of the gas generator body ; while the absence of a noticeable degree of fretting wear on these surfaces suggests that the reciprocal deflections between them and the surfaces of the gas generator body are practically absent, i.e. during movement of the gas generator body surfaces, the sealing moves with them as a whole.

It appears that the stripes detected on the sealing surface are resulted from grinding the surface of the copper sublayer by a sandpaper, while according to the design documentation, polishing of the copper coating is allowed.

Low level of adhesion of outer layers of the silver coating to the first layer demonstrates low-quality preparation of a surface of the first layer under the application of subsequent layers. It seems that application of a multilayer silver coating is a "mistake", since a single-layer coating with the same thickness is more technological and functional (Figure 5). Silver of a given thickness should be applied on copper-coated sealing surfaces polished up to the cleanliness required by the design documentation (copper is an adhesive sublayer for silver and its mechanical treatment is technological nonsense). In this case, the functionality of a silver coating will be provided and the probability of defects because of its low adhesion will significantly decrease [6].

3. Conclusions

To sum it up, unification of materials, implemented to eliminate revealed anomalies, improves plating coating adhesion and sealing element elasticity. We also optimize the process of copper-and-silver coating application.

Revealed circumstances compel us to reconsider the technology of copper-and-silver coating application, selection of surface preparation modes and of coating modes.

It is suggested to stipulate 5 μm as a minimum copper coating thickness to ensure adhesion between the base material and silver coating and, respectively, to comply with silver thickness requirements in the drawing. It is suggested to exclude copper coating polishing and, thus, prevent physical interference into the sealing surface. Absence of stripes during polishing and silver bulbs due to improved adhesion decrease a risk of leaking sealing production.

Satisfying all specified requirements, we decrease coating application time from 1 shift to 2 hours, which allows us to reduce labour costs by 4 times and further reduce production costs.

References

[1] Belousov A I 1989 *The scheme and design of rotary shaft seals in turbomachines for aircraft engines*, ed. Belousov A I and Zrelov V A, (Kuybyshev: KuAI) p 104

[2] Belousov A M, Falaev S V, Lejin D S and Vinogradov A S 2001 The experience to provide the increased resource of face seals for blade machines *The materials of the 1st scientific conf. «SINT'01»* (Voronezh) pp 215–24

[3] Belousov A I, Zrelov V A, Falaleev S V, Harlamov R V 1991 Investigation of mechanical seals with Rayleigh's chambers *Seals and vibration reliability of centrifugal machines: Materials of the 6th scientific and technical. conf.* pp. 14-9

[4] Gromiko B M, Matveev E M, Mituk Y V, Mihalev I A, Postnikov I D, Sorokin V A, Samarin V K 2006 *Works NPO Energomash named after V.P. Glushko,* **34** pp 163–79.

[5] Gromiko B M 2003 Classification and characterization of materials for metal seals *Metal science and heat treatment of metals* **6** pp 3-9

[6] Nedashkovskyi K I, Zheleznyak, Gromiko B M, Kozikov B A and Mihalev I A *Metal science and heat treatment of metals* **6** pp 33-6

Process Parameters Optimization Using Taguchi's Orthogonal Array and Grey Relational Analysis during Hard Turning of AISI D2 Steel in Forced Air-Cooled Condition

S K Rajbongshi[1]and D K Sarma[1*]

[1]National Institute of Technology Meghalaya, Shillong, 793003, India

[1*]Email: dksarma@nitm.ac.in, Phone: +919864430524, Fax: +91 364 2501113

Abstract. Optimization of process parameters play an important role in machining operations. The present work deals with a multi-objective optimization method (Grey relational analysis) and Taguchi's orthogonal array (OA) technique in machining of AISI D2 steel using coated carbide tool in forced air-cooled condition considering it to be environment friendly. The process parameters considered are cutting speed, feed and depth of cut having three levels each. The performance parameters are surface roughness, flank wear and cutting force. Taguchi's orthogonal array L_9 is considered for the experimental design. Grey relational analysis technique is used for finding the optimal settings of the process parameters to obtain the value of surface roughness, flank wear and cutting force. The optimum grey relational grade has been found out to get the optimal setting of the experimental run. Then level wise grey relational grade has been found out from the experimental run for each parameter. The optimal settings based on the rank of the grey relational grade in the experimental run and level wise optimal settings of the parameters from the grey relational grade have been compared. The improvement of performance parameters is confirmed from the grey relational prediction analysis in comparison to optimal values in the experimental run.

1. Introduction

The machining is one of the most important operations in manufacturing sector. Among the different machining operations, turning is mostly used for making different products in manufacturing areas. The turning operation using hard materials is a challenging task for the operator due to the risk of tool failure. To perform the machining operation of the hard materials, researchers are using high grade tool which is highly wear resistant and chemically inert. Researchers are mainly using ceramic and CBN tool for hard turning operations, although some of them used coated carbide tool also. The use and applications of coated carbide tools like TiAlN, TiN, TiCN, TiCON, Al2O3 and ceramic and CBN and PCBN tools in hard machining were stated by Bartarya and Choudhury [1]. The machinability of Stellite 12 alloys using coated and uncoated carbide tools were studied by Shao et al. [2]. Their study revealed that the flank wear of uncoated carbide tool was more than coated carbide tool. Strenkowski et al. used cubic boron nitride (CBN) and improved ceramic cutting tool in machining hardened steel materials to avoid the usage of coolants [3].

Due to the non-linear behaviour of machining operations, it is very difficult to obtain proper machinability by simply choosing the ranges recommended by the manufacturers. Proper setting of the machining operations is a must to obtain the optimal values of the responses. That is why, optimization technique is necessary for every types of machining operations. Taguchi's optimization technique is an important optimization tool in machining. Taguchi's orthogonal array (OA) technique helps to reduce the number of experiments, thereby saving time and cost of machining. Theses minimum number of experiments can give the optimal values of the responses. Taguchi's optimization technique is useful for optimizing single response at a time. But when multi-objective optimization criteria come, Taguchi's optimization technique cannot optimize the all responses at a time. To fulfil this discrepancy in machining, grey relational analysis is found to be an important multi-objective optimization tool which can optimize the several responses at a time. It helps to obtain optimum values of many responses at one particular setting. Gopalsamy used grey relational analysis with Taguchi's OA and ANOVA to determine the effect of process parameters viz. cutting speed, feed and depth of cut on material removal rate, surface roughness, tool wear and tool life [4]. The authors were able to optimize the machining parameters for optimal response values. Kacal and Yildrim used Taguchi's OA and grey relational analysis in turning of AISI D6 (60 HRC) steel using ceramic and cubic boron nitride cutting tool [5]. The responses studied were tool wear, surface roughness, machining force and specific cutting force. Using grey relational prediction analysis, they were able to get the optimum values of the responses as compared to the values obtained in the experimental run. Huang and Liao used a combined technique of Taguchi's OA concept and grey relational analysis to obtain optimum machining parameters in wire-EDM process [6]. Using this method, it was found that material removal rate was mainly influenced by feed rate and pulse on time had significant effect on gap width and surface roughness. Lin used Taguchi's method and grey relational analysis in multiple performance optimization of tool life, cutting force and surface roughness in machining of S45C steel bars using P20 tungsten carbide tool [7]. The authors found the improvement of response values. Grey relational technique was used by Nayak et al. in optimization of machining parameters to obtain the optimum values of material removal rate, cutting force and surface roughness in dry turning of AISI 304 austenitic stainless steel [8]. Confirmatory test using grey relational analysis gave 88.78% improvement in grey relational grade. Mishra et al. used Taguchi's L16 orthogonal array along with the grey relational analysis in optimizing process parameters in turning of AA 7075/SiC composite in an environment of dry and spray cooling [9]. The authors reported that the performance of spray cooling was better as compared to dry turning by using grey relational analysis. Sohrabpoor et al. used dry, wet and MQL as the medium of environment in turning of AISI 4340 steel using coated carbide tool [10]. Using grey relational analysis, the authors found that MQL environment method gave lowest surface roughness and tool wear. Patole et al. used grey relational analysis in optimizing machining parameters in turning AISI 4340 steel using tungsten coated carbide inserts in MQL environment [11]. Using grey relational analysis, the authors found the improvement of machinability of AISI 4340 steel using MQL.

From the above literature reviews, it has been observed that the researchers have used Taguchi's OA array along with the grey relational analysis to perform the multi-objective optimization characteristics and most of them used dry as the working environment. Also, it is observed that very few researchers have used forced air-cooling environment as the medium of operation in optimizing the machining parameters using grey relational analysis. So, in this present work, Taguchi's OA along with grey relational analysis have been used in optimizing machining parameters in turning of AISI D2 steel using coated carbide tool and force air-cooled environment as the medium of operation.

2. Experimental set up

2.1. Cutting tool & work piece material

i. Cutting Tool Material: Coated Carbide Tool (Make: Sandvik; Model: SNMG 12 04 08- KM3215).
ii. Work Piece Material (commercially available AISI D2 Steel (hardened to 48 HRC)).

3. Results and discussion

Cutting speed, feed and depth of cut are selected as process parameters and surface roughness (SR), flank wear (FW) and cutting force (CF) as performance parameters. The ranges of the cutting parameters are selected according to the recommendation of the tool manufacturers and literature. Table 1 shows the cutting parameters with their level.

Table 1. Cutting parameters with their levels.

SI No.		Level (Low)	Level (Medium)	Level (High)
1	v (m/min)	100	125	150
2	f (mm/rev)	0.05	0.10	0.15
3	d (mm)	0.25	0.40	0.55

On the basis of level and degrees of freedom (DOF) of the cutting parameters, Taguchi's OA L_9 has been chosen without any interaction effect for designing and running the experiment. Table 2 shows Taguchi's OA L_9 settings of the cutting parameters with their responses.

Table 2. Taguchi's OA L_9 settings of the cutting parameters with responses.

Expt. No.	v (m/min)	d (mm/rev)	d (mm)	SR (μm)	FW (mm)	CF (N)
1	100	0.05	0.25	1.77	0.10	48
2	100	0.10	0.40	1.61	0.19	235
3	100	0.15	0.55	1.53	0.27	299
4	125	0.05	0.55	1.57	0.29	163
5	125	0.10	0.25	1.36	0.18	99
6	125	0.15	0.40	1.64	0.28	182
7	150	0.05	0.40	1.42	0.30	79
8	150	0.10	0.55	1.93	0.40	225
9	150	0.15	0.25	2.12	0.30	96

3.1. Grey relational analysis

Grey relational analysis is a statistical technique which is used for optimizing the multi-objective characteristics phenomenon. Multi-optimization technique is necessary and helpful in the sense that it selects single setting of the process parameters which give overall optimum values of various responses considered for the study. The first thing is to normalize the experimental data in between 0 and 1 and then the next step is to find the grey relational coefficient from the normalized experimental data in grey relational analysis. After finding the experimental coefficient for each response, the grey relational grade has been obtained for each experimental run. Finally, higher grey relational grade has been selected based on the higher value. Table 3 shows the normalized values of the responses viz. surface roughness, flank wear and cutting force as obtained from Table 2. The normalized data in Table 3 follows the pattern smaller the better criteria, i.e. smaller the values of responses, better will be their performances. The grey relational formula for this is given by

$$x_i^*(j) = \frac{\max x_i(j) - x_i(j)}{\max x_i(j) - \min x_i(j)} \tag{1}$$

Here, $x_i^*(j)$ and $x_i(j)$ are the sequences after the data processing and compatibilty sequence respectively, j = 1 for SR and i = 1, 2, 3,…9 for the nine experiments from 1 to 9. After finding the normalized value of the responses, the next step is to find the deviation sequences for each experimental run. This is found out by subtracting the value of each normalized term in the experimental run from the ideal reference value. Table 3 shows the deviation sequence for the responses of each experiment.

In the Table 3, $\Delta O_i(1)$, $\Delta O_i(2)$ and $\Delta O_i(3)$ represents the deviational sequences of the responses SR, FW and CF respectively.

$\Delta_{max} = \Delta O_9(1) = \Delta O_8(2) = \Delta O_3(3) = 1$, $\Delta_{min} = \Delta O_5(1) = \Delta O_1(2) = \Delta O_1(3) = 0$

Δ_{max} and Δ_{min} represents the maximum and minimum deviation of the above three responses respectively.

Table 3. Normalized value of the responses SR, FW and CF and grey relational coefficient and grade

Expt. No.	SR	FW	CF	Expt No	SR	FW	CF	Grey relational coefficient			Grey relational grade (γ)	Rank
Reference Sequence	1.00	1.00	1.00	Deviation sequences	$\Delta O_i(1)$	$\Delta O_i(2)$	$\Delta O_i(3)$	SR $\xi_i(1)$	FW $\xi_i(2)$	CF $\xi_i(3)$		
1	0.46	1	1	1	0.54	0	0	0.48	1	1	0.826	2
2	0.67	0.7	0.25	2	0.33	0.3	0.75	0.60	0.62	0.4	0.541	5
3	0.78	0.43	0.0	3	0.22	0.57	1	0.69	0.47	0.33	0.497	7
4	0.72	0.37	0.54	4	0.28	0.63	0.46	0.64	0.44	0.52	0.533	6
5	1	0.90	0.79	5	0	0.10	0.21	1	0.83	0.70	0.843	1
6	0.63	0.67	0.47	6	0.37	0.33	0.53	0.57	0.60	0.48	0.550	4
7	0.92	0.33	0.87	7	0.08	0.67	0.13	0.86	0.42	0.79	0.690	3
8	0.25	0.0	0.29	8	0.75	1	0.71	0.4	0.33	0.41	0.380	9
9	0	0.33	0.81	9	1	0.67	0.19	0.33	0.42	0.72	0.491	8

3.1.1. Determination of grey relational coefficient and grey relational grade. Grey relational coefficient determines the values of the data after preprocessing. The grey relational coefficient correlates between the ideal and actual values of the normalized experimental data after the data preprocessing. The higher value of grey relational coefficient is an indication of better result as compared to lower value grey relational coefficient. The formula for finding the grey relational coefficient is given by

$$\xi_i(j) = \frac{\Delta min + \zeta \Delta max}{\Delta O_i(j) + \zeta \Delta max} \quad (2)$$

Where $\Delta o_i(j)$ is the difference between the ideal sequence $x_o^*(j)$ and the comparability sequence $xi^*(j)$ and ζ is the identification coefficient. For equal preference parameters ζ is taken as 0.5. The grey relational coefficient for all the nine experiments designed by Taguchi's OA L_9 can be obtained by using the Eqn. (2) and the values are shown below in Table 3. After finding the grey relational coefficient for all the three responses, the grey relational grade has been found for each experimental run by taking the average values of all the grey relational coefficients for each experiment. The formula for finding the grey relational grade is given by

$$\gamma_i = \frac{1}{n}\sum_{j=1}^{n} \xi_i(j) \quad (3)$$

Here γ_i denotes the grey relational grade for the ith experiment and n denotes the number of responses. In Table 3, γ represents the grey relational grade for each experiment, where, $\xi_i(j)$ represent the grey relational coefficients for n responses. In Table 3, grey relational grade has been shown which is obtained from the average values of grey relational coefficients of the three responses for each experimental run. The higher value of grey relational grade has been found to be 0.843 and chosen as rank1 in the Expt. No. 5. As the design of the experiment is orthogonal, so the effect of all the cutting parameters in grey relational analysis can be considered by taking the average values of all the grey relational grade for each parameter level wise. Table 4 shows the average values of grey relational grade level wise for the three cutting parameters v, f and d for the responses SR, FW and CF.

Table 4. Level wise average values of grey relational grade for SR, FW & CF in terms of cutting parameters

SI No.		d (mm/rev) Level 1	d (mm) Level 2	SR (μm) Level 3	Main effect (max-min)	Rank
2	v (m/min)	0.621	**0.642**	0.520	0.122	3
3	f (mm/rev)	**0.683**	0.568	0.512	0.171	2
4	d (mm)	**0.720**	0.593	0.472	0.248	1
	Average value of grey relational grade = 0.594					

In Table 4, the bold letter indicates the optimum value of the grey relational grade. From Table 4, it is observed that the optimum values of SR, FW and CF can be obtained at setting of v at level 2, f at level 1, and d at level 1. Again from Table 3, it is observed that the optimum values for SR, FW and CF can be obtained at a setting of v at level 2, f at level 2 and d at level 1 for Expt no. 5.

3.2. *Confirmation test*

The confirmation test for the setting v at level 2 = 125 m/min, f at level 1= 0.05 mm/rev and d at level 1= 0.25 mm has been performed according to the optimum grey relational grade from Table 4 and the corresponding response values of SR, FW and CF are found to be 1.30 μm, 0.15 mm and 84 N respectively. The estimated value of the grey relational grade $\bar{\gamma}$ can be found using the formula

$$\bar{\gamma} = \gamma_m + \sum_{i=1}^{p} (\gamma_m - \gamma_i) \quad (4)$$

So, from Table 5, it is confirmed that the grey relational grade is improved by 0.115. Comparing the settings of the Expt. No 5 and grey relational prediction it is observed that the surface roughness, flank wear and cutting force have been improved by 4.41%, 16.67% and 15.15% respectively. Hence grey relational prediction in machining AISI D2 steel using coated carbide tool has improved the machinability of AISI D2 steel. To confirm, which cutting parameters significantly affect the multiple-performance characteristics, analysis of variance (ANOVA) is also analyzed using the values of grey relational grades from Table 3. Table 6 shows the ANOVA for the grey relational grade.

Table 5. Grey relational grade improvement using the optimized machining parameters

	Optimal machining parameters	
Condition	Expt. no 5 in the OA L_9 table	Using grey relational prediction
Level	v_2-f_2-d_1	v_2-f_1-d_1
SR (μm)	1.36	1.30
FW (mm)	0.18	0.15
CF (N)	99	84
Grey relational grade	0.742	0.857
	Grey relational grade improvement = 0.115	

Table 6. ANOVA analysis of grey relational grade

Parameters	Degrees of freedom (DOF)	Sum of square (SS)	Mean square (MS)	F-Ratio	Percentage contribution (%)
v (m/min)	2	0.025	0.0125	0.641	12.51
f (mm/rev)	2	0.043	0.0215	1.10	21.52
d (mm)	2	0.093	0.0465	2.38	46.54
Error	2	0.039	0.0195		19.43
Total	8	0.1998			100

From ANOVA in Table 6, it is confirmed that depth of cut is the most influencing parameter as compared to cutting speed and feed rate. This can be confirmed from Table 4 in which the main effect of grey relational grade of depth of cut is more as compared to cutting speed and feed rate.

4. Conclusion

In this work, an experimental study has been performed in machining AISI D2 steel using coated carbide tool in forced air-cooled environment. Taguchi's OA L_9 and grey relational grade analysis have been performed for optimizing the multiple-performance characteristics of surface roughness, flank wear and cutting force. The outcome of the experimental results and grey relational analysis is summarized as follows:

The experimental results using grey relational analysis confirmed the optimum values of surface roughness, flank wear and cutting force in the Expt. No 5 of Taguchi's OA L_9. Grey relational prediction value confirmed the optimum values of surface roughness, flank wear and cutting force at a setting of cutting speed 125 m/min, feed rate 0.05 mm/rev and depth of cut 0.25 mm. Grey relational prediction value improved the responses surface roughness, flank wear and cutting force by 4.41%, 16.67% and 15.15% respectively as compared to the optimal settings of the machining parameters in the Expt. no 5. ANOVA analysis showed that the depth of cut is the most influential factor as compared to cutting speed and feed rate from the grey relational analysis.

References

[1] Bartarya G and Choudhury S K 2012 *Int. J. Mach. Tools. Manuf.* **53(1)** 1-14
[2] Shao H, Li L, Liu L J and Zhang S Z 2013 *J. Manuf. Process.* **15(4)** 673-81
[3] Strenkowski J S, Shih A J and Lin J c 2012 *Int. J. Mach. Tools. Manuf.* **42(6)** 722-31
[4] Gopalsamy B M, Mondal B and Ghosh S 2009 *Int. J. Adv. Manuf. Technol.* **45** 1068-1086
[5] Kacal A, Yildrim F 2012 *J. Mech. Eng. Sci.* **227 (7)** 1566-76
[6] Huang J T and Liao Y S 2003 *Int. J. Prod. Res.* **41 (8)** 1707-20
[7] Lin C L 2004 *Mater. Manuf. Process.* **19 (2)** 209-20
[8] Nayak S K and Potro J K 2014 *Proc. Mater. Sci.* **6** 701-08
[9] Mishra P C, Das D K, Ukamanal M, Routara B C and Sahoo A K 2015 *Int. J. Eng. Comp.* 6 445-56
[10] Sohrabpoor H, Khanghah S P and Teimori R 2009 *Int. J. Adv. Manuf. Technol.* **76 (9-12)** 2099-2116
[11] Patole P B, Kulkarni V V and Mang W 2017 *Cog. Eng.* **4(1)** 1-14.

IOP Conf. Series: Materials Science and Engineering **491** (2019) 012033 doi:10.1088/1757-899X/491/1/012033

Parametric Optimization in Photochemical Machining of Aluminium Using Taguchi Method

Rashed Mustafa Mazarbhuiya[1] and Maneswar Rahang[1, *]

[1]Department of Mechanical Engineering, National Institute of Technology, Meghalaya, India

*Corresponding Author's e-mail address: maneswar909@gmail.com, Telephone number +91 9402102355

Abstract. The objective of this study was to investigate the optimal process parameters of photochemical machining on aluminium workpiece. The input parameters considered were etching concentration, etching temperature and etching time. The study revealed their effects over Material Removal Rate, Surface Roughness and Edge Deviation. L_9 Taguchi design was used to formulate the experimental layout. ANOVA and average S/N ratio were performed for the optimal setting of the process parameters. For larger-the-better characteristics the optimal conditions for MRR was obtained at etching concentration of 600g/L, etching temperature of 60°C and etching time of 8 minutes. For, smaller-the-better characteristics, the optimal setting for surface roughness was obtained at etching concentration of 400g/L, etching temperature of 40°C and etching time of 4 minutes and for edge deviation obtained at etching concentration of 400g/L, etching temperature of 40°C and etching time of 8 minutes.

1. Introduction

Unconventional machining processes have been employed for manufacturing of micro size parts like silicon integrated circuits, microfluidic channels, decorative items and copper printed circuit board [1]. A photochemical machining is one of the emerging nonconventional machining processes that produce stress free and burr free flat complex metal parts. The process involves chemical etching through a photo-resist stencil as a method of material removal over selected areas [1]. Using ferric chloride as chemical etchant aluminium was machined at different etching temperature and found that ferric chloride was a suitable etchant for aluminium etching [2]. The selection of etchant is based on the workpiece and selection of machining parameters is based on etchant, etching temperature and etchant time [3]. This selection of efficient etchant is probably the most important parameter in chemical etching of any material. This will enhance the etching rate and produce smooth surface finish [2]. Various characteristics of ferric chloride etchants are studied which are used in industrial photo chemical machining. Although $FeCl_3$ most commonly used as etchant but there is wide variety in grades of $FeCl_3$ [4]. The cost involves for PCM defines standards for industrial etchants and methods to analyses and monitor them [5]. Many researchers had studied the effect of etching time and etchant temperature on the surface finish and rate of etching on PCM of copper, aluminium and Inconel 718 using ferric chloride and cupric chloride as etchants [6, 2, 7].

The objective of the present work is an attempt for finding the feasibility of chemical machining of aluminium by using different concentration of $FeCl_3$ at different etching temperatures and time. Taguchi design approach is used to analysis the responses material removal rate, surface roughness and edge deviation. The effect of process parameters on the machining characteristics has been investigated by ANOVA method and the optimum condition has been predicted using average S/N ratio.

2. Materials and Methods

For the experimentation, aluminium material was used as a workpiece of size 20 mm x 20 mm x 5 mm. After that photo tool was prepared in required shape and size. Photo tool is nothing but a negative film of the image to be produced. The workpiece surface was cleaned to remove dust particles to obtain good surface for adhesion of photo resist masking solution. Masking was applied by immersing into photo resist dip coater followed by drying. The coated specimen is placed under the photo tool and exposed to the ultraviolet source. As Photo resist is sensitive to U.V. radiation therefore a U.V tube exposure unit was used for exposure. Photo tools are generally used in precisely registered pairs, one on the top and one on the bottom, with the material to be machined placed in between which allow the material to be etched from both the side. After U.V. exposure the specimen is dipped into solvent-based developer solution for 60 seconds. This softens the unexposed areas of the photo resist. Then, it is dried and baked for 2 minutes followed by dipping into ferric chloride solution where, the etchant solution dissolves the unexposed masked area to get the required shape and size by removing the material. After that the work sample is cleaned by washing process. Based on the literature survey, some pilot experiments were conducted to select the controllable process parameters such as, concentration (g/L), temperature (°C) and etching time (min) as tabulated in Table 1 with their different levels.

Table 1. Control Parameters with their levels

	Control parameters	Level 1	Level 2	Level 3
A	Concentration (g/L)	400	500	600
B	Temperature(° C)	40	50	60
C	Time (min)	4	6	8

Table 2. Taguchi L_9 matrix with weights of workpiece

Sl. No.	Conc. (g/L)	Temp. (°C)	Time (min)	Trial 1		Trial 2		Trial 3	
				Initial weight	Final weight	Initial weight	Final weight	Initial weight	Final weight
1	400	40	4	1.8939	1.8926	1.9476	1.9468	1.9231	1.9225
2	400	50	6	1.7620	1.7463	1.8832	1.8730	1.8831	1.8772
3	400	60	8	1.9334	1.8790	1.9743	1.9438	2.0303	2.0044
4	500	40	6	1.9670	1.9463	1.7527	1.7490	2.0056	1.9973
5	500	50	8	1.8257	1.7814	1.8685	1.8612	1.8596	1.8473
6	500	60	4	1.9168	1.8923	2.1074	2.0974	1.9211	1.9158
7	600	40	8	1.7648	1.7557	1.9557	1.9427	1.9915	1.9846
8	600	50	4	1.9390	1.9306	1.9723	1.9654	1.8928	1.8908
9	600	60	6	1.8449	1.8184	2.0142	1.9923	1.6964	1.6807

Taguchi's design is a fractional factorial method is suitable process for parametric optimization. Taguchi's L9 orthogonal array was selected to carry out the current experimental investigation. Using ferric chloride as etchant, photochemical machining of workpiece was carried out. The effect of input parameters on the material removal rate (MRR), surface roughness (Ra) and edge deviation were studied.

The weight of the workpiece before and after machining are tabulated in table 2. Using optical microscope, edge deviation of desired shape and size were measured machining. The material removal rate was obtained by calculating the volume of material removed and then dividing it by the etching time. After the experimentation, the values of MRR, Ra and ED along with their S/N ratio were measured as tabulated in table 3. Figure 1. shows the machined specimen, measurement of edge deviation and measurement of surface roughness of work samples.

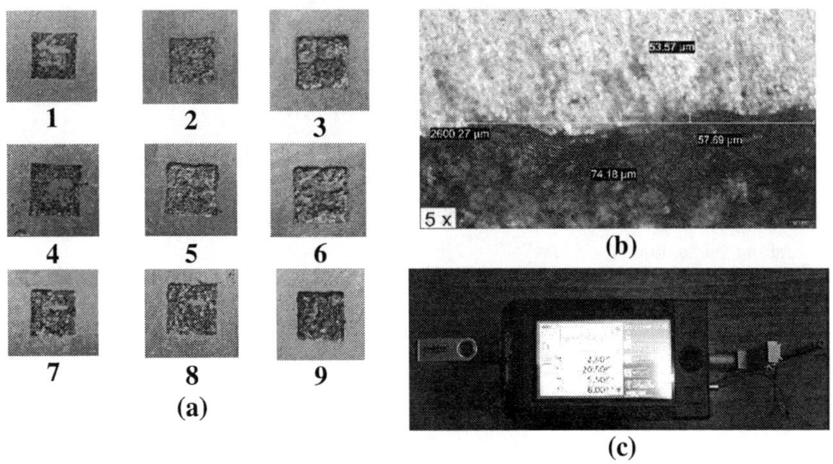

Figure 1. (a) Photo chemically machined sample specimen **(b)** Edge deviation measurement **(c)** Measurement of surface roughness

Table 3. Calculation of MRR Ra and ED along with their S/N ratios

SL. No.	Trial 1			Trial 2			Trial 3			S/N ratio		
	MRR	Ra	ED	MRR	Ra	ED	MRR	Ra	ED	MRR	Ra	ED
1	0.325	1.53	51.28	0.2	0.97	51.28	0.15	0.67	46.47	-14.201	-0.946	-33.93
2	2.617	3.1	65.25	1.57	2.9	61.58	0.983	1.43	62.96	2.766	-8.256	-36.03
3	6.8	2.63	72.12	3.812	3.4	62.04	3.237	2.93	60.89	12.078	-9.558	-36.29
4	3.45	1.37	53.57	0.617	1.3	54.28	1.383	1.57	48.76	-0.326	-3.021	-34.3
5	5.53	2.07	70.06	0.912	3.27	55.86	1.537	2.57	59.75	2.574	-8.561	-35.83
6	6.125	2.57	81.73	2.5	2.07	81.28	1.325	2.53	78.76	5.984	-7.604	-38.13
7	1.137	2.67	57.46	1.625	2.6	55.40	0.862	2.2	48.53	0.795	-7.951	-34.63
8	2.1	2.7	77.84	1.725	2.97	61.36	0.5	1.77	67.77	-1.822	-8.006	-36.82
9	4.417	3.37	79.67	3.65	3.07	62.73	2.617	0.67	64.56	10.420	-8.490	-36.83

3. Results and discussions

The experimental analysis was carried out to observe the effect of process variables on response parameters for optimal condition for high material removal rate, low surface roughness and low edge deviation. The MRR increases with increase in concentration, temperature and time. The maximum MRR was observed at concentration 600 g/L, temperature 60°C, and etching time of 8 minutes. The average S/N ratio for MRR are shown in table 4. The ANOVA is presented in table 5. Temperature is the most significant parameter for MRR followed by time. However, the etching concentration has less effect on material removal rate. Based on the average S/N ratio plot, the optimal parameters combination for MRR obtained at etching concentration at 600g/L, etching temperature at 60° and etching time at 8 minutes. The parameters that have greater influence on MRR are temperature, time and etchant concentration as shown in figure 2. The temperature has the most dominant effect amongst all other parameters.

IOP Conf. Series: Materials Science and Engineering **491** (2019) 012033 doi:10.1088/1757-899X/491/1/012033

Table 4. Ave S/N ratio MRR

Level	Etching Concentration	Etching temperature	Etching time
1	0.2143	-4.577	-3.346
2	2.744	1.173	4.287
3	**3.131**	**9.494**	**5.149**
Delta	2.9167	14.071	8.495
Rank	3	1	2
Optimal setting	3	3	3

Table 5. ANOVA for MRR

Factors	Sum of square	DOF	Mean square	F-Ratio	Percent contribution	Rank
Etching concentration	0.46	2	0.23	0.88	3.09	3
Etching temperature	11.847	2	5.923	22.655	79.53	1
Etching time	2.006	2	1.033	3.951	13.47	2
Other factor	0.522	2	0.261		3.5	
total	14.897	8			3.09	

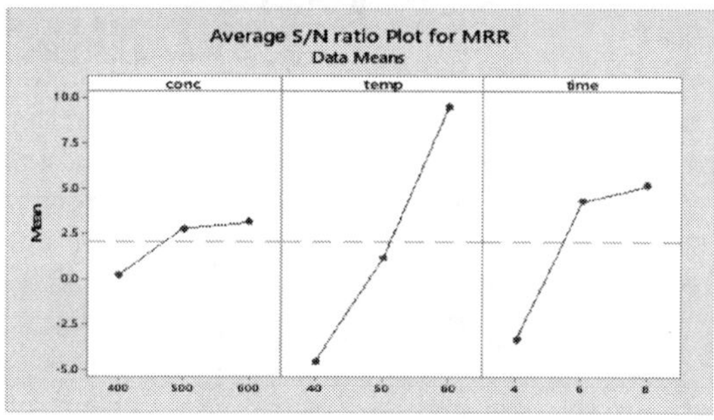

Figure 2. Average S/N ratio plot for MRR

The surface roughness increases with increase in concentration, temperature and time. This is because, as the concentration increase along with temperature then there will be increase in non-uniform material removal that results in increase in surface roughness. The minimum surface roughness observed at concentration 400 g/L, etching temperature 40°C and etching time at 4 minutes. The average S/N ratio values for surface roughness are shown in table 6. From the ANOVA table 7, was observed that etching temperature and etching time contributes more on the variability of surface roughness. However, the concentration had less effect on surface roughness value as compared to other factors. The optimal parameters combination for surface roughness obtained at etching concentration at 400g/L, etching temperature at 40° and etching time at 4 minutes. The average S/N ratio plot for surface roughness is plotted in figure 3. The parameters that have greater influence on surface roughness are same as that for MRR. The temperature has the most dominant effect amongst all other parameters.

Table 6. Ave S/N ratio **Ra**

Level	Etching Concentration	Etching temperature	Etching time
1	**-6.253**	**-3.973**	**-5.519**
2	-6.396	-8.274	-6.589
3	-8.149	-8.551	-8.69
Delta	1.896	4.578	3.171
Rank	2	1	3
Optimal setting	1	1	1

Table 7. ANOVA FOR **Ra**

Factors	Sum of square	DOF	Mean square	F-Ratio	Percent contribution	Rank
Etching concentration	0.164	2	0.082	0.572	5.44	3
Etching temperature	1.636	2	0.818	5.703	54.24	1
Etching time	0.928	2	0.464	3.234	30.77	2
Other factor	0.286	2	0.143			
total	3.016	8				

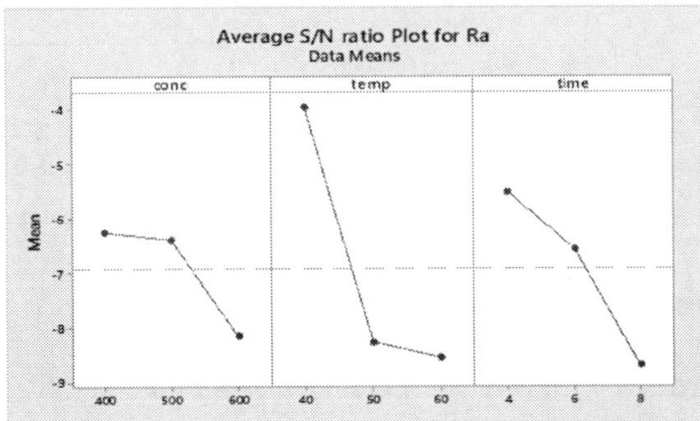

Figure 3. Average S/N ratio plot for Ra

The edge deviation increases with increase in etching concentration and etching temperature but, decreases with increase in etching time. The minimum edge deviation observed at etching concentration 400 g/L, etching temperature 40°C and etching time of 8 minutes. The average S/N ratio value for ED are shown in table 8. From the ANOVA table 9 was observed that etching temperature is the most significant parameter on the variability of edge deviation. However, the etching time and etching concentration had less effect on edge deviation. The optimal parameters combination for surface roughness was obtained at etching concentration at 400g/L, etching temperature at 40° and etching time at 8 minutes. The average S/N ratio plot for edge deviation is plotted in figure 4. The parameters that have greater influence on edge deviation in the decreasing order of importance is same as that for MRR and surface roughness. The temperature has the most dominant effect amongst all other parameters.

Table 8. Ave S/N ratio **ED**

Level	Etching Concentration	Etching temperature	Etching time
1	**-35.42**	**-34.287**	-36.293
2	-36.087	-36.227	-35.72
3	-36.093	-37.083	**-35.583**
Delta	0.673	2.796	0.71
Rank	3	1	2
Optimal setting	1	1	3

Table 9. ANOVA for **ED**

Factors	Sum of square	DOF	Mean square	F-Ratio	Percent contribution	Rank
Etching concentration	51.058	2	25.529	0.914	6.78	3
Etching temperature	579.082	2	289.541	10.371	76.92	1
Etching time	66.841	2	33.42	1.197	8.88	2
Other factor	55.832	2	27.916			
total	752.815	8				

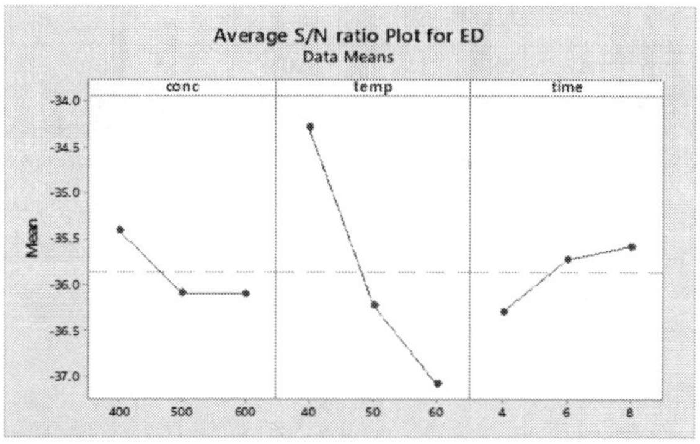

Figure 4. Average S/N ratio plot for ED

4. Conclusions

Based on the results obtained, the following inferences were drawn.

- MRR increases with increase in concentration, temperature and time respectively.
- Surface roughness increases with increase in concentration, temperature and time respectively.
- The edge deviation increases with increase in concentration and temperature but decreases with increase in etching time.
- The optimal parameters combination for MRR obtained at etching concentration at 600g/L, etching temperature at 60° and etching time at 8 minutes.
- The optimal parameters combination for surface roughness obtained at etching concentration at 400g/L, etching temperature at 40° and etching time at 4 minutes.
- The optimal parameters combination for edge deviation obtained at etching concentration at 400g/L, etching temperature at 40° and etching time at 8 minutes.
- From ANOVA it was observed that for MRR, surface roughness and ED temperature is the most significant factor followed by time while machining Aluminium work piece.

References

[1] Allen D. M., Photochemical Machining: From manufacturing's best kept secret to a $6 billion per annum rapid manufacture process *CIRP journal of Manufacturing systems,* **53**(2) 559-572 (2004).

[2] Cakir O., Chemical etching of aluminum, Journal of Materials Processing Technology **199** 337–340. (2008)

[3] Yadav R.P., Teli S.N., A Review of issues in photochemical 450 machining, *International Journal of Modern Engineering Research* **4** 49–53 (2014)

[4] Allen D.M., Almond H.J., Characterization of aqueous ferric chloride etchant used in industrial photo chemical machining, *Journal of Materials Processing Technology* **149** 238-245 (2004)

[5] Roy R., Allen D.M., Zamora O., Cost of photochemical machining, *Journal of Materials Processing Technology,* **149** 460–465 (2004)

[6] Cakır O., Temel H., Kiyak M., Chemical etching of Cu-ETP copper, *Journal of Materials Processing Technology* **162–163** 275–279 (2005)

[7] Wagh D.V., Dolas D. R., Dhagate M. D., Experimental investigation of photochemical machining on Inconel 600 using ferric chloride, *International Journal of Engineering Research & Technology* **4**(2), 289–293 (2014)

IOP Conf. Series: Materials Science and Engineering **491** (2019) 012034 doi:10.1088/1757-899X/491/1/012034

Influence of Brazing Process Parameters on the Strength of Liquid Rocket Engine Brazed Structures

A.B. Aminov[1,*] and K.E. Dubrovsky[2]

[1] Post-graduate student, Chief welder, NPO Energomash Named after Academician V.Glushko, JSC, Russian Federation, Khimki, Burdenko st., bldg. 1
[2] Main specialist of Brazing and Welding Department, NPO Energomash Named after Academician V.Glushko, JSC, Russian Federation, Khimki, Burdenko st., bldg. 1

[*]E-mail: aminov_ab@npoem.ru

Abstract: The article addresses the issues of brazing parameters optimization: mediums; pressure; rate of temperature change; transition from one rate of temperature change to another one; the excess pressure in the cavity of the installation with respect to the medium pressure in the brazing cavity. Impact of the parameters is demonstrated by the examples of the engine gimbal flange, booster pump unit rotor and guide vanes of the turbo-pump unit pump. Improvement in the quality of the soldered joints of the flange can be achieved by introducing homogenizing annealing after brazing. A theoretical calculation of the nonstationary temperature distribution in the structural elements during soldering was preliminarily carried out to solve the problem of soldering the guide apparatus. The calculation resulted in that a reduction in the rate of heating and cooling, and temperature exclusions were excluded. The problem of an impregnated rotor of the booster aggregate was solved by analyzing the thermal interaction of the brazed components, which determined the need for a slow rate of the initial stage of heating the rotor and introducing a soak in temperature.

1. Introduction

Nowadays, brazing, or hard soldering, is successfully used in many engineering branches. The significant progress in its development has been achieved in rocket engineering, particularly, in the production of liquid rocket engines. The prime advantage of brazing in comparison with an alternative method of producing a fixed joint of materials, i.e. welding, is that the process is performed at a temperature below their fusing point. This enables selecting the process temperature depending on the requirement to maintain the brazed materials properties. Brazing, however, also has disadvantages that include the change of structural and phase state of some alloys at high brazing temperatures with the probability of crack formation, defects in the joint area such as pores, microporosity, dry joints, formation of brittle layers, etc. The process of producing the brazed joint is performed in special brazing machines and represents heating to the brazing temperature, soaking at the brazing temperature and cooling. At the same time, the experience of NPO Energomash, JSC in brazing liquid rocket engine structures has proven that the following brazing process parameters are of great importance: atmosphere in the brazed structure (vacuum or shielding gas); gaseous atmosphere pressure in the brazing machine working cavity; rate of temperature variation during heating and cooling; transition from one temperature variation rate to another, overpressure in the machine cavity with respect to the pressure of the atmosphere in the brazing cavity. The optimum combination of these process parameters enables producing a high-quality, defect-free, sound brazed joint and the strength of the brazed structure in general.

2. Pros and cons of brazing in vacuum

Brazing may be performed both in neutral atmosphere (argon, helium) and in vacuum, i.e. in the rarefied gas state created by pumping gas out from the brazing cavity. Brazing in vacuum prevents the brazed part from oxidation providing for a high-quality brazed joint, with no use of flux as required by the gaseous atmosphere. It also prevents embrittlement of stainless steels and heat-resistant alloys. [1]

Moreover, brazing in vacuum allows to provide the contact force between the brazed parts required to produce the brazed joint. A higher contact force is provided by creating the shielding gas increased pressure in the brazing machine working cavity.

Despite all the positive features, vacuum has a certain disadvantage, namely that an intensive evaporation of the brazed materials is possible at high temperatures, particularly, of brazes containing high vapor pressure components. As a result, the braze becomes enriched with infusible components that results in increase of the braze fusing temperature and reduction of its fluidity that may lead to defects and the brazed joint strength reduction.

Mn-base braze is the braze containing a high vapor pressure component that is, in the meantime, widely used in NPO Energomash when brazing structures [2].

3. Influence of the degree of vacuum and homogenizing annealing during the engine gimbal flange brazing

The engine gimbal flange is one of the liquid rocket engine assemblies where Mn-base braze is used (Figure 1).

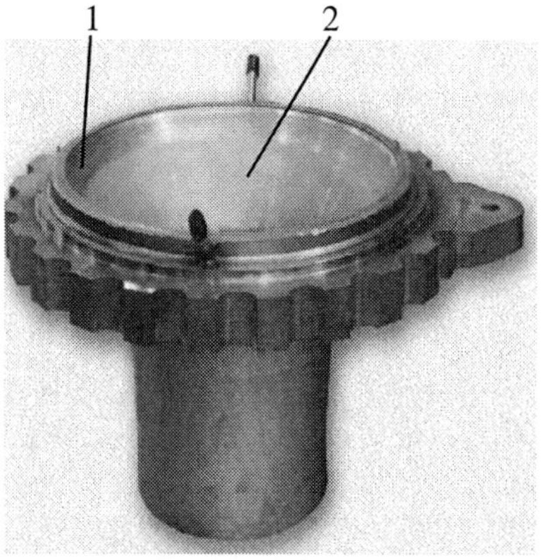

Figure 1. Engine gimbal flange
1 - outer wall; 2 - inner wall

NPO Energomash brazing experience has shown that the brazed joints of the flange with stainless steel parts and nickel alloy parts have sufficient, but yet relatively low strength and brittle fracture pattern.

The flange is brazed in vacuum created in a cavity between the inner and outer walls, under an increased pressure of the shielding gas, argon, in the machine working cavity int order to compress the walls together. The inner wall has ribs on the tops of which Mn-base braze strips are attached with the help of spot welding during the assembly for brazing.

An experimental operation with brazing the flanges in two machines with various degrees of medium vacuum in the brazing cavity has been carried out in order to check the influence of the vacuum degree and homogenizing annealing after brazing to increase the brazed joint strength using Mn-base braze. These are HR-64x80VC (10^{-2} mm of mercury vacuum) and VCA (10^{-1} mm of mercury vacuum) machines.

After homogenizing annealing and standard heat treatment (cold treatment and tempering) metallographic specimen to study the brazed joint quality, as well as samples for tensile strength tests have been cut out from the flange brazed joints.

Metallographic study of the brazed seam showed no defects in both flanges . The brazed seams have a thickness of ~ 50 μm and are formed with double fillets (Figure 2).

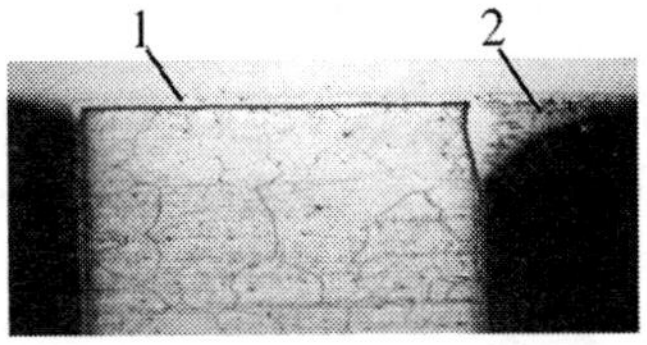

Figure 2. Flange brazed joint
1 - brazed seam; 2 - fillet

The test results have demonstrated that the tensile strength of flange samples brazed with subsequent homogenizing annealing in both machines with various degrees of vacuum is sufficiently high and exceeds the strength of brazed joints without homogenizing annealing by approximately 1,5 times. The fracture surface of brazed seams is matt, typical for ductile fracture. The fractures show no defects (Figure 3).

Figure 3. Brazed seam fracture

4. Regarding the rate of temperature variation in the course of brazing

4.1 Temperature distribution in structural elements

In order to determine the approach to select the rate of brazing temperature variation during heating and cooling, a theoretical analytical treatment of the non-stationary temperature distribution in the structural elements during brazing has been carried out, assuming that the heating is performed via a radiant heat flux, as it is implemented in the brazing machines in NPO Energomash, JSC [3].

A heat transfer model for an unlimited plate with a linear temperature variation over time on one plate surface (in our case it is the wall outer surface) and with no heat flow on the other surface (wall inner surface), as implemented in the process of heating for brazing, has been taken as the analytical model.

The heating rate and wall thickness have been varied in a range corresponding to the existing brazing practice, particularly, the rate has been varied from 0,1 to 1 K/s and the wall thickness - from 5 to 40 mm.

The following characteristics have been received as the result:

- difference of temperatures across the entire wall thickness upon attainment of the brazing temperature by the outer surface;
- wall thickness warm-up time;
- rate of wall outer surface heating due to the incident radiant flux and wall thickness value;

The obtained results have shown that the heating rate is inversely related to the wall thickness. Based on this we can get an idea of the wall heating nature with sections of various thicknesses. Indeed, the wall thin part will heat up faster than the thick one that will result in a temperature gradient between them. Consequently, conditions for thermal stresses will occur that will increase manifold under certain conditions (low material plasticity, stress concentrators) and result in crack formation.

The obtained calculated data of the temperature distribution in the structure in the course of brazing allow selecting optimal brazing conditions and finding design solutions to reduce the probability of structural failure due to the thermal stresses.

4.2 Solution to the problem of crack formation in the guide vanes of the turbo-pump unit pump

The obtained data have been used to solve the problem of crack formation during brazing of the guide vanes of the turbo-pump unit pump (Figure 4). An assumption on the thermal stresses in the housing material due to the difference of temperatures between the nozzle and the header during brazing has been made as the reason for crack formation. The presence of the stress concentrator in the form of a surface transitional section with a small bending radius in the area of the nozzle interface with the header enhances this effect. Attention has been also paid to the following feature related to the presence of soaking in the brazing temperature conditions, both during heating and cooling stages.

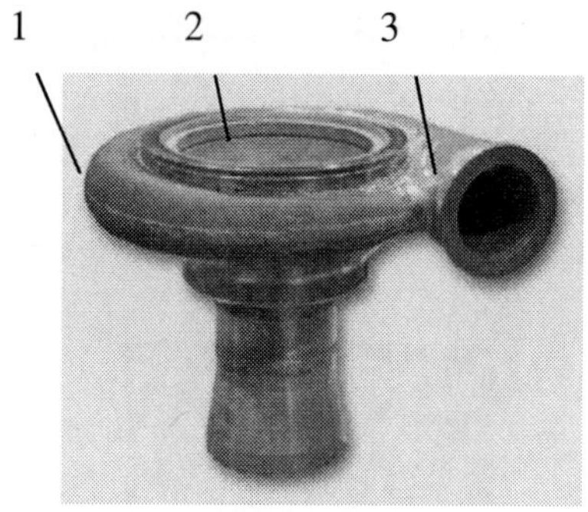

Figure 4. Guide vanes of the turbo-pump unit pump:
1 - housing (header); 2 - inner wall;
3 - housing (header nozzle)

Such soaking is provided to equalize the temperature between the brazed parts that have different weight. However, the soaking assumes a relative jump from the heating rate set by the brazing method to zero rate (constant temperature). In this case, the structural element with smaller weight is heated faster before the soaking, and during soaking its temperature will be approaching the temperature of the element with greater weight and, thus, a steplike temperature excess of the element with a smaller weight over the element with a larger weight takes place.

Additional thermocouples in addition to the standard ones were installed on several assembly units to identify the temperature difference in the structural elements. With the help of these thermocouples it was really possible to register the presence of the steplike temperature difference between the nozzle and the header within the soaking temperature ranges . Subsequent strength calculations have demonstrated the presence of tensile stresses [4].

In accordance with the calculation results it has been decided to exclude soaking from the brazing cycle during heating and cooling in order to reduce stresses, having left the same brazing time that led to a significant decrease of both heating and cooling rates. In the result of the measures taken, the crack formation in the guide vanes was no longer observed.

4.3. Solution to the problem of dry joints in the booster pump unit rotor

The decrease of the heating rate was also required for another liquid rocket engine structural element - the booster pump unit rotor, upon brazing of the rotor shroud blades (Figure 5), however, for a different reason. This was due to a specific feature determining the main difficulties of selecting the rotor brazing conditions that consist in the fact that the material of the both braced parts is (steel of a transitional class with structural transformations upon its temperature variation defining a significantly nonmonotonic dependence of the linear thermal expansion coefficient from temperature.

Following the standard rise with the temperature increase to 600°C, it then drops sharply to a minimum at 700 °C, then the rise repeats with the subsequent temperature increase.

Thus, as the temperature increases, the part increases in size, then decreases, then increases again that may be interpreted as the distortion "hole".

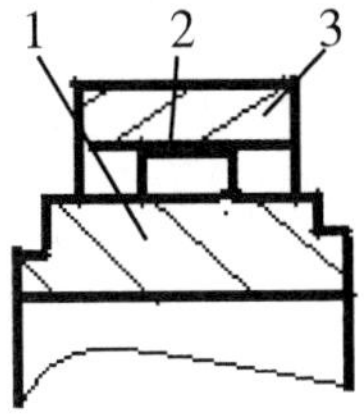

Figure 5. Booster pump unit rotor: 1 - rotor with blades; 2 - braze; 3 - shroud

This results in the fact that in case of the difference of the brazed rotor parts temperature and mismatch of the distortion "holes" there is a danger of their counter thermal deformation that may lead to the shroud plastic deformation with subsequent unacceptable increase of the brazed clearance upon the brazed parts temperature equalizing at the brazing temperature.

To obtain high-quality brazing of parts made of such material, it is required to provide the temperature difference between them not exceeding a certain maximum permissible value that may only be selected by test.

Based on the shroud - rotor thermal interaction analysis results [5], a conclusion has been made that following requirements have to be unconditionally followed. Firstly: slow rate of the rotor initial heating ensuring the lowest possible temperature difference between the shroud and rotor. Secondly: introduction of soaking for equalizing the brazed parts temperature before they enter the "hole" by the coefficient α, as the factors ensuring the shroud plastic deformation absence.

The appropriateness of the above described thermophysical scenario of the shroud - rotor interaction has been confirmed by subsequent successful and sound rotor brazing.

5. Conclusions

The following conclusions may be made as the result:

1. Homogenizing annealing of the engine gimbal flange made after brazing with Mn-base braze under the conditions of various medium vacuum values enables significant increase of the strength of the brazed joint produced without homogenizing annealing.
2. The conducted analytical and experimental studies on the temperature distribution in the structural elements during brazing allowed to exclude the excessive thermal stresses in the guide vanes of the turbo-pump unit pump, thus avoiding crack formation in it during brazing that was achieved by means of the slow heating and cooling, as well as excluding soaking in the course of brazing.
3. Based on the thermophysical design analysis, the booster pumping unit rotor brazing conditions have been defined as follows: decrease of the heating rate and introduction of soaking that allowed to ensure defect-free rotor brazing.

References

[1] Petrunin I.E. Physicochemical processes during soldering, Moscow: Higher School, 1972. 280 p.
[2] Khorunov V.F. Fundamentals of soldering thin-walled structures from high-alloy steels, Kiev: Naukova Dumka, 2008. 240 p.
[3] A.B. Aminov, K.E. Dubrovsky On some features of temperature distribution in the liquid rocket engine structures during brazing // NPO Energomash works. M., 2015. No. 27. P. 198-210.
[4] A.B. Aminov, K.E. Dubrovsky On technological advancement of the brazing method for the liquid rocket engine turbo-pump unit pump housing. // NPO Energomash works. M., 2016. No. 33. P.255– 269.
[5] A.B. Aminov, V.V. Dmitriev, K.E. Dubrovsky, V.I. Petrov, V.N. Semyonov. Brazing features fo the liquid rocket engine turbo-pump unit rotor // NPO Energomash works. M., 2010. No. 27. P.217-222.